D1526203

MAGNESIUM PRODUCTS DESIGN

MECHANICAL ENGINEERING

A Series of Textbooks and Reference Books

EDITORS

L. L. FAULKNER

Columbus Division
Battelle Memorial Institute

and

Department of Mechanical Engineering
The Ohio State University
Columbus, Ohio

S. B. MENKES

Department of Mechanical Engineering
The City College of the
City University of New York
New York, New York

1. Spring Designer's Handbook, *by Harold Carlson*
2. Computer-Aided Graphics and Design, *by Daniel L. Ryan*
3. Lubrication Fundamentals, *by J. George Wills*
4. Solar Engineering for Domestic Buildings, *by William A. Himmelman*
5. Applied Engineering Mechanics: Statics and Dynamics, *by G. Boothroyd and C. Poli*
6. Centrifugal Pump Clinic, *by Igor J. Karassik*
7. Computer-Aided Kinetics for Machine Design, *by Daniel L. Ryan*
8. Plastics Products Design Handbook, Part A: Materials and Components; Part B: Processes and Design for Processes, *edited by Edward Miller*
9. Turbomachinery: Basic Theory and Applications, *by Earl Logan, Jr.*
10. Vibrations of Shells and Plates, *by Werner Soedel*
11. Flat and Corrugated Diaphragm Design Handbook, *by Mario Di Giovanni*
12. Practical Stress Analysis in Engineering Design, *by Alexander Blake*
13. An Introduction to the Design and Behavior of Bolted Joints, *by John H. Bickford*
14. Optimal Engineering Design: Principles and Applications, *by James N. Siddall*
15. Spring Manufacturing Handbook, *by Harold Carlson*
16. Industrial Noise Control: Fundamentals and Applications, *edited by Lewis H. Bell*
17. Gears and Their Vibration: A Basic Approach to Understanding Gear Noise, *by J. Derek Smith*

Contents

to select the correct process, design the part in an optimum way, and fabricate the part itself. The book is divided into four sections in order to address these needs.

Background information on the history and production of magnesium and a discussion of specifications and units used are found in the introductory Section I. More importantly, selection and design criteria for using magnesium in structural parts are given, together with many illustrations which show successful applications.

Design principles for basic shapes such as castings or wrought products are given in Section II, Fabrication. The technology for machining, forming and joining is covered in detail.

Section III, Properties, contains extensive tables and graphs of physical and mechanical properties for all the current commercial alloys in cast and wrought form.

The final section, Corrosion and Finishing, contains a discussion of the principles of magnesium corrosion in a variety of media, together with quantitative data on the corrosion performance of alloys when exposed to many different environments. Finishing procedures for applying protective chemical, organic, electrolytic, and metallic coatings are given.

The book could not have been written without the support and the extensive help provided by the members of the International Magnesium Association. Special recognition and thanks are due to the individuals on the Editorial Board. In addition, David L. Hawker, Materials Consultant, was of great assistance to J. S. Viland of the Editorial Board.

Robert S. Busk

Preface

There are three excellent books on magnesium technology: The classic *Technology of Magnesium and Its Alloys*, edited by Adolph Beck and published in the 1930s; *Magnesium and Its Alloys* by C. Sheldon Roberts, published in 1960; and the monumental *Principles of Magnesium Technology* by E. F. Emley, published in 1966. The latter in particular is a tower of strength for all those engaged in the fabrication of magnesium.

However, there is no single publication addressed particularly to those who wish to use magnesium as the material for constructing a part. Magnesium alloys are now becoming more widely used and the industry is poised on the threshold of significant expansion. It is timely to gather the widely scattered information to be found in producers' trade literature, in various specifications, and in the general literature into a single publication addressed specifically to the potential user of magnesium.

Magnesium is broadly used in widely divergent nonstructural applications such as alloying in aluminum, desulfurizing iron and steel, producing ductile iron, and producing pharmaceuticals and fine chemicals via the Grignard reaction. While these applications are important to the industry, this book is addressed specifically to those wishing to make a serviceable part using magnesium as the material of construction.

The point of view adopted is that the user of this book will buy suitable basic shapes from the magnesium industry, such as castings, sheet, plate, extrusions, or forgings. While not needing to know how to make the shapes purchased, he or she will need to know how

Editorial Board

Terje Kr. Aune

Norsk Hydro a.s
Porsgrunn, Norway

Allan Froats

Chromasco–A division of Timminco Ltd.
Haley, Ontario, Canada

S. B. Hirst

Magnesium Elektron Limited
Manchester, England

Shigeru Nemoto

Japan Light Metal Association
Tokyo, Japan

Henry J. Proffitt

Haley Industries, Ltd.
Haley, Ontario, Canada

J. S. Viland

AMAX Magnesium
Salt Lake City, Utah

Joseph Waibel

Dow Chemical Company
Freeport, Texas

Library of Congress Cataloging-in-Publication Data

Busk, Robert S.
Magnesium products design.

(Mechanical engineering ; 53)
Includes bibliographies and index.
1. Magnesium 2. Magnesium alloys. 3. Design Industrial. I. International Magnesium Association. II. Title. III. Series
TA480.M3B78 1986 669'.723 86-19783
ISBN 0-8247-7576-7

The guidelines and recommendations in this book are based on information believed to be reliable and are offered in good faith but without guarantee. The operational conditions which exist in individual plants and facilities vary widely. Users of this information should adapt it, as appropriate, to the precise conditions of the individual facility and should always exercise independent discretion in establishing plant or facility operating procedures. No warranty, express or implied, is made of this information by the publisher, author, the International Magnesium Association or by any of its member companies and no responsibility of liability is assumed for the completeness of the data or the general applicability of the guidelines and recommendations herein, which are based on state-of-art knowledge and may not be appropriate in all situations.

Nothing is to be construed as granting any right, by implication or otherwise, for manufacture, sale or use in connection with any method, apparatus or product covered by letters patent, nor as insuring anyone against liability for infringement of letters patent.

COPYRIGHT © 1987 by MARCEL DEKKER, INC. All Rights Reserved

Neither this book nor any part may be reproduced or transmitted in any form or by any means, electronic or mechanical, including photocopying, microfilming, and recording, or by any information storage and retrieval system, without permission in writing from the publisher.

MARCEL DEKKER, INC.
270 Madison Avenue, New York, New York 10016

Current printing (last digit):
10 9 8 7 6 5 4 3 2 1

PRINTED IN THE UNITED STATES OF AMERICA

MAGNESIUM PRODUCTS DESIGN

ROBERT S. BUSK

Sponsored by the
International Magnesium Association

MARCEL DEKKER, INC. New York and Basel

44. Machinery Adhesives for Locking, Retaining, and Sealing, *by Girard S. Haviland*
45. Couplings and Joints: Design, Selection, and Application, *by Jon R. Mancuso*
46. Shaft Alignment Handbook, *by John Piotrowski*
47. BASIC Programs for Steam Plant Engineers: Boilers, Combustion, Fluid Flow, and Heat Transfer, *by V. Ganapathy*
48. Solving Mechanical Design Problems with Computer Graphics, *by Jerome C. Lange*
49. Plastics Gearing: Selection and Application, *by Clifford E. Adams*
50. Clutches and Brakes: Design and Selection, *by William C. Orthwein*
51. Transducers in Mechanical and Electronic Design, *by Harry L. Trietley*
52. Metallurgical Applications of Shock-Wave and High-Strain-Rate Phenomena, *edited by Lawrence E. Murr, Karl P. Staudhammer, and Marc A. Meyers*
53. Magnesium Products Design, *by Robert S. Busk*

ADDITIONAL VOLUMES IN PREPARATION

Mechanical Engineering Software

Spring Design with an IBM PC, *by Al Dietrich*

Mechanical Design Failure Analysis: With Failure Analysis System Software for the IBM PC, *by David G. Ullman*

18. Chains for Power Transmission and Material Handling: Design and Applications Handbook, *by the American Chain Association*
19. Corrosion and Corrosion Protection Handbook, *edited by Philip A. Schweitzer*
20. Gear Drive Systems: Design and Application, *by Peter Lynwander*
21. Controlling In-Plant Airborne Contaminants: Systems Design and Calculations, *by John D. Constance*
22. CAD/CAM Systems Planning and Implementation, *by Charles S. Knox*
23. Probabilistic Engineering Design: Principles and Applications, *by James N. Siddall*
24. Traction Drives: Selection and Application, *by Frederick W. Heilich III and Eugene E. Shube*
25. Finite Element Methods: An Introduction, *by Ronald L. Huston and Chris E. Passerello*
26. Mechanical Fastening of Plastics: An Engineering Handbook, *by Brayton Lincoln, Kenneth J. Gomes, and James F. Braden*
27. Lubrication in Practice, Second Edition, *edited by W. S. Robertson*
28. Principles of Automated Drafting, *by Daniel L. Ryan*
29. Practical Seal Design, *edited by Leonard J. Martini*
30. Engineering Documentation for CAD/CAM Applications, *by Charles S. Knox*
31. Design Dimensioning with Computer Graphics Applications, *by Jerome C. Lange*
32. Mechanism Analysis: Simplified Graphical and Analytical Techniques, *by Lyndon O. Barton*
33. CAD/CAM Systems: Justification, Implementation, Productivity Measurement, *by Edward J. Preston, George W. Crawford, and Mark E. Coticchia*
34. Steam Plant Calculations Manual, *by V. Ganapathy*
35. Design Assurance for Engineers and Managers, *by John A. Burgess*
36. Heat Transfer Fluids and Systems for Process and Energy Applications, *by Jasbir Singh*
37. Potential Flows: Computer Graphic Solutions, *by Robert H. Kirchhoff*
38. Computer-Aided Graphics and Design, Second Edition, *by Daniel L. Ryan*
39. Electronically Controlled Proportional Valves: Selection and Application, *by Michael J. Tonyan, edited by Tobi Goldoftas*
40. Pressure Gauge Handbook, *by AMETEK, U.S. Gauge Division, edited by Philip W. Harland*
41. Fabric Filtration for Combustion Sources: Fundamentals and Basic Technology, *by R. P. Donovan*
42. Design of Mechanical Joints, *by Alexander Blake*
43. CAD/CAM Dictionary, *by Edward J. Preston, George W. Crawford, and Mark E. Coticchia*

MAGNESIUM PRODUCTS DESIGN

I

Introduction

Chapter 1 contains a very brief description of the history and production methods used for making magnesium. More detailed information is contained in the references cited.

Chapter 2 has a qualitative description of the application principles for magnesium, together with many illustrations of successful applications. It is worth careful reading before proceeding with the further chapters covering quantitative design information.

Details of the nomenclature used for alloys and tempers, lists of specifications for magnesium alloys and processes, the compositions of alloys cited in the book, and units of measurement used throughout this book are given in Chapter 3.

1

Background

1.1 OCCURRENCE

Magnesium is the sixth most abundant element on the earth's surface, amounting to about 2.5% of its composition. Magnesium-bearing minerals are widespread and concentrated enough so that there will never be a shortage of ore supply anywhere in the world.

1.2 PRODUCTION METHODS

Two major methods of producing the metal have survived the test of time: electrolytic reduction of magnesium chloride, and high-temperature chemical reduction of magnesium oxide by silicon. Both methods are used to produce significant quantities, although currently amounts from electrolytic sources exceed those from the silicon reduction technique.

There are two basic cell types for the electrolytic production of the metal, from which all modern cells are derived. One of these, based upon the Dow Chemical cell design [1], uses slightly hydrous magnesium chloride as a feed material for the cell; the other, based on the I. G. Farben cell [2] design (IG), uses anhydrous magnesium chloride. Magnesium chloride for either process is obtained from the ocean, from magnesium-chloride-rich brines, from buried salt deposits, from dolomite, or from magnesium oxide ores.

Over the years many design changes have occcurred in both types so that the descendants of the two original cells now vary considerably from the originals. In particular, those cells based on the IG process are now much larger, closed, and fully automatic [3]. The cells based on the Dow Chemical design have become much larger and more efficient. Energy efficiency for both types has been greatly improved over the past decade.

When magnesium chloride is electrolyzed, metallic magnesium and chlorine are produced. These are separated, the magnesium being collected in the molten state for casting into ingots, and the chlorine in the gaseous state. Since until recently it has not been practical to produce good quality magnesium chloride without using chlorine in the process, that produced in the electrolytic cell was recycled back to the making of magnesium chloride. Production methods have now been developed to enable the production of anhydrous magnesium chloride by directly drying brines containing the compound [3]. In those plants taking advantage of the new technology, the chlorine produced in the electrolytic cell is available for marketing.

Two different technologies have also been developed for the reduction of magnesium oxide by silicon. The original process, developed by L. M. Pidgeon [4], uses ferrosilicon in the solid state to contact magnesium oxide, also in the solid state. The reaction takes place at a high enough temperature and a low enough pressure so that the magnesium produced by the reduction of the magnesium oxide is in the gaseous state, and condenses as solid magnesium crystals in a cooler part of the reactor. The crystals are then melted and cast into ingots. An alternate process, developed by Pechiney-Ugine-Kuhlman in France [5], uses a molten solution of aluminum oxide, slag, and magnesium oxide, causing the ferrosilicon to contact a liquid rather than a solid. The magnesium produced in this process, also in the gaseous state, is condensed to a liquid and then frozen to a solid in a cooler part of the reactor. It is then melted and cast into ingots.

The molten magnesium produced by any of the processes can be cast directly into ingots, or can be alloyed with other metals and then cast. Alloys are cast into ingots which are then remelted to produce shaped castings, or into billets to produce wrought products such as sheet, extrusion, and forgings.

1.3 APPLICATIONS HISTORY

Commercial production of both aluminum and magnesium began in 1886 [6]. However, the technical difficulties in the production of magnesium associated with the necessity for using anhydrous magnesium chloride and with the greater reactivity of molten magnesium as compared to molten aluminum resulted in a slower development for magnesium. In fact, until World War I the only producer was in Germany, and production capacity was very small.

With the advent of World War I, it became necessary for the Allied countries to have a source of magnesium, and many plants were started. Although there was severe attrition immediately after the war, by 1939 plants were in operation in Germany, England, France, Japan, Russia, Switzerland, and the United States. Total production in that year was 32,850 tons, with Germany accounting for 20,000 tons; England, 5,000; the United States, 3,350; France, 2,500; Japan, 1,000; and Russia and Switzerland, each 500 [7, p. 48; 8].

The demand for the metal was large during World War II, requiring great expansion of production throughout the world. In the United States alone, production in the peak year of 1943 attained 184,000 tons, an increase of a little over five-fold from the 1939 figure [8]. Structural applications, limited to military uses during the war, were predominately in aircraft. Typical applications were parts for engines, air frames, and landing wheels.

Following the war, the military uses largely disappeared, and civilian uses had to be found to absorb the production capacity created to meet the war needs. A valuable legacy of the war effort, however, was the large body of knowledge created in fabrication and design technology. This was put to use to seek commercial outlets for the metal, an effort that was world-wide in extent, and intensive. While some applications were unsuccessful from a technical point of view, most were viable; failures were more commonly for economic or other market-place reasons.

Overall, the magnesium industry has grown well in the years since World War II, and capacity has kept pace with growing consumption. Data on the production, capacity, and consumption of the metal for the past few years are given in Table 1.1.

TABLE 1.1 Capacity, Production, and Consumption of Magnesium in the Western World (1000 Metric Tons)

Year	Capacity	Production	Consumption
1970		169	179
1971		180	178
1972		179	191
1973		181	205
1974		189	219
1975		176	167
1976		183	217
1977		187	197

TABLE 1.1 (*Continued*)

Year	Capacity	Production	Consumption
1978		209	210
1979	247	226	225
1980	256	236	219
1981	260	209	199
1982	260	157	169
1983	261	166	204
1984	262	232	211

Source: Refs. 9, 10, 11, 12, 13, and 14.

REFERENCES

1. R. M. Hunter, *Transactions of the Electrochemical Society 86*:21 (1944).
2. C. J. P. Ball, *Transactions of the AIME 159*:28 (1944).
3. Nils Høy-Petersen, "Magnesium Production at the Porsgrunn Plant of Norsk Hydro," *Proceedings of the International Magnesium Association 32* (1979).
4. L. M. Pidgeon and W. A. Alexander, *Transactions of the AIME 159*:315 (1944).
5. F. Trocmé, "The Development of the 'Magnetherm' Process," *Transactions of the AIME* (1974).
6. W. H. Gross, "The Story of Magnesium," American Society for Metals, Metals Park, Ohio, 1949.
7. Adolph Beck, "The Technology of Magnesium and its Alloys," F. A. Hughes and Co., Limited, London, 1940.
8. Magnesium Association, "Selected Magnesium Statistics," 1962.
9. J. Laboree, "World Review of Magnesium Supply and Demand," *Proceedings of the International Magnesium Association 1* (1978).
10. R. Hansen, "World Review of Magnesium Supply and Demand," *Proceedings of the International Magnesium Association 7* (1980).
11. Tor Ness, "World Review of Magnesium Supply and Demand," *Proceedings of the International Magnesium Association 1* (1981).
12. J. Thomas Timmins, "World Review of Magnesium Supply and Demand," *Proceedings of the International Magnesium Association 23* (1982).

13. P. Cohen, "Magnesium Supply and Demand Overview," *Proceedings of the International Magnesium Association 110* (1983).
14. J. S. Viland, "Magnesium Supply & Demand Report," *Proceedings of the International Magnesium Association 1* (1984).

2

Preliminary Design Considerations

2.1 TECHNOLOGY BASE

The foundations for magnesium technology were laid in the years between the first and second world wars. While important work was done in the United States and England, I. G. Farben A. G., in Germany, was the leader in these efforts. Their work culminated in the classic book compiled by Beck [1], published in Germany in 1939. It summarizes the information up to that time on the production and handling of magnesium, its physical, mechanical, and chemical properties, and includes some discussion of applications. These efforts were thorough and comprehensive, resulting in a good base for the very large expansion in the use of magnesium that occurred during World War II.

The extreme service requirements characteristic of military use necessitated a great deal of development work, leading to new alloys, new fabrication techniques, and a good understanding of engineering principles required for the design of a successful part. Problems with fatigue, stress corrosion, notch sensitivity, creep, and general and galvanic corrosion were all met in practice and solutions found for the specific applications that were current. The natural sequel to the work reported by Beck [1] was compressed into a very short time, and provided in its turn a good base for expansion in the use of magnesium in civilian applications after the end of the war.

The disappearance of military applications after World War II and the resulting need to develop civilian uses, required a further expansion of the technology base gained during the war. For the military applications, technical suitability was sufficient for use, with very little attention paid to cost. With civilian usage, however, not only must the part perform a function, but it must do so at a competitive cost. The major centers for the work to develop civilian markets were initially in the United States and in England, with important work also being done in France. Today, new developments are widespread, occurring not only in the United States, England, and France, but in Norway, Italy, Germany, Japan and Russia.

After about forty years of application developments around the world, there is now a good understanding of what constitutes a sound application for magnesium, from both performance and cost standpoints. New alloys and new improved fabrication procedures have been developed; optimum methods for exploiting especially useful characteristics have been determined; ways of minimizing those characteristics that pose serviceability problems have been worked out. The information exists today for selecting magnesium for applications where it can serve well, and for designing properly for optimum serviceability in those applications.

2.2 FORMS AVAILABLE

In general, magnesium shapes are produced by all of the methods used for other metals. Some forms are peculiarly suited to the inherent properties of magnesium, while others are peculiarly unsuited. Thus selection of the proper form for optimum utilization of magnesium for a specific part is an important part of the overall design process. As a rule of thumb, the less complicated the path from the ingot to the finished part, the more competitive is magnesium for that part. Thus, high-pressure die casting and extrusion are good ways to form magnesium. Thin sheet is not. Table 2.1 shows the forms in which magnesium is customarily produced, together with some comments on the characteristics of each.

2.2.1 Castings

Sand and permanent-mold castings can be of extremely intricate shapes, including tubeless passageways of complicated design and thin walls. Quality is good, and, in some alloys, mechanical properties in castings are equal to or better than those of separately-cast test bars. Specific mechanical properties (the property divided by density) of magnesium are competitive with other materials.

The high-pressure die casting process is peculiarly suited to magnesium, because of two inherent characteristics: a low heat content, and no reactivity with iron. Because of the low heat content, less heat needs to be extracted during the cooling portion of the casting cycle and castings can be made at a faster rate than is the case with aluminum. Because of the lack of reactivity with iron, simple tools can be used, there is little sticking in the die, long die life is common, and equipment for metering accurate quantities of molten metal to the die-casting machine can be constructed of steel. Hot-chamber die casting machines, a low-cost technology, are being used for magnesium, but have not yet been successful for aluminum. Parts can be made with very little or no draft. Mechanical properties are competitive with those for either aluminum or zinc. These factors often combine to result in a part that is less expensive when die cast in magnesium than when die cast in either aluminum or zinc.

Low-pressure casting is a relatively new technique for magnesium that is now coming into more general use. The same inherent advantages that are important for high-pressure die casting apply also to low-pressure casting. Large castings, of intricate design, are produced by this process. The quality is generally higher than that of high-pressure castings.

Investment casting is a process that permits the casting of shapes not possible by other means, since there is the ability to cast blind holes and re-entrant angles. Thin walls and exceptionally high mechanical properties for the metal in the casting are characteristic of the process.

2.2.2 Wrought Products

The extrusion process, with ability to produce complicated shapes, including hollow cavities, lends itself to flexibility of design. Almost any shape that can be machined into a die can be extruded. The quality of the metal is high, and mechanical properties are consistent.

Forgings are readily produced in magnesium, and have good mechanical properties.

Thick plate can be readily produced in magnesium at a reasonable cost. Dimensional stability is exceptional for the usual alloy produced in thick plate.

Thin sheet is available in magnesium, but cost tends to be high. This is inherent in the fundamental properties of the metal, and will probably always be true, although new methods may reduce the costs significantly.

A characteristic peculiar to wrought products made of magnesium is that, with the exception of a few alloys, the compressive yield strength is lower than the tensile yield strength. This must be kept in mind and accounted for in design.

TABLE 2.1 Forms Available

Major category	Minor category	Remarks
Cast products	Sand	Intricate shapes, suitable for low volume, good quality, high cost, very large parts possible
	Permanent mold	Intricate shapes, suitable for high volume, good quality, high cost, large parts possible
	Low-pressure die	Intricate shapes, high quality, intermediate size parts possible, low cost
	High-pressure die	Accurate dimensions, requires high volume, intermediate size parts possible, intermediate quality, lowest cost
	Investment	Extremely accurate dimensions, high quality, very intricate shapes, high cost
Wrought products	Sheet	Low properties, high cost
	Plate	Low properties, low cost, good dimensional stability
	Extrusions	Intermediate properties, low cost, intricate shapes
	Forgings	High quality, intermediate properties, medium cost
	Impact extrusion	Good quality, simple shapes, low cost, good properties

2.3 CRITERIA FOR SELECTION OF MAGNESIUM

The basis for selecting any material for constructing a specific part is that the part will serve with the best compromise between function and cost when made of that material.

Magnesium is superior to other materials for a number of properties, and inferior for others. A knowledge of what these properties are, and the full implications of each for the service of a part is required for proper selection of the material. After selection, design must maximize exploitation of the favorable properties and minimize the unfavorable.

The density of pure magnesium is 1.74 g/cc, less than that for any other structural metal. For a specific part, a lower density can be translated into lower weight, lower cost, improved ruggedness, or even all three.

Whether lower density will result in lower weight for a part depends upon the mechanical properties of competing materials, upon ability to manufacture the part with dimensions that will utilize mechanical properties fully, and upon the design of the part. For some forms, for example high-pressure die casting, the properties of magnesium are equal to those of competing materials. In this case, of course, the part will be lighter if made of magnesium than of other metals.

For other forms, for example most wrought products, some properties of magnesium are less than those of some competing materials. If the part design is such that the critical property is the elastic modulus (as in buckling), the part will be lighter if made of magnesium. If yield and tensile strength are critical and if the part design is such that the mechanical properties of the competing materials are fully utilized, then the part may be heavier when made of magnesium. However, if the part design is such that the mechanical properties of the competing material are not fully utilized, then the part may be, even in this case, lighter when made of magnesium. The total weight saving is often not only the direct saving by using magnesium rather than another metal, but also further savings if other parts of the structure can be made lighter because the specific part made of magnesium is light.

Low density can lead to lower costs for the finished part or structure in several ways. If the part is lighter when made of magnesium, less material is used to make the part and this is a cost saving. If the use of magnesium permits making other components of the structure lighter, a cost saving in those materials is realized. Because walls can be thicker without weight penalty when using a material of low density, it is sometimes possible to use monocoque construction, eliminating the need for stiffeners with their accompanying fasteners. Simplicity of construction and lower cost results.

Because thicker wall design can be used without weight penalty, other properties improve. Thus, a lawn mower housing made of magnesium resists stones thrown by the rotor much more than will a housing made of steel simply because the magnesium housing can be so much thicker and still be lighter than the steel housing. A part, even though of light weight when made of magnesium will often be much more rugged and serviceable against abuse than the same part made of another, more dense, material. There is an extra margin of safety because of the low density.

Damping capacity is a measure of the rate at which unforced vibrations disappear due to internal mechanisms in the material. The high damping capacity of magnesium [2] means that vibrations due to sound or mechanical stress diminish rapidly and are not transmitted readily throughout the structure. A high damping capacity is directly useful, for example, when the material is used as a platform for objects sensitive to vibration damage or when sound absorption is important. Magnesium is used as a platform for electronic devices in missiles because of its good damping capacity. The indirect benefits of a high damping capacity are often even more important. Since vibrations due to stress are not transmitted efficiently, alternating stress applied to one part of a structure is lessened as it reaches a further part of the same structure, if the vibrations must be transmitted by magnesium. Because the vibrations due to alternating stress are dampened within the magnesium itself, the stress level actually experienced by a part made of magnesium is less than if the part were made of another material. Both of these effects result in a greater fatigue life for the structure than would be predicted from laboratory data.

While the energy required to form a crack in magnesium is usually less than is required for other metals, the energy to create a permanent dent is more than that for other metals [3]. There are many applications where failure is better defined by denting than by destruction. A high resistance to denting is important to luggage or to shipping containers, for example, to prevent a battered look.

Bare magnesium will not "crock" when handled. There are many applications where the discoloration picked up by hands or by clothing from a bare piece of metal requires that the metal be finished before use. If the environment is such that corrsion is not a problem, this is not required for magnesium. Thus, hospital racks for bedding and clothing, paper spools, drafting tools, or printing equipment can be fabricated from unfinished magnesium, resulting in a lowered cost.

As with crocking, the galling tendency of magnesium is low compared to other metals. Thus, sliding surfaces will often operate more smoothly and with fewer problems than if galling occurs readily.

Magnesium has an exceptionally low thermal neutron cross section [4]. While this is a specialty property and not often a design consideration, it is the reason for the use of magnesium as cladding for the fuel in gas-cooled nuclear reactors [5].

Magnesium is the most machineable of all metals, in the sense of less power requirements for metal removal and less tool wear. A fine surface is obtained with relatively deep cuts, eliminating the need for a final, low-feed cut in order to get a fine finish. Deeper small-diameter holes can be drilled, sometimes holes that can not be drilled in any other material. Fewer cuts, higher tool speeds, less power demand, and less down time for tool resharpening, translate into less cost for any machining steps required for producing a part.

The metallurgy of magnesium is such that the stability of both properties and dimensions is excellent. Thus parts such as tooling jigs retain their dimensions over long usage.

The essentials of the corrosion of magnesium can be summarized by three statements. It is stable in alkaline and fluoride environments. It is unstable in most acidic or chloride environments. The solution potential is higher than that for any other structural metal. Since the earth we live on is generally acidic and since chloride is far more prevalent than fluoride, the corrosion resistance of magnesium is less than that of aluminum (although better than that of mild steel). Occasionally there are specialty applications that make good use of resistance to alkali. For example, bulk shipping containers for paint were made of magnesium because of its good resistance to alkaline cleaning after shipping. In general, for magnesium, finishing is required if salt water will be contacted or if there will be standing fresh water in contact with the part that will pick up carbon dioxide from the atmosphere and thus become acidic. If the part will be exposed only to dry conditions, finishing will not be required except for decorative purposes.

Because of its position in the galvanic series, magnesium is subject to galvanic corrosion if it is electrically connected to any other structural metal and there is an electrolyte present. This is the most serious corrosion problem for magnesium, and must be guarded against in any design. Removal of either the metallic connection or the electrolyte eliminates galvanic corrosion.

Thus, the major points to be considered when deciding whether to select magnesium for a specific part are:

1. Is light weight important in the use of the part? Are the mechanical properties of magnesium, the manufacturing requirements, or the design of the part such that it will be lighter if made of magnesium than of something else?

2. Does low density lead to any subsidiary advantages, such as simpler construction, lower cost, or a more rugged part less subject to damage in service?

3. Is the part subject to vibrating stress that may lead to fatigue failure? Will improved damping capacity be an advantage?

4. Does the part produce noise that should be blocked as much as possible? Will good damping capacity be useful?

5. Will the part need to be finished for protective or decorative reasons? If not, will non-crocking be useful?

6. Are there sliding surfaces present during the use of the part so that a non-galling tendency will be useful?

7. Will the service environment be such that corrosion is a problem? If so, can magnesium be finished satisfactorily at a reasonable cost?

8. Does making the part require extensive or particularly difficult machining? Will ease of machining, or low cost for machining, be important?

9. Can the part be made using methods that are especially suited to magnesium?

10. Taking all factors into account, is the cost of the part, including the costs during service, competitive when produced in magnesium?

Table 2-2, with its referenced figures, has been constructed to aid in visualizing how the various properties of magnesium have been managed in many successful applications.

TABLE 2.2 Representative Applications

Application	Properties significant to the application	Figure number
Automotive		
Cylinder head cover	Light weight, ease of machining, dimensional stability, easily pressure die cast, overall low cost	2.1
Clutch and transmission housing	Light weight, ease of machining, easily pressure die cast, overall low cost	2.2
Carburetor	Light weight, hot-chamber die cast, low cost	2.3
Fan and fan housing	Light weight, low vibration	2.4
Air intake grill center panel	Light weight	2.5
Brake and clutch pedal bracket	Light weight	2.6

TABLE 2.2 (*Continued*)

Application	Properties significant to the application	Figure number
Steering column lock housing	Light weight	2.7
Clutch housing	Light weight	2.8
Gear housing	Light weight	2.9
Truck bodies	Light weight, ability to use monocoque construction	2.10
Wheels	Light weight, toughness	2.11
Aerospace		
Gear casing	Light weight, ease of casting complex parts	2.12
Gear casing	Light weight	2.13
Gear casing	Light weight	2.14
Mounts	Light weight, high damping capacity	2.15
Tools		
Vibration fixtures	Ease of machining, high damping capacity, dimensional stability	2.16
Chain saw	Light weight, high damping capacity, ease of die casting	2.17
concrete tools	Alkali resistance	2.18
Materials handling		
Hand trucks	Light weight, ease of fabricating	2.19
Bakery racks	Light weight, ease of fabricating, noncrocking	2.20
Sporting goods		
Ski binding	Light weight	2.21
Handle and middle section of archery bow	Light weight, high damping	2.22
Fly reel	Light weight	2.23
Baseball bat	Light weight, high damping	2.24
Tennis racquet	Light weight, high stiffness	2.25
Deep sea diving suit	Light weight, ability to protect against severe environment	2.26

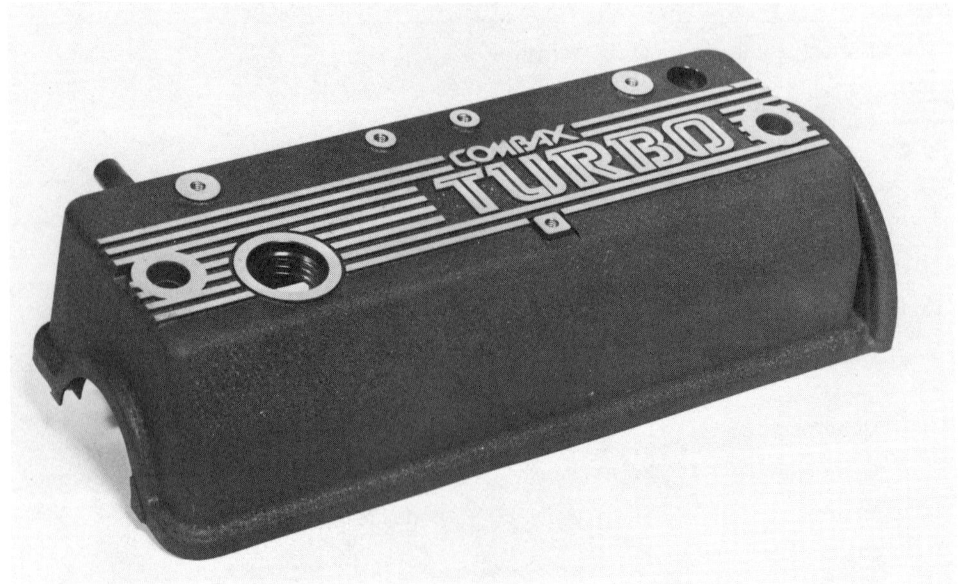

FIGURE 2.1 Cylinder head cover for the Honda City Turbo Engine. Made by the high-pressure die-casting process by Teisan Diecasting, Japan. AZ91B.

FIGURE 2.2 Clutch and transmission housing for Volkswagon. Made by the high-pressure die-casting process by Volkswagen, Germany. AZ81A.

FIGURE 2.3 Carburetor for Renault. Made by the high-pressure die-casting process by Solex, France. AZ91B.

FIGURE 2.4 Fan and fan housing for the Porsche 911. Made by the high-pressure die-casting process by Mahle, Germany. AZ81/91B.

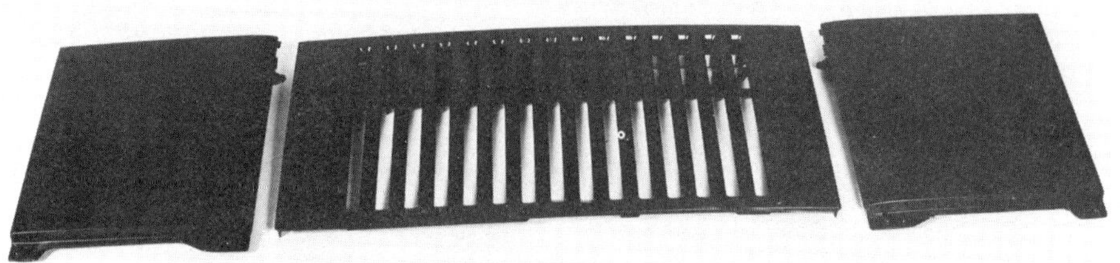

FIGURE 2.5 Air intake grill center panel for Pontiac Fiero. Made by the high-pressure die-casting process by Webster Mfg., Canada. AZ91D.

FIGURE 2.6 Brake and clutch pedal bracket for the Ford Ranger pickup truck. Made by the high-pressure die-casting process by Diemakers, Inc., U.S.A. AZ91D.

FIGURE 2.7 Steering column lock housing for the Ford Ranger pickup truck. Made by the high-pressure die-casting process by Webster Mfg., Canada. AZ91B.

FIGURE 2.8 Clutch housing for the Ford Ranger and Bronco pickup trucks. Made by the high-pressure die-casting process by Webster Mfg., Canada. AZ91D.

FIGURE 2.9 Gear housing for the Alfa Romea GTV 2,5-6. Made by the high-pressure die-casting process by Leibfried, Germany. AZ91B.

FIGURE 2.10 Delivery truck built by Metropolitan Body Company, a subsidiary of International Harvester. Monocoque construction of AZ31B sheet and extrusions. No longer in production.

FIGURE 2.11 Automobile wheel made for Ferrari by Speedline, Italy using the low-pressure casting process.

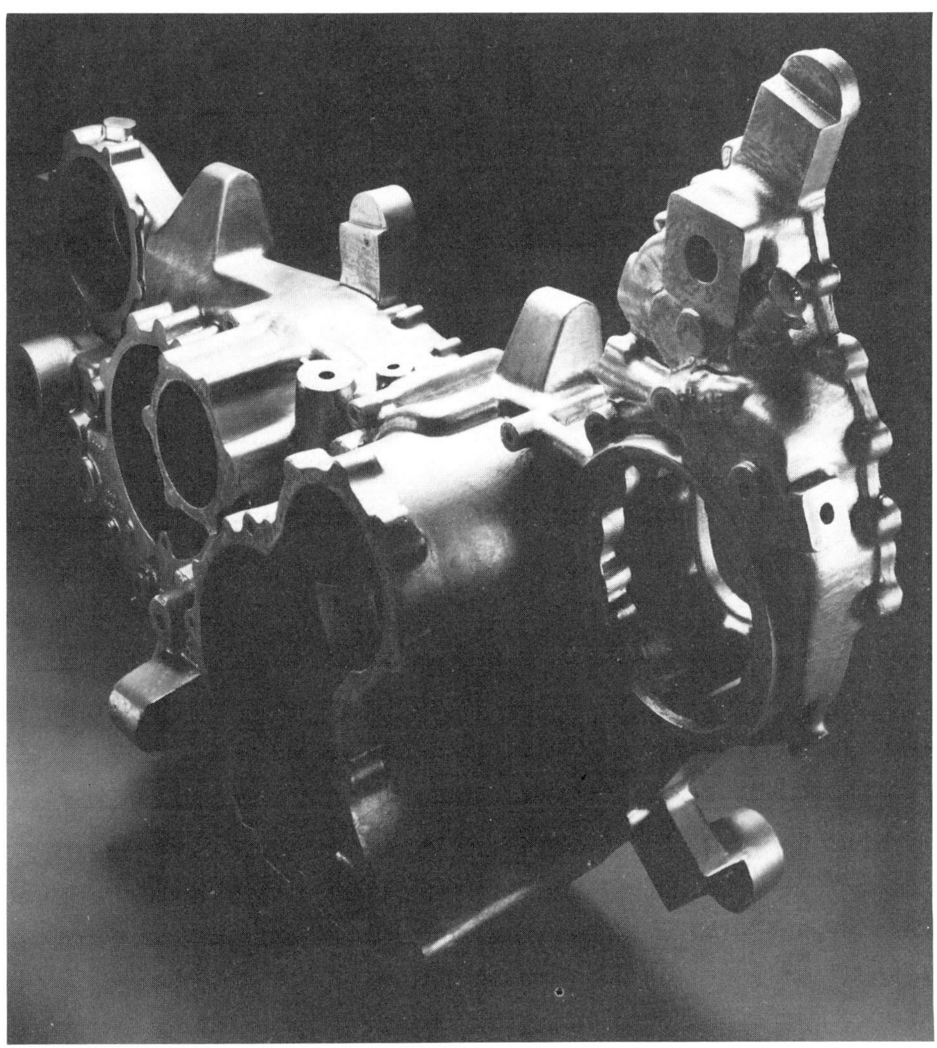

FIGURE 2.12 Gearbox housing for a fighter aircraft. Made by the sand-casting process by Haley Industries, Canada. ZE41A.

FIGURE 2.13 Gearbox for the Boeing CH47C and D helicopters. Made by the sand-casting process by Haley Industries, Canada. ZE41A.

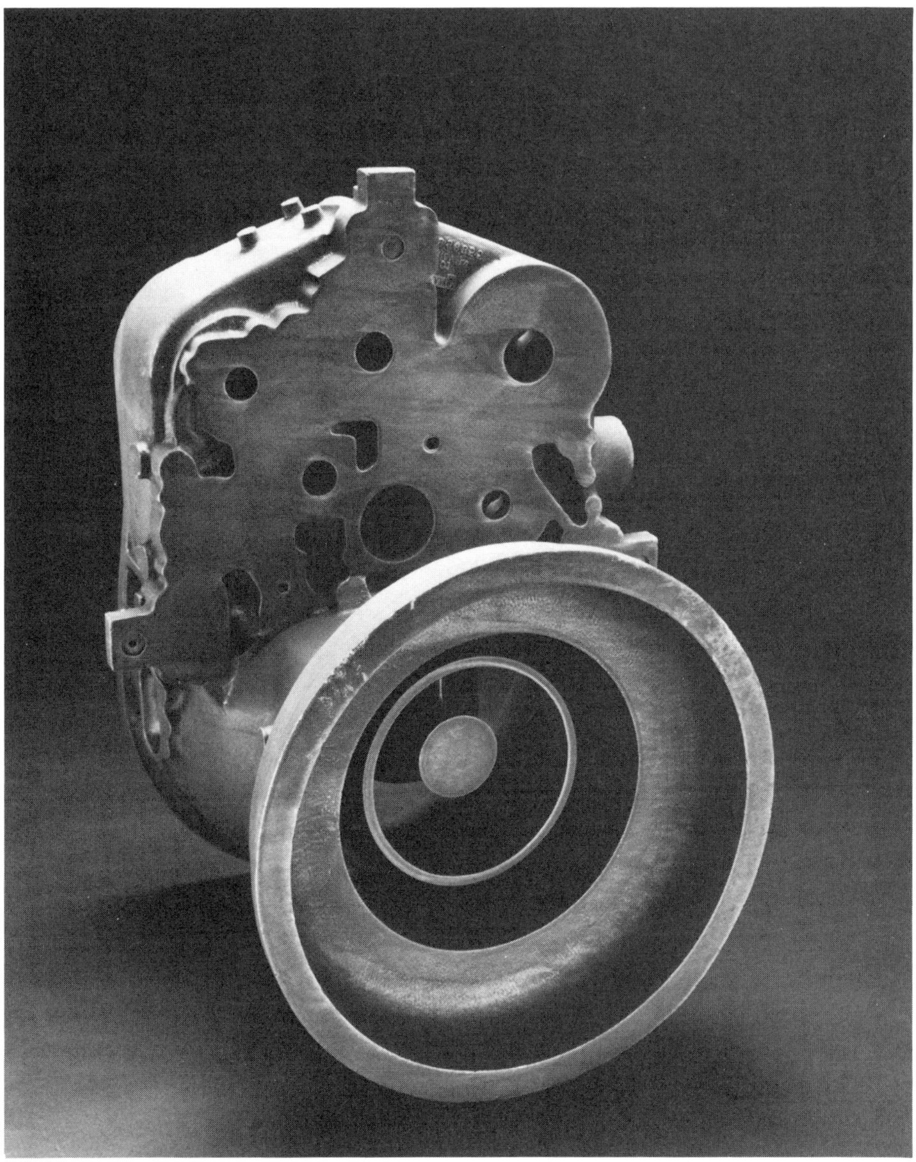

FIGURE 2.14 Gearbox for the Garrett TPE331 F-16 Turboprop engine. Made by the sand-casting process by Haley Industries, Canada. QE22A.

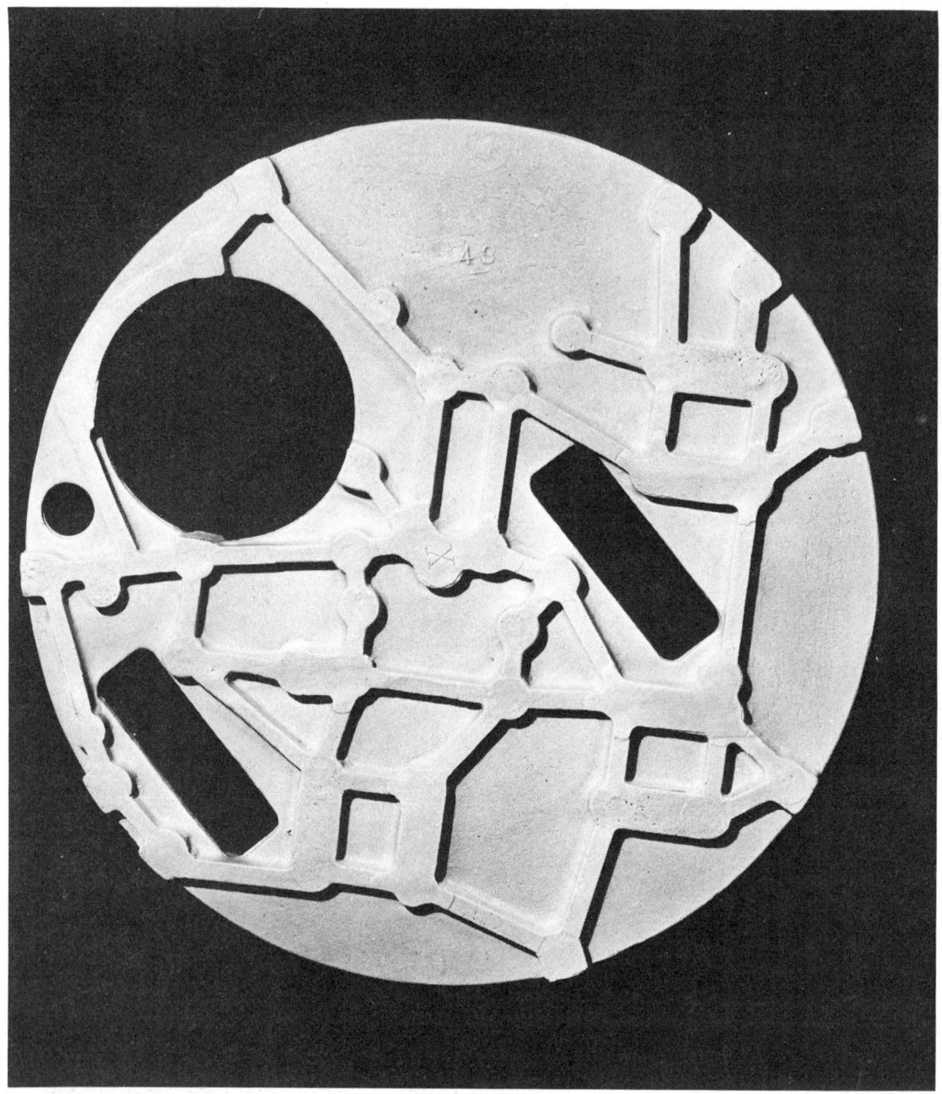

FIGURE 2.15 Bracket for mounting electronic equipment in a guided missile. Made by the sand-casting process. K1A.

FIGURE 2.16 Vibration fixtures.

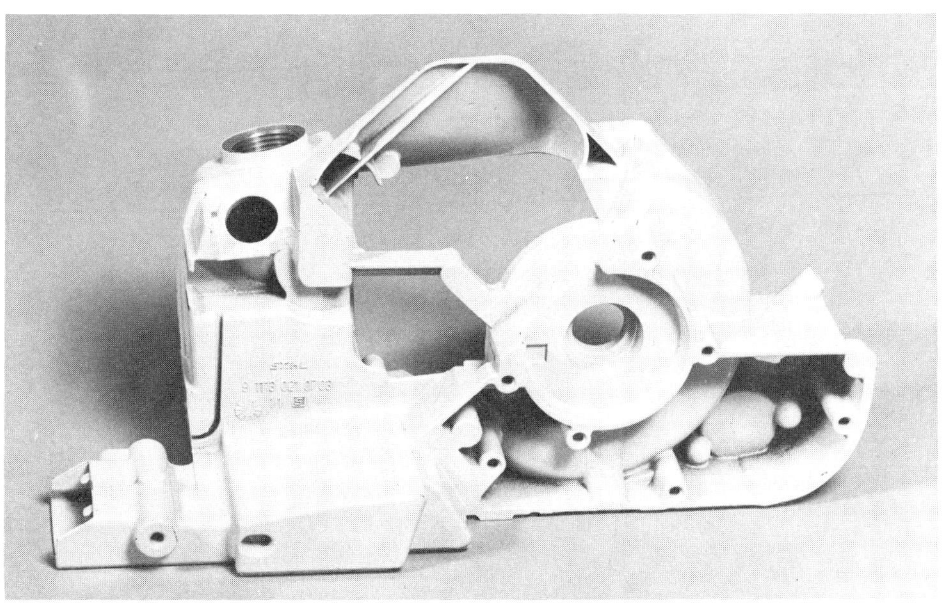

FIGURE 2.17 Combined gas tank, oil tank, and crankcase housing for a chain saw. Made by the high-pressure die-casting process by Stihl, Germany. AZ19B.

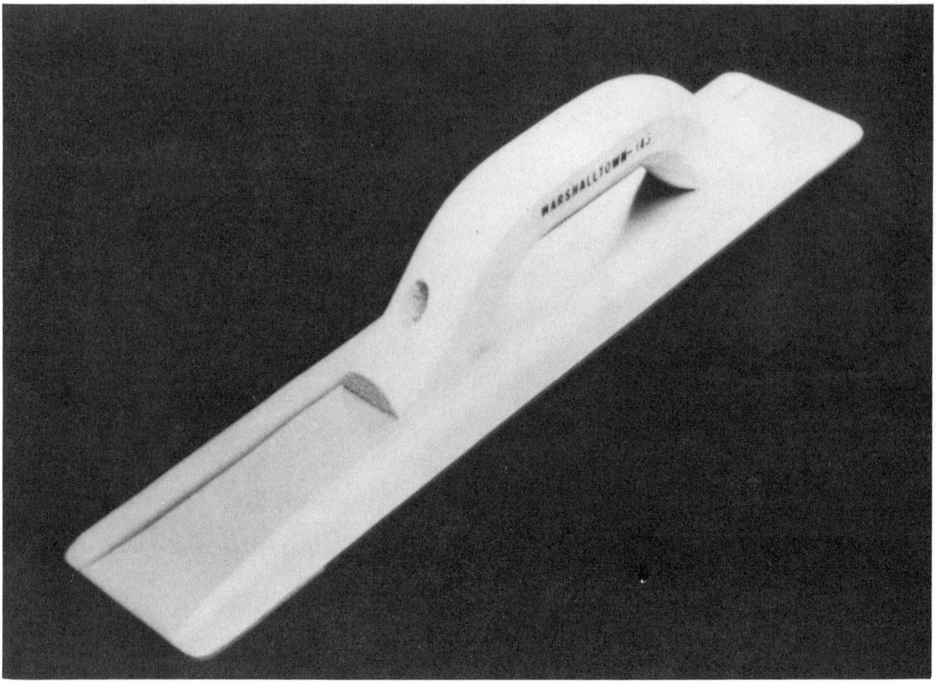

FIGURE 2.18 Concrete finishing tool.

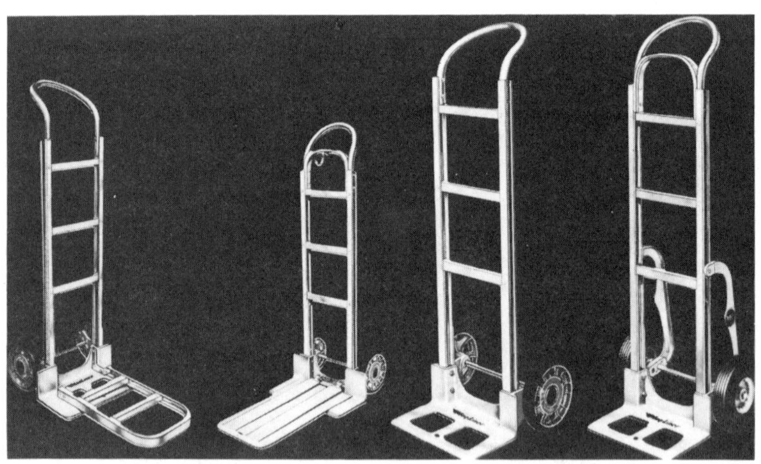

FIGURE 2.19 Hand trucks. Made by Magline, Inc., U.S.A. of AZ31B extrusions and AZ91C sand castings.

FIGURE 2.20 Bakery racks. Made by Kaiser Magnesium, U.S.A., of AZ31B extrusions.

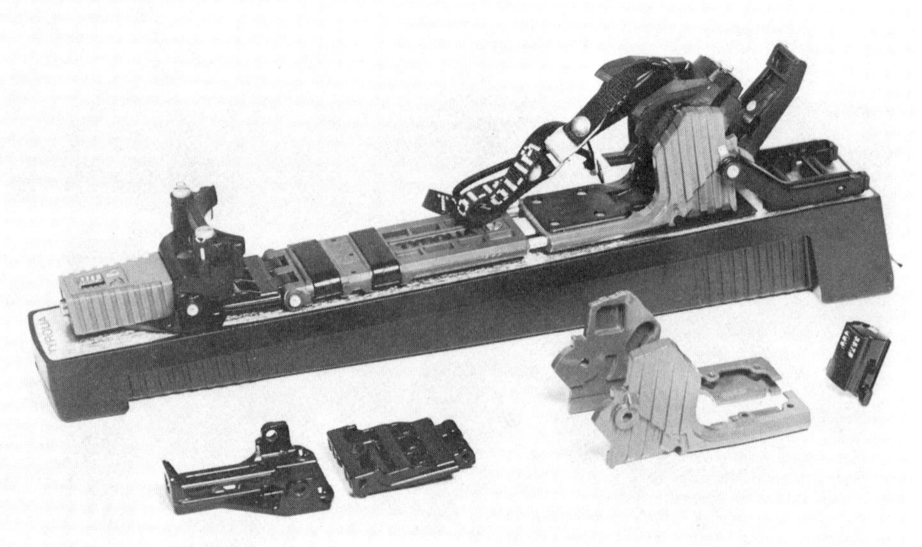

FIGURE 2.21 Ski binding for Tyrolia. Made by the high-pressure die-casting process by Peter Brockhaus, Austria. AZ81/91B.

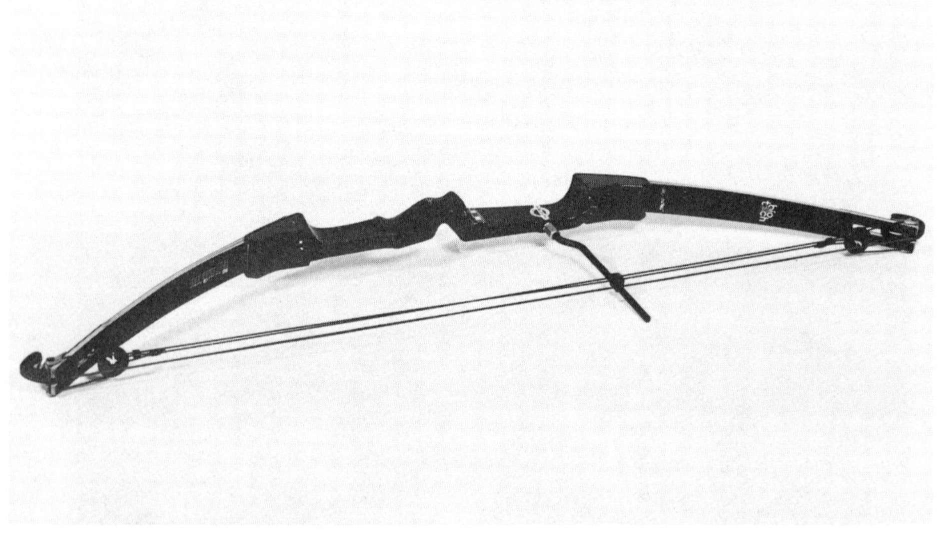

FIGURE 2.22 Handle and middle section of a bow for Bear Archery, U.S.A. made by the high-pressure die-casting process. AM60A.

FIGURE 2.23 Fly reel for Hardy Bros. Ltd., England. Made by the high-pressure die-casting process by Magnesium Castings Ltd., England. AZ81/91B.

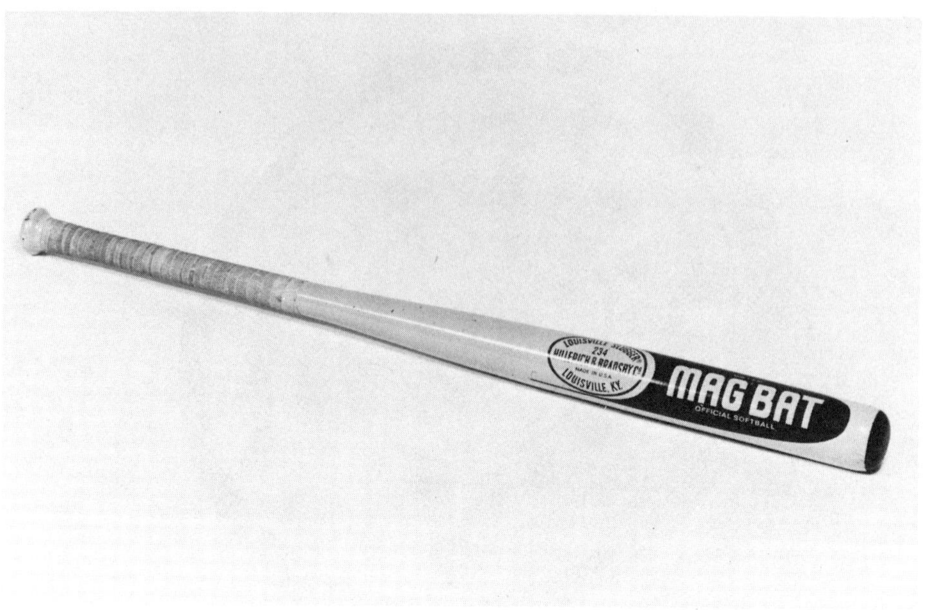

FIGURE 2.24 Baseball bat for Hillerich & Bradsby Co., U.S.A. Made by the high-pressure die-casting process. AM60A. No longer in production.

FIGURE 2.25 Tennis racquet made by Prince Manufacturing, Inc., U.S.A. Made of ZK60A extrusions.

FIGURE 2.26 Deep-sea diving suit, constructed entirely of magnesium Light weight was essential and proper finishing enabled excellent service in a very hostile environment.

2.4 SAFETY

Magnesium can be ignited, upon which it will burn with a brilliant white light at a temperature of about 2800°C. However, it is only magnesium vapor that will burn. Thus, the metal must be heated to a temperature high enough to produce a sufficient quantity of vapor to support combustion. As a practical matter, this means the metal must be melted. Pure magnesium melts at 650°C. Because the heat conductivity of magnesium is high, all of a massive piece must be raised to the melting point for the piece to burn freely. Thus it is very difficult to start a fire with a massive piece such as a casting. Even if a torch is used to raise a part of the casting to the ignition temperature and burning starts, it will cease when the torch is removed, by heat conduction to other parts of the casting and consequent lowering of the temperature below the ignition temperature. On the other hand, finely divided magnesium, such as some kinds of machining scrap, and very thin ribbon, cannot conduct the heat away from a source and burning can be initiated with a match, which has a flame temperature of about 900°C. Continuous burning of the chip or ribbon will be sustained after the match is removed. Fire hazard is therefore principally associated with finely divided magnesium, and is most apt to be encountered in machining or grinding operations, rather than in the service of a part.

Once started, a magnesium fire can be extinguished by using the normal strategies of cooling, removing oxygen, or letting the magnesium be totally consumed. Machining chips that have started to burn can be cooled effectively with cast iron chips, and this technique is used quite successfully. Since magnesium is an active chemical, it will react with oxygen preferentially to many other materials. Hence, any extinguisher that contains oxygen as a part of its chemical structure will probably support the combustion of magnesium rather than stop it. This is true of water, and one should not use water to cool the magnesium without being aware of the fact that, during cooling with water, reaction will occur, producing heat and hydrogen. The hydrogen, in turn, may react with the air to produce explosions. Water can be used, but only in large quantities in order to cool the mass of magnesium below the melting point, with recognition of the hydrogen explosion hazards that are being introduced. Water should never be used for extinguishing a fire of finely divided material, since the vigorous reaction with water will scatter the chips and spread the fire.

Oxygen can be excluded by covering with a non-reactive material, such as melting fluxes or other proprietary materials. Any oxygen-containing material, such as sand, should be avoided. In a confined volume, such as a heat-treatment furnace, boron trifluoride is an extremely effective fire extinguisher through the formation of magnesium fluoride, which excludes oxygen.

As with all oxydizable materials, an air suspension of fine magnesium powder can explode, and this is the most serious hazard associated with the burning of magnesium. Explosion will occur only if the powder is fine enough to remain suspended in the air for an appreciable period of time. As a practical matter, this means that the powder must be at least as fine as about 200 mesh, which has an average particle size of 74 μm. Thus, the only time that an engineer is apt to encounter an explosion hazard with magnesium is in handling dust created by a grinding operation.

Since magnesium will react very slowly with water, even at room temperature, to produce hydrogen, the large surface area associated with finely-divided machining chips can, when wet, produce sufficient hydrogen to cause a hydrogen explosion hazard. The worst case is a large amount of damp powder since the temperature will then rise, leading to still greater hydrogen production and even to ignition of the magnesium itself. Powder should be stored dry. If it must be wet, the amount of water should be copious and means provided for hydrogen dispersal.

In summary, fire is not a significant hazard for structures made of magnesium, since the magnesium is too massive to support combustion under service, or even abusive, conditions. It is a hazard mainly during the preparation of structures, such as during grinding, machining, or heat treatment.

REFERENCES

1. Adolph Beck, "The Technology of Magnesium and its Alloys," F. A. Hughes and Co., Limited, London, 1940.
2. Stephen C. Erickson, "Magnesium's High Damping Capacity for Automotive Noise and Vibration Attenuation," *Proceedings of the International Magnesium Association*, 1975.
3. J. A. Gusack and D. E. Ritzema, "Magnesium in Focus," *Proceedings of the Magnesium Association*, 1961.
4. J. W. Fredrickson, "Pure Magnesium" in *Metals Handbook*, American Society for Metals, Metals Park, Ohio, 1949.
5. K. G. Sumner, "Magnesium in the British Nuclear Industry," *Proceedings of the International Magnesium Association*, 1980.

3

Nomenclature and Specifications

3.1 NOMENCLATURE

3.1.1 Composition

There are many systems of nomenclature for alloys used in the world. However, because of its simplicity and ready identification with composition, the ASTM system will be used as the basis for naming alloys throughout this book [1]. In the ASTM system, one or at most two letters are used to indicate the principal alloyed elements followed by the weight percent of each of those elements present in the alloy to the nearest whole percentage. The letters representing the elements are listed in order of the amount of the element present, the highest being first. In the event that both elements are present in the same amount, the letters are listed alphabetically. Following the two letters and the numbers is a serial letter assigned chronologically in order to differentiate among otherwise identical names. The letters assigned to elements are:

A: Aluminum
B: Bismuth
C: Copper
D: Cadmium
E: Rare Earths (Mischmetal)
F: Iron
G: Magnesium
H: Thorium
K: Zirconium
L: Lithium
M: Manganese
N: Nickel
P: Lead
Q: Silver
R: Chromium
S: Silicon
T: Tin
W: Yttrium
Y: Antimony
Z: Zinc

Because the system is designed for use with both aluminum and magnesium alloys, there are elements in the list, such as iron, that are not used as alloying elements in magnesium.

In this book, the full ASTM designation, consisting of one or two letters, numbers representing amount of addition, and the serial letter will be used for specific alloys. If the specific alloy in question is in an ASTM specification, then the letter used will be the same as used by the ASTM. However, if the alloy is not in any ASTM specification, an arbitrary final letter will be assigned. When a generic alloy is being discussed, the serial number will be omitted. For example, there are four variants of the magnesium alloy containing 9% Al and 1% Zn as principal alloying additions: AZ91A, AZ91B, AZ91C, and AZ91D. If the generic alloy is meant, it will be named as AZ91. The names and nominal compositions for all of the magnesium alloys cited in the book are listed in Table 3.1.

3.1.2 Temper

The temper of an alloy refers to the state of heat treatment or cold work that is present. For castings, only variations of heat treatment are used to affect properties; for wrought products, variations of cold work, annealing, and heat treatment are used alone or in combination.

Since the ASTM system is being used for composition designation, the ASTM method for designating temper will also be used in this book [2]. The designations used and their meanings are shown in Table 3.2.

There are other combinations used, but those in Table 3.2 are the most common and will appear in this book. Reference [2] should be consulted for full details.

The full name of an alloy consists of both the composition and the temper designation. Thus the designation ZK60A-T5 refers to the first variant of an alloy containing 6% Zn + 0-0.5% Zr which has been stabilized after cooling from a shaping temperature. If the temper is not important, and only the composition is being referred to, the temper designation is omitted.

TABLE 3.1 Nominal Compositions of Alloys (Weight Percent)

Name	Al	Cu	Mn	Rare earths	Si	Ag	Th	Y	Zn	Zr	Remarks
AM60A	6.0		0.15								
AM100A	9.9		0.10								
AS41A	4.2				1.0						
AZ31B	3.0		0.20						1.0		Fe, Cu, Ni low
AZ31C	3.0		0.20						1.0		
AZ61A	6.5		0.15						1.0		Fe, Cu, Ni low
AZ61B	6.5		0.15						1.0		
AZ61C	6.0		0.3						1.0		
AZ63A	6.0		0.15						3.0		
AZ80A	8.5		0.12						0.5		Fe, Ni low
AZ81A	7.5		0.13						0.7		
AZ81B	8.0		0.3						0.6		
AZ81C	8.4		0.5						0.90		
AZ81D	8.0		0.4						0.6		Cu, Fe, Ni, Si low
AZ81E	8.4		0.2						0.6		Fe, Mo low
AZ91A	9.0		0.13						0.68		Cu low
AZ91B	9.0		0.13						0.68		
AZ91C	8.7		0.13						0.70		Cu low

TABLE 3.1 (*Continued*)

Name	Al	Cu	Mn	Rare earths	Si	Ag	Th	Y	Zn	Zr	Remarks
AZ91D	9.0		0.13						0.68		Fe, Ni, Cu low
AZ92A	9.0		0.10						2.0		
AZ101A	9.8								0.6		
EQ21A[c]				2.2[a]		1.5				0.7	
EZ33A				2.8					2.6	0.7	
HK31A							3.2			0.7	
HM21A			0.8				2.0				
HZ11A							0.8		0.6	0.6	
HZ32A							3.2		2.1	0.7	
K1A										0.7	
M1A			1.2								
QE22A				2.2[a]		2.5				0.7	
QH21A				1.0[a]		2.0	1.1			0.7	

WE54A[c]			3.5[b]		5.25		0.5
ZC61A	1.2	0.7				6.5	
ZE10A			0.17			1.2	
ZE41A			1.2			4.2	0.7
ZE63A			2.6			5.8	0.7
ZH11A				0.75		0.5	0.6
ZH62A				1.8		5.7	0.7
ZK10A						1.3	0.6
ZK30A						3.0	0.6
ZK40A						4.0	0.7
ZK51A						4.6	0.7
ZK60A						5.5	0.4
ZK61A						6.0	0.8
ZM21A		1.2				2.2	

[a]Rare earth metals present as Didymium (normally present as Mischmetal).
[b]Present as Neodymium plus heavy rare earths.
[c]Alloy patented by Magnesium Elektron, Ltd.

TABLE 3.2 ASTM Temper Designations and Their Meanings

Temper designation	Meaning
F	As fabricated, whether cast or wrought.
O	Annealed, recrystallized wrought products. The softest temper.
H	Strain-hardened wrought products. Always followed by two or more digits to indicate the degree of strain hardening present.
T	Thermally treated to produce stable tempers other than F, O, or H. Always followed by one or more digits to indicate the specific treatment.
Subdivisions	
H1	Strain hardened only. Followed by a digit to indicate the amount, from 0 (annealed) to 8 (full hard).
H2	Strain hardened and then partially annealed. Followed by a digit to indicate the amount, from 0 (annealed) to 8 (full hard).
H3	Strain hardened and then stabilized. Followed by a digit to indicate the amount of strain remaining, from 0 (annealed) to 8 (full hard).
T4	Solution heat treated and aged at room temperature to a substantially stable condition. Most magnesium alloys remain in the metastable, single-phase, fully solution heat treated state.
T5	Cooled from an elevated-temperature shaping process (e.g., casting or extrusion) and then artificially aged at an elevated temperature to improve properties or stability.
T6	Solution heat treated and then artificially aged at an elevated temperature.
T7	Solution heat treated and then stabilized at an elevated temperature.
T8	Solution heat treated, cold worked, and then artificially aged at an elevated temperature.

3.2 SPECIFICATIONS

There are many agencies that have written specifications dealing with magnesium. The major ones in the western world are:

Germany
- DIN—Deutsche Norman

Great Britain
- BSI—British Standards Institute
- L Series—Aerospace Specifications

International
- ISO—International Standards Organization

Japan
- JIS—Japanese Standards

United States
- ASTM—American Society for Testing and Materials
- SAE—Society of Automotive Engineers
- AMS—Aerospace Material Specifications of the Aerospace Materials Division of the SAE
- MIL—U. S. Military
- Federal—U. S. Government agencies (civil)

Tables 3.3, 3.4, and 3.5 are tabulations of current specifications for magnesium products and processes.

TABLE 3.3 Specifications Covering Magnesium Products

Product	Alloy	AMS	ASTM	BSI	DIN	Federal	ISO	L series	JIS	Military	SAE
Primary ingot and stick	9980A		B92				114				
	9980B		B92								
	9990A		B92		17800						
	9995A		B92		17800						
	9998A		B92								
Alloy ingot	AM100A		B93						2221		
	AM60A		B93								
	AS41A		B93								
	AZ63A		B93				121		2221		
	AZ81A		B93								
	AZ81B			2970			121				
	AZ81C			2970			121				
	AZ81D			2970							
	AZ91A		B93						2221 2222		
	AZ91B		B93					3L125	2222		
	AZ91C		B93				121				
	AZ91D		B93								
	AZ92A		B93				121		2221		
	AZ101A			2970							
	EZ32A			2970							
	HZ32A			2970							
	ZE41A			2970							
	ZH62A			2970							
	ZK414			2970							

Castings: sand	AM60A				1729(2)						
	AM100A		B80						5203		J465
	AZ63A	4420 4422 4424	B80			QQ-M-56	121		5203		J465
	AZ81A		B80		1729(2)	QQ-M-56					J465
	AZ81B			2970			121				
	AZ81C			2970			121				
	AZ81D			2970							
	AZ91C	4437	B80		1729(2)	QQ-M-56	121		5203	MIL-M-46062	J465
	AZ92A	4434	B80			QQ-M-56	121		5203	MIL-M-46062	J465
	AZ101A			2970							
	EK41A	4434 4440 4441									
	EZ32A			2970			3115	2L126			
	EZ33A	4442	B80		1729(2)	QQ-M-56			5203		J465
	HK31A	4445	B80			QQ-M-56				MIL-M-46062	J465
	HZ32A	4447	B80	2970	1729(2)	QQ-M-56	3115				J465
	K1A		B80								J465
	QE22A	4418	B80		1729(2)	QQ-M-56	3115			MIL-M-46062	J465
	QH21A		B80								
	ZE41A	4439	B80	2970	1729(2)			2L128			J465
	ZE63A	4425	B80							MIL-M-46062	J465
	ZH62A	4438	B80	2970	1729(2)	QQ-M-56	3115			MIL-M-46062	J465
	ZK41A			2970			3115				
	ZK51A	4443	B80			QQ-M-56			5203	MIL-M-46062	J465
	ZK61A						3115		5203		
Castings: gravity die	AM100A	4483	B199			QQ-M-55				MIL-M-46062	J465
	AZ63A					QQ-M-55					

TABLE 3.3 *(Continued)*

Product	Alloy	AMS	ASTM	BSI	DIN	Federal	ISO	L series	JIS	Military	SAE
	AZ81A		B199		1729(2)	QQ-M-55					J465
	AZ91C		B199		1729(2)	QQ-M-55				MIL-M-46062	J465
	AZ92A	4484	B199			QQ-M-55				MIL-M-46062	J465
	EZ33A		B199			QQ-M-55					J465
	HK31A		B199			QQ-M-55					J465
	HZ32A					QQ-M-55					J465
	QE22A		B199			QQ-M-55					J465
Castings: pressure die	AZ91A	4490	B94			QQ-M-38			5303		J465
	AZ91B		B94						5303		J465
	AZ91D		B94								
	AM60A		B94		1729(2)						J465
	AS41A		B94		1729(2)						J465
	AZ61A				1729(2)						
Castings: investment	AM100A	4455	B403								J465
	AZ81A		B403								J465
	AZ91C	4452	B403								J465
	AZ92A	4453	B403								J465
	EZ33A		B403								J465
	HK31A		B403								J465
	K1A		B403								J465
	QE22A		B403								J465
	ZE63A		B403								J465
	ZK61A		B403								J465

Sheet and plate	AZ31B	4375	B90		9715	QQ-M-44			4201		J466
		4376									
		4377									
	AZ31B[a]									MIL-F-46048	
	AZ31B[b]	4382									
	AZ31C		B90	3370			3116				
	AZ61B						3116				
	AZ81E						3116				
	HK31A	4384	B90							MIL-M-26075	J466
		4385									
	HM21A	4383	B90							MIL-M-8917	J466
		4390									
	M2A		3370								
	ZE10A		B90							MIL-M-46037	J466
	ZM21A			3370							
	ZK11A			3370			3116				
	ZK31A			3370			3116				
	ZK61A						3116				
Extruded rods and shapes	AZ10A					QQ-M-31					
	AZ31B		B107			QQ-M-31			4203		J466
									4204		
	AZ31C		B107	3373	1729(1)		3116				
	AZ61A	4350	B107			QQ-M-31		L.51283	4203		J466
								2L.503	4204		
	AZ61B						3116				
	AZ61C			3373	1729(1)						
	AZ80A		B107			QQ-M-31			4203		J466
									4204		
	AZ81E				1729(1)		3116				

TABLE 3.3 (*Continued*)

Product	Alloy	AMS	ASTM	BSI	DIN	Federal	ISO	L series	JIS	Military	SAE
	HM31A	4388 4389								MIL-M-8916	J466
	M1A		B107		1729(1)	QQ-M-31					J466
	M2A			3373							
	ZK11A			3373			3116		4203 4204		
	ZK21A	4387						2L.505		MIL-M-46039	
	ZK31A			3373			3116	L.514	4203 4204		
	ZK40A		B107								J466
	ZK60A	4352	B107			QQ-M-31					J466
	ZK61A			3373			3116		4203 4204		
	ZM21A			3373							
Tubes	AZ31B		B107			WW-T-825			4202		J466
	AZ31C		B107	3373	1729(1)		3116				
	AZ61A	4350	B107		1729(1)	WW-T-825			4202		J466
	AZ61B						3116				
	AZ61C			3373							
	AZ80A		B107								
	AZ81E				1729(1)		3116				
	M1A		B107		1729(1)	WW-T-825					J466
	M2A			3373							
	ZK11A			3373			3116		4202		
	ZK21A	4387									

	ZK31A			3373			3116		
	ZK40A		B107						J466
	ZK60A	4352	B107			WW-T-825			J466
	ZK61			3373			3116		
	ZM21A			3373					
Forgings	AZ31B		B91			QQ-M-40			J466
	AZ31C				1729(1)		3116		
	AZ61A	4358	B91			QQ-M-40			J466
	AZ61C			3372	1729(1)		3116		
	AZ80A	4360	B91			QQ-M-40			J466
	AZ81E				1729(1)		3116		
	HM21A	4363	B91			QQ-M-40			J466
	M1A					QQ-M-40			
	M2A			3372					
	ZK11A			3372			3116		
	ZK31A			3372			3116		
	ZK61A			3372			3116		
	ZM21A			3372					
Welding rod	AZ61A							MIL-R-6944	J466
	AZ92A	4395						MIL-R-6944	
	AZ101A							MIL-R-6944	
	EZ33A	4396						MIL-R-6944	
Brazing rod	AZ91A					QQ-B-655			
	AZ125A					QQ-B-655			

[a] Tooling plate.
[b] Tread plate.

TABLE 3.4 Process Specifications

Process	AMS	Federal	JIS	Military
Finishing and painting				
Cleaning, weapons				MIL-S-5002
Chemical treatment	2475		8651	MIL-M-3171
HAE coating	2476		8651	MIL-M-45202
Plating to solder	2421			
Treatment and painting				MIL-T-704
Dow #17	2478		8651	MIL-M-45202
Anodic cleaning				MIL-M-3171
Selenious acid			8651	
Manganese phosphate			8651	
Inspection				
General specifications		151		
Quality assurance	2355			
Quality control				MIL-Q-9858
Marking, shipment, and Storage				
Identification marking		184 123		MIL-STD-129
Identification	2815			
Welding wire	2816			
Shipment and atorage				MIL-STD-649
Tolerances				
Extrusions	2205	245		
Sheet and plate	2202	245		
Forgings		245		
Welding				
Arc welding				MIL-W-18326
Gas welding				MIL-W-18326
Radiographic inspection				MIL-R-45774

TABLE 3.5 Miscellaneous Specifications

Subject	AMS	ASTM	BSI	DIN	Federal	ISO	JIS	Military	SAE
Codification		E527 B275							
Temper designation		B296							
Chemical analysis		E35							
Sampling: wrought		E55				2142			
Sampling: cast		E88							
Reference radiographs		E155 E505							
Tension testing		B557 B557M		50125 50145 50148					
Heat treatment		B661						MIL-M-6857	
Reference test bar and castings						2377			

3.3 UNITS

Systems of units for measurements of physical properties proliferated over the centuries, with very little consistency either with one another, or even internally. In 1960, a universal internally consistent system known as Le Système International d'Unités (SI) was adopted and is being maintained by the Conférence Générale des Poids et Mesures. The SI system of units has been widely adopted and will therefore be the basis for the units used in this book. However, adoption of SI units in the United States has been slow, with most engineers still more comfortable with the British inch-pound units. Because of the importance of the technology in the United States, inch-pound units for stress (kilopound-force per square inch, or ksi) will be used in conjunction with SI units where feasible. For convenience, Table 3.6 gives conversion factors for those units used most frequently in the book [3]. In those tables where both units are not used, the pertinent conversion factors are given in the tables.

Only the centigrade scale is used for temperature.

TABLE 3.6 Conversion Factors

To convert from	To	Multiply by
Atmosphere (standard)	Pascal (Pa)	1.013×10^5
Atmosphere (technical)	Pascal (Pa)	9.806×10^4
Btu (thermochemical)	Joule (J)	1.054×10^3
Btu (thermochemical)/lb.	Joule per kilogram (J/kg)	2.324×10^3
Btu (thermochemical)/(lb.°F)	Joule per kilogram kelvin (J/(kg. K))	4.184×10^3
Btu. ft/(h.ft^2.°F) (thermal conductivity)	Watt per metre kelvin (W/(m. K))	1.730
Calorie (thermochemical)	Joule	4.184
Dyne	Newton (N)	1.000×10^{-5}
Foot	Metre (m)	3.048×10^{-1}
Ft. lbf	Joule (J)	1.356
Free fall (standard g)	Metre per second squared (m/s^2)	9.806
Gallon (U.S. liquid)	Cubic meter (m^3)	3.785×10^{-3}
Horsepower (550 ft. lbf/s)	Watt (W)	7.457×10^2

TABLE 3.6 (*Continued*)

To convert from	To	Multiply by
Horsepower (electric)	Watt (W)	7.460×10^{2}
Inch	Metre	2.540×10^{-2}
kgf/cm^2	Pascal (Pa)	9.807×10^{4}
kW. h	Joule (J)	3.600×10^{6}
kip/in^2	Pascal (Pa)	6.895×10^{6}
kip/in^2	Megapascal (MPa)	6.895
Pound (avoirdupois)	Kilogram (kg)	4.536×10^{-1}
Pound-force (lbf)	Newton (N)	4.448
lbf/in^2 (psi)	Pascal (Pa)	6.895×10^{3}

Note: Symbols of SI units are given in parentheses.

REFERENCES

1. "Standard Practice for Codification of Certain Nonferrous Metals and Alloys, Cast and Wrought," ASTM B275 (1980).
2. "Standard Recommended Practice for Temper Designations of Magnesium Alloys, Cast and Wrought," ASTM B296 (1983).
3. "Standard for Metric Practice, " ASTM E380 (1979).

II
Fabrication

The operations of machining, joining, and forming are used to put a part into its final form. The designer must be familiar with these operations in order to design the best and most economical part for his purposes.

Magnesium is the easiest of metals to machine, and the machining step in the fabrication of a part is often important enough to result in a lower cost for the part if made in magnesium than in another material. Chapter 4 gives details on machining practice for magnesium alloys.

Magnesium can be joined to itself or to other materials by all of the common methods. Welding, in particular, is well suited to the properties of magnesium alloys, and high-quality welds of good properties are common. Chapter 5 gives information important to the designer on the many methods of joining parts.

Forming by bending, drawing, or forging are commonly applied. techniques, and details of magnesium practice are given in Chapter 6.

4

Machining

The information in this chapter is derived in large part from The American Society for Metals [1], The Dow Chemical Company [2], and Norsk Hydro [3].

The machineability of magnesium is superior to that of any other metal. More metal can be removed per unit of time or per unit of power; surface finish is smoother for any given set of conditions; deeper holes can be drilled; and tools retain their sharpness for a longer time.

4.1 TOLERANCES

If very close tolerances are required, then the high coefficient of thermal expansion (25.0 microunit strain/°C at 100°C) must be taken into account. It may be necessary to use low cutting speeds and small feeds to minimize any temperature increase.

Because magnesium has a low modulus of elasticity (4.48 Gpa; 6,500 ksi), elastic distortion during clamping can occur more easily than with other metals. This must be guarded against by, for example, using light clamping pressures, clamping on heavy sections, and using shims between the part and the tool bed.

When heavy cuts are taken, the surface, to a depth of as much as 0.5 mm (20 mils), can be heavily cold worked. Cold work results in elastic stresses which will be relieved with time through conversion to plastic strain. The time required for relief is proportional to the temperature, but can take place over days or weeks even at room temperature, resulting in distortion of the part. If the distortion that occurs is too great, it may be necessary to insert a stress-relief treatment between the last rough cut and a final finishing cut, since the latter will be free of significant cold work. The time and temperatures required for proper stress relief depend upon the alloy, form, and temper. Later chapters in the book should be consulted for the correct treatments for each case. This problem is not as acute for magnesium as for many other metals because the ease of chip formation does not lead to large stresses.

4.2 LUBRICATION

While magnesium is usually machined dry, there are times when a cooland or lubricant is useful. If such is the case, water-soluble oils, animal, or vegetable oils should not be used. Only mineral oils are suitable for use on magnesium, with properties as given in Table 4.1. Chips that are wet with water are a particular hazard during storage because the slow reaction between the magnesium and the water can produce both heat and hydrogen. The heat buildup can even be great enough to cause ignition; if this occurs, and the hydrogen has been trapped, an explosion can result.

TABLE 4.1 Properties of Recommended Coolants

Specific gravity	0.79 to 0.86
Viscosity (Saybolt) at 38°C	Up to 55 sec
Flash point (closed cup), min	135°C
Saponification No., max	16
Free acid, max	0.2%

4.3 COMPARISON WITH OTHER METALS

As a background for the development of tool and part design in later sections of this chapter, comparisons of the machining characteristics of magnesium with other metals are helpful in understanding the properties specific to magnesium.

The single-point tool used for turning will be used as the basis of comparison, and the standard conventions for describing such a tool will be used as shown in Figure 4.1. Specification of the six angles and the radius of curvature at the nose is sufficient to describe any tool completely.

Table 4.2 compares the speed of machining and the power requirements for steel, aluminum, and magnesium. Aluminum can be machined considerably more rapidly than steel, and magnesium as much more rapidly than aluminum. If machined at the same speed as aluminum, the metal removed is far larger for magnesium than for aluminum. Power requirements are lower for magnesium than for steel or aluminum. The important implications of Table 4-2 are that during the machining of magnesium, copious quantities of chips are produced in a given time, and that only low forces are required for machining. Both factors enter into tool design.

In Table 4.3, there is a comparison of nominal designs for a single-point tool to be used for roughing or finishing work for the three metals: mild steel, aluminum, and magnesium. The numbers given are for "average" conditions and would be modified in detail for any given machining job for any of the metals. However, they illustrate very well the differences among the metals.

Magnesium differs from either steel or aluminum in the low power required for generating chips, in the large volume of chips produced in a unit time, in the good surface finish produced even with relatively rough cuts, and in the necessity for preventing friction and heat build-up in order to minimize fire hazard. All of these differences are illustrated by the figures in Table 4.3.

The purpose of back rake is to turn the chips away from the work. Both magnesium and aluminum use a larger angle than does steel, but do not differ greatly from each other. With no back rake, any ribbon produced is spiralled very tightly. As the angle increases, the helix produced is longer. Since magnesium characteristically produces chips rather than coils, the high back rake angle promotes freer chipping.

Side rake controls the strength of the cutting edge, with a small angle being stronger than a large angle. At the same time, the smaller the angle, the larger are the forces required for machining. Since the forces required for machining magnesium are low, the side rake angle can be low for maximum strength without undue power penalty.

The end relief angle provides clearance between the tool and the finished surface. If this angle is too small, rubbing of the tool on the part occurs. If the angle is too large, digging can result. Since rubbing, with its accompanying heat generation and potential fire hazard, is to be avoided with magnesium, and since digging is less likely than with other metals, the end relief angle is kept large.

The side relief angle provides clearance between the cut surface and the tool flank. If the angle is too small, rubbing takes place; if too large, the cutting edge is weak. Since the forces required for machining magnesium are low, this angle can be quite large to prevent rubbing, and yet retain sufficient strength at the cutting edge for machining. During finishing operations, heat generated is relatively low and the angle for magnesium can be lower.

The side cutting edge acts to turn the chip away from the finished surface. As the angle increases, the chip width increases and the chip thickness decreases. Magnesium chips should be as large as practical to minimize fire hazard, so this angle is kept large.

A large nose radius can cause chatter, but is desirable for good finish. Since the finish on magnesium after machining is excellent, even under poor machining conditions, the nose radius can be kept low to prevent any possibility of chatter, with no loss of smoothness. This is true both during roughing and during finishing operations.

The general principles illustrated in Table 4.3 are utilized in all tool designs for the machining of magnesium. The objective is to minimize rubbing by providing adequate clearance, to maximize chip size, to provide for the rapid accumulation of chips, to take advantage of the low forces required for machining, and to capitalize on the smooth surface produced during machining.

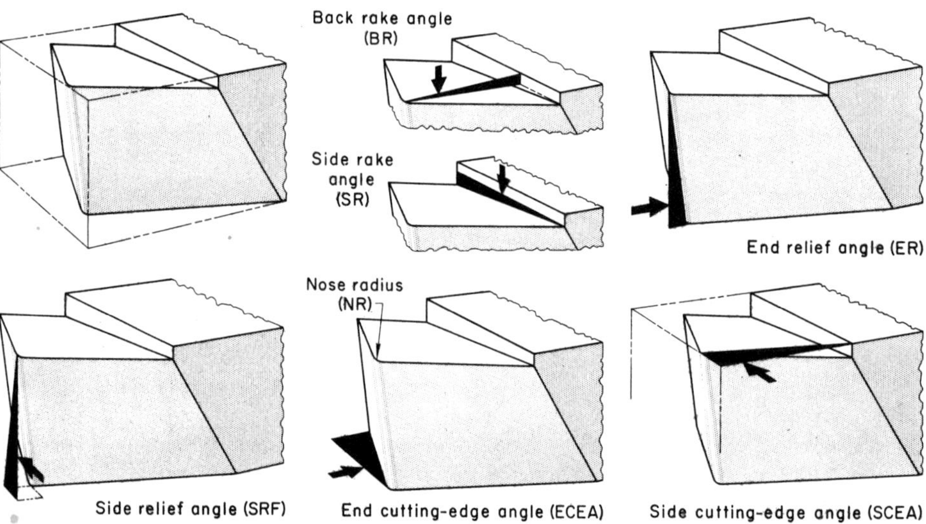

FIGURE 4.1 Standard nomenclature and abbreviations for tool angles. Back rake angle (BR) is the angle between the cutting face of the tool and the shank or holder, measured parallel to the side of the shank or holder. The angle is positive if, as in the sketch above, it slopes from the cutting point downward toward the shank, and negative if it slopes upward toward the shank. Side rake angle (SR) is the angle between the cutting face of the tool and the shank or holder, measured perpendicular to the side of the shank or holder. The angle is positive if, as in sketch above, it slopes downward away from the cutting edge to the opposite side of the shank, and negative if it slopes upward.

TABLE 4.2a Nominal Feeds and Speeds for Machining with a Single-Point Turning Tool (Roughing)

	Cut depth		Feed		Speed (surface distance/min)	
Material	mm	in	mm/rev	in/rev	Meters	Feet
Steel						
100 Brinnell	3.81	0.150	0.38	0.015	40	130
250 Brinnell	3.81	0.150	0.38	0.015	23	75
Aluminum						
Simple cast	3.81	0.150	0.38	0.015	230	755
All other	3.81	0.150	0.38	0.015	180	590
Magnesium	3.81	0.150	0.50	0.020	1100	3600
	12.70	0.500	2.50	0.098	180	590

TABLE 4.2b Volume Removable by Unit Horsepower/Minute

Material	Cubic centimeters	Cubic inches
Mild steel	18	1.1
Aluminum	60	3.7
Magnesium	110	6.7

FIGURE 4.1 (*Continued*)
End relief angle (ER) is the angle between the end face of the tool and a line drawn from the cutting edge perpendicular to the base of the shank or holder, and usually is measured at right angles to the end cutting edge. Side relief angle (SRF) is the angle between the side flank immediately below the side cutting edge and a line drawn through the side cutting edge perpendicular to the base of the tool or tool holder, and usually is measured at right angles to the side flank. End cutting-edge angle (ECEA) is the angle between the end cutting edge of the tool and a line perpendicular to the side of the shank. Side cutting-edge angle (SCEA), called also "lead angle," is the angle between the side cutting edge and the projected side of the shank or holder. Nose radius (NR) is the radius on the tool between the end and the side cutting edges. From Ref. 1.

TABLE 4.3 Typical Design for Single-Point Turning Tools

Material	Degrees						Nose radius	
	Back rake	Side rake	End relief	Side relief	End cutting edge	Side Cutting edge	mm	in
Roughing:								
Steel	8	14	6	6	6	0	1.63	0.064
Aluminum	20	20	10	10	5	10	1.63	0.064
Magnesium	15	5	15	15	30	30	1.00	0.040
Finishing:								
Steel	8	14	6	6	6	0	1.63	0.064
Aluminum	20	20	10	10	5	10	5.10	0.200
Magnesium	15	8	15	15	15	30	1.00	0.040

4.4 TOOL DESIGN

4.4.1 Material and Finish

Any tool material, including ordinary carbon steel, can be used successfully for machining magnesium. However, high-speed tool steels maintain sharpness enough longer to justify the larger cost even for relatively short machining periods. Carbide-tipped tools are preferred for lengthy machining operations, because of the still longer life. While diamond-tipped tools are seldom necessary, they are occasionally used for cases requiring extremely fine finish.

The cutting edge of the tool must be ground to as sharp a finish as possible, free of burrs, scratches, or wire edges. If the tool has been used for the machining of any other metal, it should be reground and rehoned before being used for magnesium.

It is important that all frictional surfaces and surfaces over which chips pass be very smooth to prevent any buildup of chips and accumulation of heat. After grinding on fine wheels (at least 100-grit aluminum oxide for high-speed steels and 200- to 320-grit silicon carbide or diamond for carbide-tipped), the cutting edges and the surfaces should be hand honed to remove all grinding marks. The finer the surface on the tool, the better will be the surface finish on the part and the longer will the tool last. With proper preparation, the time between sharpening can be as much as five to ten times longer than that required when machining aluminum.

4.4.2 Turning and Boring

The range of tool geometries which are commonly used are shown in Table 4.4. Speeds, feeds, and depths of cut that can be used are shown in Table 4.5. Using these conditions, surface finishes of 250 μmm (10 μin) or less can be obtained readily in production. Significantly finer finishes of 50-100 μmm (2-4 μin) require careful attention to tool sharpness, use of the lower speeds and feeds, larger nose radii, and perhaps the use of diamond-tipped tools.

Parting or grooving operations need good clearance along the side of the tool to prevent rubbing on the cut surfaces, a rake angle of about 5° being adequate. When parting or grooving, the tool should be advanced into the work as rapidly as possible, and removed promptly and rapidly after the cut is made.

4.4.3 Planing and Shaping

Single-point tools are used for planing and shaping, and the geometry of the tool is the same as is used for turning and boring. Because speeds are necessarily lower, both depths of cut and feeds can be high, commonly as much as 12.5 and 25 mm (0.5 to 1 in) per stroke respectively. The potential effects of severe cold work as discussed in Section 4.1 should be taken into account when these heavy depths and feeds are used.

4.4.4 Drilling

The ease of machining of magnesium is especially noticeable when drilling. Holes with a depth of 20 times the hole diameter are readily produced without removing the drill from the hole to remove chips. Since drilling speeds are also high, this means that chips are produced rapidly and in large quantity. It is critical that the flutes be highly polished to assure that the chips will be removed readily as drilling proceeds.

The general design of a twist drill is shown in Figure 4.2. The nomenclature of this figure will be used in describing drills in this section.

The design features for three general classes of twist drills for magnesium are given in Table 4.6, depending upon whether the hole to be drilled is shallow (up to 5 times the hole diameter), deep (over 5 times the hole diameter), or to be in thin sheet. As with all tools, it is important to keep the cutting edge sharp, to avoid excessive burring, undersize holes, and too much heating. The corners of the cutting lips are rounded to improve the surface finish.

The general parameters of speeds and feeds are given in Table 4.7. When tools are properly maintained, and the conditions given in Table 4.7 are followed, as much as 1300 m (4,250 ft) can be drilled before sharpening. More usually, resharpening is needed after 150 to 400 m (500 to 1300 ft) of drilling.

4.4.5 Milling

The nomenclature for milling cutters and for the tooth design of milling cutters is shown in Figure 4.3. Values for magnesium are given in Table 4.8. Milling speeds and feeds are given in Table 4.9. As can be seen, the general principles of good clearances and provision for chip flow as described in Section 4.3 are followed for the design of milling cutters. Speeds and feeds should be as high as is consistent with a good surface finish. Typically, both rough and finish milling is done at speeds of 300 to 900 surface m/min (1000 to 3000 ft/min).

The high speed at which magnesium is milled results in copious quantities of chips. For this reason, milling cutters for magnesium typically have only one third to one half the number of teeth as is usual for steel. Milling is usually done dry, but if a coolant is needed for any reason, a mineral oil can be used as described in Section 4.2. 4.2.

4.4.6 Reaming

Consistent with the principle of allowing ample space for chip removal, reamers for magnesium should have fewer flutes than is usual for other metals, typically four to six for reamers under 25 mm (1 in) in diameter. There should be an even number, with opposing cutting edges spaced 180° apart. If chatter is a problem, opposing pairs of cutting edges can be spaced a few degrees unequally. The helix angle of the flutes can vary from 0 to -10°.

The design of a reamer for magnesium does not differ greatly from that for other metals. The chamfer is 45°; the margin, 0.1 to 0.3 mm (25 to 75 mils); the primary relief angle, 5 to 8°; and the secondary relief angle, about 20°.

As with all machining operations for magnesium, the hole being reamed should be sufficiently undersize so that a definite cut will be taken by the reamer. The optimum range of stock to be removed is 0.25 to 0.4 mm (60 to 100 mils). Less leads to burnishing, more can produce enough chips to clog the flutes.

Speeds and feeds are given in Table 4.10. While the speed is often limited by rigidity or the machine being used, high speeds and medium feeds produce the best surface finish. Tool life should be between 15,000 and 20,000 holes.

4.4.7 Countersinking and Counterboring

Typical designs for countersinking and counterboring are shown in Figure 4.4. All cutting edges and margins should be kept sharp, and the tools should be highly polished. Speeds are from 30 to 100 m/min (100 to 325 ft/min), and feeds from 0.2 to 0.4 mm/revolution (50 to 100 mils/revolution). An important added precaution when counterboring is to round or chamfer the corners at the ends of the cutting edges. The rounded interior corner in the bottom of the hole that is produced in this way minimizes stress concentration in that area.

4.4.8 Tapping and Threading

The standard nomenclature for solid taps is shown in Figure 4.4. The recommendations for taps to be used for magnesium are given in Table 4.11, where the eccentric design is the one used most commonly. If there is trouble with chips jamming or if much closer tolerances are needed, then the concentric type should be used. The latter cuts during the backing out of the tap, resulting in a clean, accurately-tapped hole. If the tap cuts oversize, the rake angle should be decreased. Conversely, if the tap cuts undersize, the rake angle should be increased.

As has been pointed out earlier, the flutes should be highly polished to facilitate the removal of chips. Magnesium holes can be tapped dry, but the use of a coolant is recommended for better chip removal, longer tool life, improved surface finish, and more accurate threads. Typical cutting speeds are given in Table 4.12.

4.4.9 Die Threading

Threading dies should have approximately the same cutting angles as taps. A cutting edge relief of 0.1 mm (4 mils) is recommended; if a part, however, requires extremely fine surfaces and close tolerances, a 3° positive heel rake will clean up the thread when the die is removed. Self-opening dies help to provide smooth threads.

The cutting angles of thread chasers should be about the same as those used on single-point turning tools, except that the rake angle should be somewhat larger. Threads may be chased at speeds up to 150 m/min (600 ft/min) or higher.

Thread rolling is not recommended for magnesium because of its limited cold workability.

4.4.10 Sawing

Because the power requirement for sawing magnesium is only about one-tenth that required for steel, speeds are great and copious quantities of chips are produced. Adequate provision for chip removal through a large tooth pitch and large chip spaces is a requirement.

Both high-speed steel and carbide-tipped blades are used. The carbide-tipped blades can be operated at much higher speeds and have longer life than the high-speed steel blades.

Recommended designs and speeds for saws are given in Table 4.13. Feeds are best determined for individual cases, but are always high. As an example, a 300-mm (75-in) carbide-tipped circular saw can cut a 25-mm (6.4-in) thick plate at a rate of 6500 linear mm/min (1600 in/min.

4.4.11 Grinding

Magnesium is seldom ground because the surface finish obtained by machining is excellent. However, if grinding is desired, the conditions given in Table 4.14 are a guide to good practice.

Since the fine dust produced by grinding is extremely hazardous, good safety practice is necessary. Nonsparking equipment, spark-proof motors, and wet collection of the dust must be used. Grinding equipment used for magnesium must not be used for any other material. Good housekeeping must be practiced through daily inspection and cleaning of all surfaces on which dust can collect. Chrome-pickled magnesium surfaces spark when ground, and any grinding of a part that will receive a chrome pickle for surface protection should be done before rather than after such treatment.

The dust should be collected in a good wet collector such as is recommended by the National Fire Protection Association [4], or by the United Kingdom Health and Safety Executive [5]. These or similar publications should be consulted and their recommendations followed before any grinding is undertaken. The regulations given in [5] are mandatory in the United Kingdom.

TABLE 4.4 Normal Variations of Single-Point Tool Design

Tool signature	Range of values
Back rake	10-20°
Side rake	0-10°
End relief	10-20°
Side relief	10-20°
End cutting edge	15-45°
Side cutting edge	0.60°
Nose radius	0.5-1.5 mm (0.020-0.060 in)

TABLE 4.5 Conditions for Turning and Boring

	Speed		Feed		Max depth	
Operation	m/min	ft/min	mm/rev	in/rev	mm	in.
Roughing	90-180	295-590	0.8-2.5	0.032-0.100	12.7	0.50
	180-300	590-980	0.5-2.0	0.020-0.078	10.2	0.40
	300-460	980-1500	0.2-1.5	0.008-0.064	7.6	0.30
	460-610	1500-2000	0.2-1.0	0.008-0.040	5.0	0.20
	610-1500	2000-4900	0.2-0.8	0.008-0.032	3.8	0.15
Finishing	90-180	295-590	0.1-0.6	0.004-0.024	2.5	0.10
	180-300	590-980	0.1-0.5	0.004-0.020	2.0	0.08
	300-1500	980-4900	0.1-0.4	0.004-0.016	1.3	0.05

TABLE 4.6 Twist Drill Design

Feature	Deep holes	Shallow holes	Sheet
Point angle	118°	118°	60°
Helix angle	40-50°	10-30°	10°
Chisel-edge angle	135-150°	120-135°	120-135°
Relief angle	24°	12°	
Web	Constant thickness	Constant thickness	Thinned at point
Corners	Rounded	Rounded	Rounded
Flutes	Polished	Polished	Polished

TABLE 4.7 Speeds and Feeds for Drilling

		Feed		
Drill diameter in	Speed fpm	Sheet	Shallow holes in/rev	Deep holes
0.25	300 to	0.005-0.030	0.004-0.030	0.004-0.008
0.50	2000	0.010-0.030	0.015-0.040	0.012-0.020
1.0		0.010-0.030	0.020-0.050	0.015-0.030
mm	m/min		mm/rev	
5	90 to	0.13-0.75	0.10-0.75	0.10-0.20
13	600	0.25-0.75	0.38-1.02	0.30-0.50
25		0.25-0.75	0.50-1.3	0.38-0.75

TABLE 4.8 Typical Milling Cutter Design

	Angle in degrees			
Cutter type	Tooth	Radial rake	Primary clearance	Secondary clearance
Face		20	10	20
Plain mill		25	10	25
Form		8	10	
Shell end		10	10	20
Inserted blade	7	15	10	20

TABLE 4.9 Speeds and Feeds for Milling

Operation	Speed fpm	Feed in/min	Feed in/tooth	Depth of cut in
Roughing	Up to 900	10-20	0.005-0.025	Up to 0.500
	900-1500	10-60	0.005-0.020	Up to 0.375
	1500-3000	15-75	0.005-0.010	Up to 0.200
Finishing	Up to 1000	10-50	0.005-0.015	Up to 0.075
	1000-3000	10-70	0.004-0.008	0.005-0.050
	3000-5000	10-90	0.003-0.006	0.003-0.030
	5000-9000	10-120	0.002-0.005	0.003-0.030
	meters/min	mm/min	mm tooth	mm
Roughing	Up to 275	254-508	0.13-0.63	Up to 13
	275-450	254-1525	0.13-0.51	Up to 10
	450-900	380-1900	0.13-0.25	Up to 5
Finishing	Up to 300	254-1270	0.13-0.38	Up to 2
	300-900	254-1780	0.10-0.20	0.13-1.25
	900-1500	254-2290	0.08-0.15	0.08-0.75
	1500-2750	254-3050	0.05-0.13	0.08-0.75

Source: Ref. 2.

TABLE 4.10 Conditions for Reaming

Reamer diam.		Feed		Speed (steel)		Speed (carbide)	
mm	in	mm/rev	in/rev	m/min	ft/min	m/min	ft/min
5-12	0.2-0.5	0.5-1.0	0.02-0.04	15-50	50-165	25-100	80-330
12-25	0.5-1.0	0.1-3.0	0.04-0.12	15-50	50-165	25-100	80-330
25-50	1.0-2.0	2.5-4.0	0.10-0.16	15-50	50-165	25-100	80-330

Source: Ref. 3.

TABLE 4.11 Tap Design Parameters

Feature	Eccentric tap	Concentric tap
Total land width	30% of tap circumference	30% of tap circumference
Number of flutes	2 for up to 5 mm (0.2 in) 3 for up to 18 mm (0.7 in) 4 for above 18 mm (0.7 in)	2 for up to 5 mm 3 for up to 18 mm 4 for above 18 mm
Land relief	0.1 mm (0.004 in)	0
Rake angle	20°	10-20°
Heel rake angle	0°	3-5°
Core diameter	30% of tap diameter	30% of tap diameter
Flutes	Straight or helix	Straight or helix

TABLE 4.12 Cutting Speeds for Tapping

Tap diameter		Speed	
mm	in	mm/min	in/min
3-6	0.12-0.24	15-25	0.60-0.10
6-12	0.24-0.48	20-30	0.80-1.2
12-25	0.24-1.0	20-40	0.80-1.6
25-50	1.0-2.0	30-50	1.2-2.0

Source: Ref. 3.

TABLE 4.13 Designs for Saws

Design feature	Circular saw	Band saw	Power hacksaw	Hand hacksaw
Pitch (teeth/cm)	0.2-1.6	1.6-2.4	0.8-2.4	4.7-7
(teeth/in)	0.5-4.0	4.0-6.1	2.0-6.1	12-18
Tooth set (mm)	None	0.5-1.3	0.4-0.8	
(in)	None	0.02-0.05	0.016-0.032	
End relief angle	9 to 10°	10 to 12°		
Side relief (HSS)	1 to 1.5°			
Clearance angle	10 to 30°	20 to 30°	20 to 30°	20 to 30°
Face bevel angle	0 to 5°			
Kerf (mm)	2.0-15			
(in)	0.08-0.6			
Surface (m/min)				
HSS	90-600			
Carbide	90-3000			
Surface (ft/min)				
HSS	295-1970			
Carbide	295-9850			

TABLE 4.14 Typical Wheels and Grinding Conditions

Surface grinding	
Wheel type	Aluminum oxide, 46 mesh
Wheel speed	1700-2000 surface meters/min (5500-6500 ft/min)
Table speed	15-30 surface meters/min (50-100 ft/min)
Downfeed/pass, roughing	0.08 mm (0.003 in)
Downfeed/pass, finishing	0.02 mm (0.001 in)
Crossfeed/pass	1/3 wheel width
Cylindrical grinding	
Wheel type	Silicon carbide, 46 mesh
Wheel speed	1700-2000 surface meters/min (5500-6500 ft/min)
Work speed	45 surface meters/min (150 ft/min)
Infeed/pass, roughing	0.05 mm (0.002 in)
Infeed/pass, finishing	0.04 mm (0.002 in)
Traverse/work revolution	
Roughing	1/3 wheel width
Finishing	1/6 wheel width
Centerless grinding	
Wheel type	Silicon carbide or aluminum oxide, 60 mesh
Wheel speed	1700-2000 surface meters/min (5500-6500 ft/min)
Infeed/pass, roughing	0.13 mm (0.005 in)
Infeed/pass, finishing	0.04 mm (0.002 in)
Work feed	130 cm/min (50 in/min)
Regulating wheel	30 rpm rotation: 3° angle
Internal grinding	
Wheel type	Aluminum oxide, 36 mesh
Wheel speed	1500-2000 surface meters/min (4900-6500 ft/min)
Work speed	25-50 surface meters/min (80-165 ft/min)
Infeed/pass, roughing	0.08 mm (0.003 in)
Infeed/pass, finishing	0.005 mm (0.0002 in)
Traverse/work revolution	
Roughing	1/3 wheel width
Finishing	1/6 wheel width

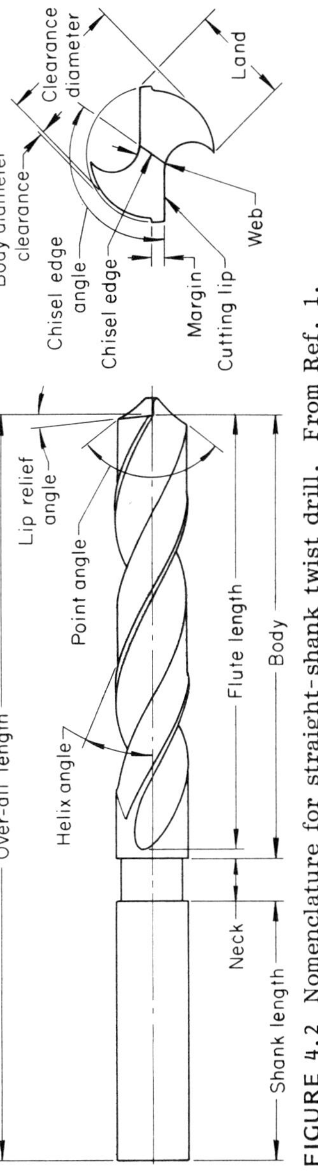

FIGURE 4.2 Nomenclature for straight-shank twist drill. From Ref. 1.

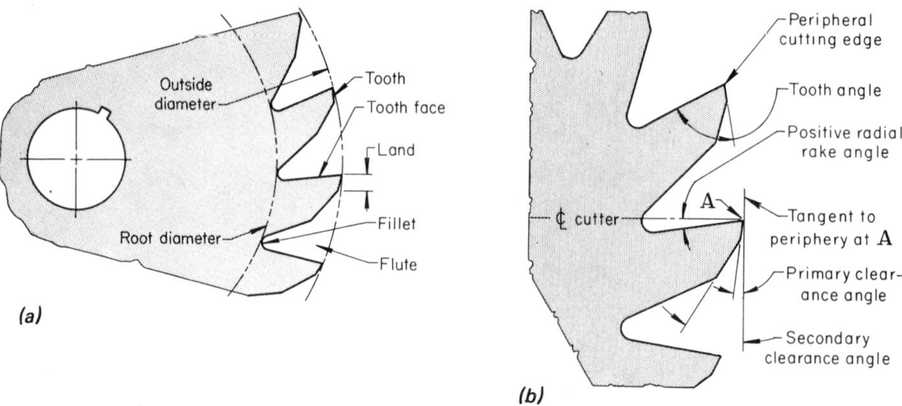

FIGURE 4.3 (a) Nomenclature for essential features of milling cutters. (b) Nomenclature of tooth design of milling cutters. From Ref. 1.

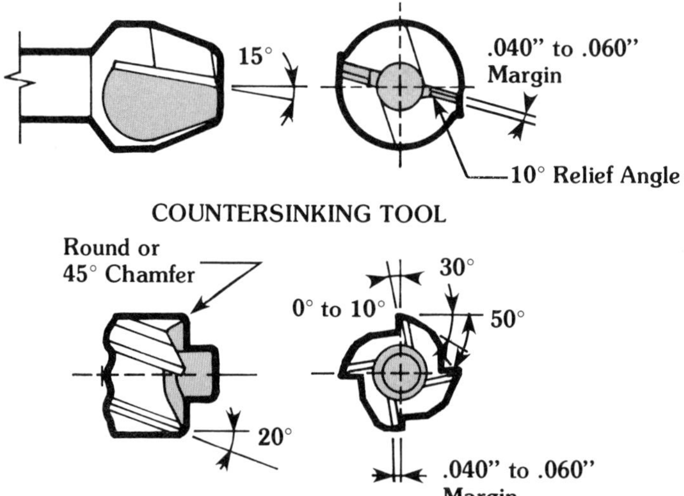

FIGURE 4.4 Typical design for (top) countersinking and (bottom) counterboring.

4.5 SAFETY

As pointed out in Chapter 2, chips produced during machining can be raised to the ignition temperature with a small source of heat. Once some chips become ignited, the fire can spread to adjacent chips. This necessitates certain precautions that must be followed carefully.

The generation of heat during machining should be kept to a minimum. Thus, tools must be kept very sharp so that they truly cut rather than tear; rubbing of the tool on the machined surface must be avoided to minimize frictional heat; the tool must not be allowed to dwell on the work when a cut is finished. The work place must be kept clean of accumulated chips.

A very careful and detailed study by Peloubet [6] has shown the relation between speeds and feeds of cutting, relative humidity, and alloy composition on the probability of heating a chip to the flash point during machining. The work was done using high-speed steel or carbide tips for tools, and tools were properly sharpened. The observations made were the frequency of flashes as machining proceeded, using a grooving tool on a turning cylinder. His results are summarized in Figures 4.5 through 4.8. The salient points to note are:

1. For any given set of conditions, there is a regime of feed and speed that may result in flashing. Flashing can be eliminated if this regime is avoided. As feeds become higher, the chip becomes thick enough that it cannot be raised to the ignition temperature. If the speed is low enough, not enough heat is generated for any chip size. If the speed is high enough, the chip, regardless of size, is not in contact with the tool long enough to be heated to the ignition temperature.

2. Relative humidity has a strong influence on the size of the speed-feed regime that may produce flashing. The higher the relative humidity, the more likely is flashing. It was noted in the work that the relevant variable is relative humidity, not the absolute amount of water vapor in the air.

3. A single-phase alloy (AZ31 vs. AZ92 and AZ92-T4 vs. AZ92-F) is less likely to exhibit flashing than a multiple-phase alloy.

While the work was limited to a study of alloys containing aluminum as an alloying element and the quantitative information in the figures applies only to those alloys and conditions studied, the general conclusions should apply to any alloy system. If tools are sharp, there is no dwell time, there are no external sources of ignition, and yet flashing is a problem, changing the conditions of speed and feed in the directions suggested by Figures 4.5 through 4.8, and assuring that the relative humidity is low should solve the problem.

Any external sources of ignition should be minimized. Thus, the striking of a steel insert in the magnesium during machining can result in sparking. Tools that have been used for machining other metals can have surfaces that will cause excessive heating if then used to machine magnesium, and should be resharpened before use on magnesium. Some sand-cast surfaces, particularly if the casting has been chrome pickled before machining, will spark when machined. If sparking during machining is too severe and cannot be corrected by changing conditions, a coolant, as discussed in Section 4.2, should be used.

Smoking should not be allowed while machining, or while in an area where magnesium fines are present.

Chips should not be allowed to accumulate at the machine. As they are produced, chips should be kept cleared and put in covered steel containers for safe storage. The containers should be vented to allow the escape of any hydrogen that may be generated. Machinists should not wear clothing to which dust and fine filings will cling. For example, fuzzy cloth, pockets, and cuffs should be avoided.

If a chip fire does occur, it can be extinguished by using Met-L-X, G-1, or melting fluxes to cover the fire and exclude oxygen. Since melting fluxes are hygroscopic, they should be used only as a last resort in a machine shop. Dry cast-iron chips can be used to extinguish the fire by cooling. If the burning chips are on a combustible floor, such as wood, they should be covered with the extinguisher and then shoveled on to a non-combustible surface such as a steel plate. Burning chips should also be removed from a concrete floor since the water in the concrete can be removed rapidly enough to cause spalling with consequent scattering of the burning chips. Water, or other oxygen-containing materials, should never be used to extinguish a magnesium chip fire.

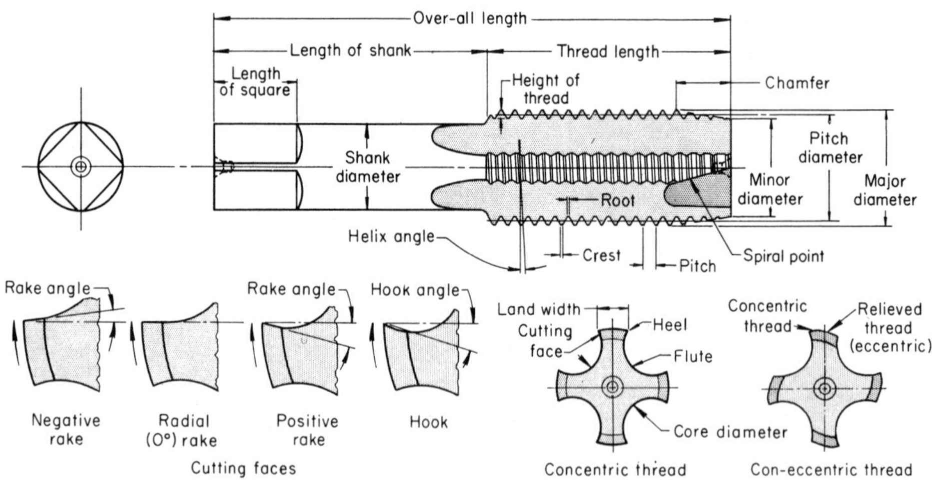

FIGURE 4.5 Standard nomenclature for design details of solid taps. From Ref. 1.

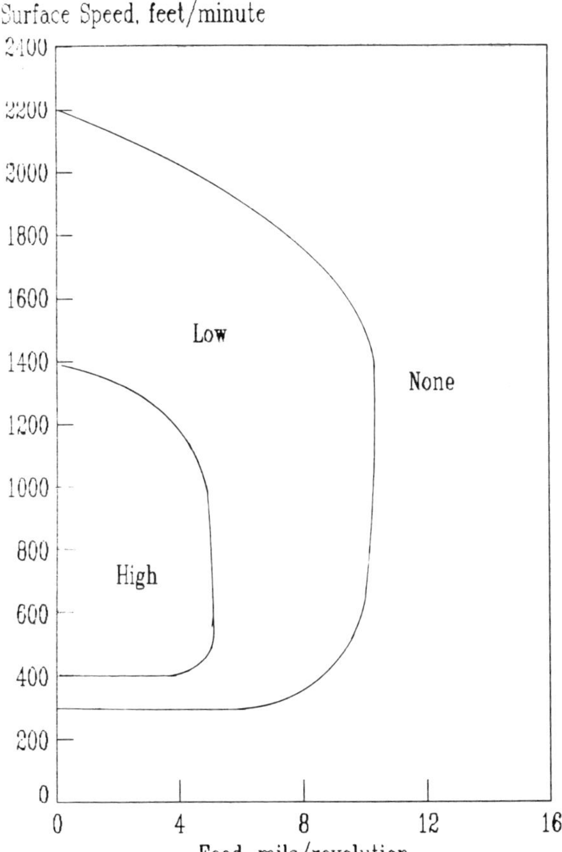

FIGURE 4.6 Degree of flashing as related to speed and feed. Extruded billet of AZ92A. 45% relative humidity. Conversions: 0.0254 times mils/revolution equals mm/revolution: 0.3048 times ft/min equals m/min. From Ref. 6.

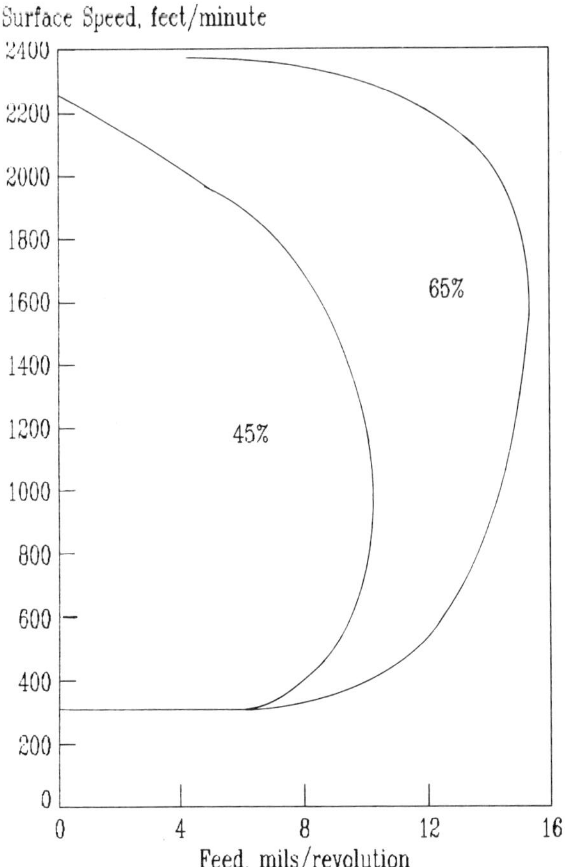

FIGURE 4.7 Envelope of low flash frequency as affected by relative humidity. Extruded billet of AZ92A. Conversions: 0.0254 times mils/revolution equals mm/revolution; 0.3048 times ft/min equals m/min. From Ref. 6.

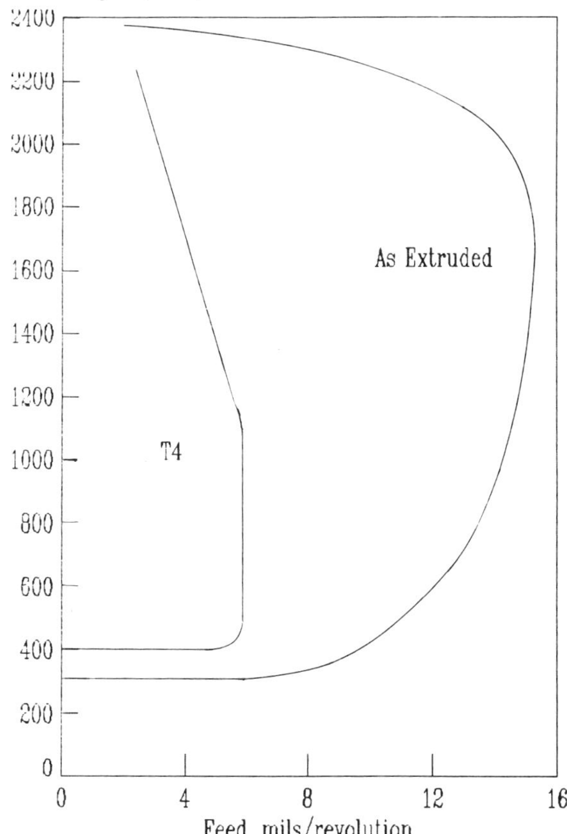

FIGURE 4.8 Envelope of low flash frequency at 65% relative humidity as affected by state of homogenization. Conversions: 0.0254 times mils/revolution equals mm/revolution; 0.3048 times ft/min equals m/min. From Ref. 6.

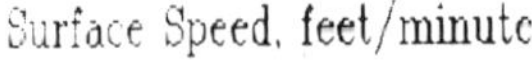

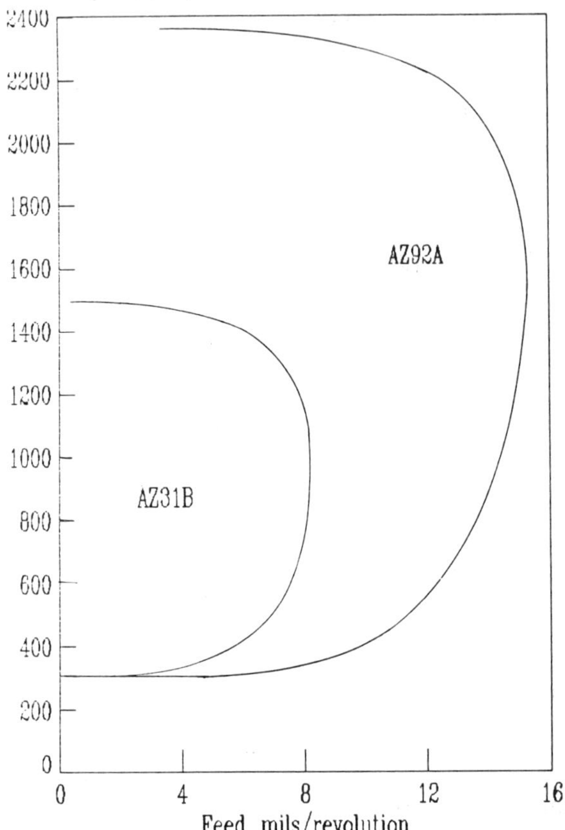

FIGURE 4.9 Effect of alloy composition on envelope of low flash frequency at 65% relative humidity. Conversions: 0.0254 times mils/revolution equals mm/revolution: 0.3048 times ft/min equals m/min. From Ref. 6.

REFERENCES

1. American Society for Metals, "Machining," *Metals Handbook*, Vol. 3, 8th ed., American Society for Metals, Metals Park, Ohio, 1967.
2. Dow Chemical Company, "Machining Magnesium," Bulletin 141-480-82, 1982.
3. Norsk Hydro, a.s., "Magnesium Pure and Alloys; Machining," 1968.
4. National Fire Protection Association, "Storage, Handling, and Processing of Magnesium," Bulletin 48, 1982.
5. U.K. Health and Safety Executive, "Magnesium (Grinding of Castings and Other Articles) Special Regulations," 1947,
6. John A. Peloubet, "Machining Magnesium—A Study of Ignition Factors," *Fire Technology, 1*:5-14 (1965).

5

Joining

Magnesium can be joined to itself or other materials by all of the usual methods. There are, however, two properties of magnesium which need to be considered carefully during the design of joints.

The high coefficient of thermal expansion of magnesium compared to other metals can result in greater distortion or induced stress, either during preparation of the joint—as in welding—or during use.

The high solution potential of magnesium compared to other metals results in a higher risk of galvanic corrosion when other metals are joined to magnesium.

5.1 LIQUID METAL METHODS

5.1.1 Arc Welding

Magnesium in general is readily welded. The low heat capacity and low heat of fusion both mean that power inputs can be low and welding speeds high. The metallurgy of magnesium is such that excellent mechanical properties are obtained. Thus good welds can be made easily at low cost.

The reactivity of molten magnesium requires protection from the atmosphere during welding. At one time, welds were made using gas welding, with flux protection. However, the difficulty of thorough removal of flux from the weld and the severe corrosion penalty for failure to clean completely have resulted in the gas welding of magnesium almost never being used. If the circumstances are such that gas welding must be used, such as in field repair, the welding is straightforward and simple, but great care must then be taken to completely remove the flux. Arc welding is universally used, with protection supplied by the use of an argon or helium gas shield. The protective gas cannot be nitrogen or carbon dioxide, since both of the latter react readily with molten magnesium.

There are two major classifications of arc welding used for magnesium. The first, known as gas tungsten-arc welding, uses a non-consumable tungsten electrode to supply the power for melting by means of an arc between the tungsten and the metal being welded; if any filler metal is used, it is added as a separate rod fed into the molten pool. The second, known as gas metal-arc welding, uses the welding rod itself as the electrode which, while being consumed, is fed continuously into the molten pool. There are three variations used for the gas tungsten-arc welding method: the current can be either ac or dc; if dc, the tungsten electrode can be either positive or negative. With the gas metal-arc welding process, the current is always dc, and the metal electrode is always positive. However, there are several modes of metal transfer from the metal electrode to the weld bead, depending upon the current and wire feeding speed. These modes and the dependence on current and speed are shown in Figure 5.1 [1]. Transfer must be by pulsed-arc in the region of globular transfer. Table 5.1 lists some of the characteristics and differences for these variations [2].

While, in general, magnesium alloys are easily welded, there are differences among the different compositions, and there are a few alloys considered unweldable. Table 5.2 [3] shows the relative weldability of the commercial magnesium alloys. The rating is based on sensitivity to cracking in the weld, with the best ratings being the least susceptible to cracking.

Four alloys, AZ61A, AZ92A, AZ101A, and EZ33A, are used for general welding rods. The selection of the rod to use depends upon the alloys being welded, the kind of weld being made, and the cost of the rod. When both pieces being welded are of alloys containing Zr, EZ33A is the proper welding rod. If either of the pieces contains aluminum, a rod containing aluminum should be used. Alloys ZE10, M1A, and K1A are preferably welded with the aluminum-containing rods. Welding is easier and more trouble-free with rods made of alloys AZ101A and AZ92A than of AZ61A when welding cast alloys; AZ61A is better for most wrought alloys and is less expensive. In all cases, rods of the same composition as the base metal being welded can be used, a selection which is almost always made when welding is for the purpose of repairing sand or permanent-mold castings.

Joint Design

A good weld joint is one that has full penetration of the weld metal into the joint, is free of cracks and porosity, supports the full design loads, and is produced at minimum cost by optimizing weld preparation and amount of weld metal used. A good design is one that assures that the welded joint will be good, and, in addition, provides unimpeded access to the joint during welding. A specific consideration for magnesium welds is that overall cost will most often be least when the least weld metal is used even if weld preparation is more expensive. The high coefficient of thermal expansion of magnesium results in a high risk of distortion; the smaller the amount of weld metal that is used, the lower is that risk. Thus, for example, double-beveled joints are often more desirable than single-beveled.

Table 5.3 [3] illustrates some typical joint dimensions that are used for magnesium sheet, plate, and extrusions. Sand and permanent-mold castings can be joined to other castings or to wrought products, using these or similar joints. However, almost all welding on castings is for the purpose of repair, for which see *Welding of Castings*, below.

Welding Operations

All surfaces and edges must be cleaned just prior to welding. Mechanical cleaning with aluminum oxide abrasive cloth, with stainless steel brushes, or aluminum or steel wool is satisfactory. Chemical cleaning is also possible, using a solution of 180 gms/liter of chromic acid, 40 gms/liter of ferric nitrate, 0.45 gms/liter of potassium fluoride, and water to make one liter. Parts should be dipped in this solution, held at 20 to 35°C, for about 3 min, rinsed thoroughly in hot water, and then dried in air. Welding rod should also be clean. The separate filler rods used in the tungsten-arc process can be cleaned in the same way as the surfaces to be welded. Coils used in the metal-arc process must be stored in a way to stay clean.

The important welding variables for estimating the cost of making a welded joint are the speed of welding, the amount of weld rod required, the amount of shielding gas required, and the power consumption. Tables 5.4 and 5.5 [3] give some typical conditions for welding various thicknesses of magnesium with the four types of arc welding. Any specific joint design will differ from these typical conditions to some degree, but the numbers can serve as a guide.

Backup plates, made of steel, aluminum, or magnesium, are used when welding sheet in order to support the liquid weld metal at the bottom of the joint, and to aid in heat removal. Gas shielding for the bottom of the weld is provided by holes in the plate. The plates have a groove running parallel to the seam of the weld, with Table 5.6 [3] providing a guide to the depth of the groove for various conditions. Weld fixtures must be rigid to prevent movement of the pieces during the welding operation.

Almost any weld will have sufficient restraint so that residual stresses remain after welding. These will sometimes be self-relieved immediately resulting in distortion which then requires a straightening operation. Many times distortion does not occur, leaving stresses in the piece that can be as high as the proportional limit. For magnesium alloys containing more than about 1.5% aluminum, it is necessary to stress relieve in order to avoid stress-corrosion cracking during service. For other alloys, it may be desirable to stress relieve, particularly if machining will be performed after welding. The removal of metal from a piece containing residual stresses can result in relief of the stresses with consequent distortion after machining. Time and temperatures which will relieve about 90% of the stresses in the material are shown in Table 5.7 [3-5].

Welding of Castings

Castings can be welded to other castings or to wrought products for assembling structures. In this case, the welded joints and procedures described above can be used. In most cases, however, castings are welded to repair defects such as cracks, misruns, or blowholes that have appeared in the casting. Such repair is efficient and should be permitted, except where porosity, flux, or oxide may be present.

Because castings often are of complex design, containing both thick and thin sections, the restraint on a weld can be high and preheating may be necessary to prevent weld cracking. This is particularly true for alloys that have a long freezing range, such as those high in zinc. Preheating can sometimes be done by using a torch to heat a local area, but it is preferable to put the entire casting in a furnace and raise it to the proper temperature. It is important not to overheat, staying within the maximum temperatures for various alloys and tempers as given in Table 5.8. Note that these are maximum temperatures. It will often be possible to obtain sound welds by preheating at lower temperatures than are given in Table 5.8 [3].

Castings made from some alloys are usually heat treated to the T4 or T6 tempers. Because the weld metal freezes very rapidly, the metal in the bead is fine grained and the time required for complete solution of the second phases in the weld metal is short compared to the time required for the as-cast metal in the casting itself. If heating is continued after solution is complete, there will be excessive grain enlargement in the weld metal and mechanical properties will suffer. If the casting that has been welded is in the as-cast condition before welding, it is not possible to solution heat treat the entire welded casting without causing either excessive grain growth in the weld bead or insufficient heat treatment of the unwelded metal. Therefore, it is necessary to weld only on castings that have been already put into either the T4 or T6 temper. Then, after welding, both the casting and the weld metal can be put into the solution-heat-treated state with only a short time at the heat-treating temperature. Table 5.8 [3] shows the recommended procedures, and it is important that these be followed.

Mechanical Properties of Welds

The mechanical properties of a welded joint depend upon the properties developed in the weld bead itself, the properties of the zone of metal immediately adjacent to the molten metal that has been heated to a high temperature, the properties of the base metal, and the geometry of the joint itself.

The weld bead itself is cast metal that has been frozen at a rapid rate, resulting in both a fine grain size and a fine second-phase structure. Fenn and Lockwood [6] measured the tensile properties of bars cut from weld-bead metal which demonstrate that the freezing rate of weld beads is between that of a sand casting and that of a high-pressure die casting. Their data are given in Table 5.9 [6]. Note that the strength is considerably improved over the properties obtained by casting in sand and that for alloy AZ61A the ductility is also markedly increased. While not demonstrated in Table 5.9, the metal in the weld-bead is much sounder than that in a high-pressure die casting. When weld-bead metal is heat treated according to the schedules in Table 5.8, no blistering occurs. By contrast, it is seldom possible to heat treat high-pressure die-cast metal to the T4 temper without blistering. The properties in the weld bead itself are therefore those characteristic of sound, fine-grain cast metal. If the alloy used for the rod is heat treatable, then properties can be further improved by heat treatment.

That portion of the base metal that is immediately adjacent to the weld bead is known as the heat-affected zone. The excessive heat experienced by the metal in this zone can have several effects. If the base metal has been work hardened, as by cold rolling, the strengthening due to the cold work will be removed. If the alloy is susceptible to grain growth, the grains will enlarge in the heat-affected zone with consequent lowering of mechanical properties. If the alloy is in a fully tempered state, such as T6, the metal in the heat-affected zone may be returned to the T4 condition.

There is little that can be done to remedy the lowering of properties by removal of cold work, although hot hammering of the welds has been used to strengthen the joint [7]. If, for example, AZ31B-H24 sheet is welded, the metal in the heat-affected zone will be in the -0 temper and, this being the weakest metal in the structure, failure will occur there. Metal that is in the T4, T5, or T6 condition can be returned to that condition by following the schedules given in Table 5.8. Any lowering of properties will be related to grain growth in the heat-affected zone. This is generally rather small, so that the overall lowering of properties is minor.

Mechanical properties of butt-welded alloys are given in Table 5.10 [8]. While yield strength and %E as well as tensile strength are shown, it should be noted that values for elongation and yield may be false.

Because both yield strength and elongation were measured by reference to a 2-in gauge length, any local yielding distorted the data. Since the gauge length extended over the base metal, the heat-affected zone, and the weld bead, the metal was not uniform and local yielding may have occurred. Weld efficiency is therefore better measured by comparing the tensile strength of the welded to the unwelded metal. As can be seen by the data in Table 5.10, the weld efficiencies for all the alloys are generally between 85 and 95%.

Because the cast metal characteristic of the weld bead is of high quality, and because castings are generally heat treated for full development of properties, the weld efficiency of repaired casting sections is close to 100%. This also applies to fatigue strength [9], which is unaffected by welding of castings.

Quality Control

Improper welding technique can result in a number of defects, and it is important to inspect the finished joint carefully for their presence. A common defect is cracking, which can be in the form of crater, longitudinal, or underbead cracking. These can often be found simply by visual inspection, but die-penetrant fluorescent inspection shows them clearly. Undercutting, porosity, incomplete fusion, incomplete penetration, gas cavities, and tungsten inclusions can be seen by radiography. All of these defects can be eliminated by careful attention to the welding procedures.

5.1.2 Resistance Spot Welding

While magnesium can be resistance welded by all of the usual techniques, such as spot, seam, projection, or flash welding, the most common is spot welding.

The design of a spot-welded joint must include the spot size desired, the spacing of the spots, both from each other, and from the edges of the work pieces, and good access for the machine to the joint. The spot size is governed by the thickness of the work piece, and the welding parameters such as electrode size, current, and electrode force applied. The penetration of the spot should be at least 20% of the work piece thickness, but not more than 80%. Too little penetration results in a weak spot; too much, in metal expulsion, warpage of the work piece, and too much penetration of the electrode into the spot.

Spot spacing is governed by both current and strength requirements. Spots that are too close together result in too much current being shunted through the spot already made, leaving insufficient to produce a good new spot. Too large a spacing results in a weak joint. The spacing should be so designed that consistently good welds are made and that the joint, while strong, will be the point of failure if the structure is overstressed during service. It is easier to repair a joint than a torn work piece.

Spot welding takes place in a number of distinct steps, each of which must be controlled accurately:

1. Squeeze: During this first interval, the upper electrode is brought into contact with the work pieces and the full electrode force is applied. This serves to bring the work pieces together and to hold them tightly enough together to assure good contact. No current flows during the squeeze period.
2. Weld: Once good contact is obtained and the full force applied, the current is switched on, resulting in heating of the volume between the electrodes to above the melting point. The liquid nugget forms. The full electrode force is continuously applied during the weld part of the cycle.
3. Hold: The current is discontinued, but the electrode force is maintained. During this period, the liquid weld nugget freezes, and the weld is completed.
4. Off: The electrode force is discontinued; the electrodes are separated to allow the work piece to be removed, and the machine is ready for the start of the next cycle.

The total cycle takes about a half second. Since the timing is critical to the production of good welds, it is mandatory that automatic controls be used. There are many variants of the basic procedures that can be introduced. The current can be increased to its full value slowly, and then allowed to taper off rather than stopping suddenly. The current can be interrupted to allow for brief holding periods between application of short current bursts. A "forging" force can be applied during the hold period, which is an increase of electrode force during the freezing cycle to assure defect-free weld metal.

Machines and Techniques

Either ac or dc current can be used, although ac is somewhat more satisfactory for welding magnesium. The ac machine can be single-phase, or frequency-converter three-phase. An electrostatic, condenser-discharge type is used for dc welding. As stated above, automatic controls are mandatory.

Before welding, the work pieces must be cleaned thoroughly enough to result in a surface resistance no higher than about 50 microhms. This can be accomplished by brushing with steel wool, by acid cleaning in dilute mineral acid, in a chromic-sulfuric pickle, or in dilute chromic acid pickle. See Chapter 14 on finishing for details for these solutions. The electrodes must also be cleaned frequently enough to prevent copper pickup on the spots. Inattention to this can lead to galvanic corrosion problems during use. With wire-brush cleaning, from 10 to 50 spots can be made between cleanings; with pickling, from 50 to several hundred, depending in both cases on the alloy being welded.

The electrodes used must be RMWA (Resistance Welding Manufacturers Association) Group A, either Class 1 or Class 2. Class 1 has a lower strength and a higher conductivity than Class 2 [11]:

Class 1: 120 MPa (17 ksi) proportional limit; 80% IACS
Class 2: 240 MPa (35 ksi) proportional limit; 75% IACS

The selection will depend upon the relative importance of strength and conductivity for the specific work being done.

The exact welding schedule must be determined for each job, taking into account the major variables of electrode force, current, current duration, hold time, and whether or not such additional steps as a forging force or interrupted current are used. The schedule is derived by starting with those suggested in Tables 5.11 and 5.12 [10] and adjusting until the desired size and quality of spot is obtained. The characteristics governing spot quality are the degree of penetration, the spot diameter, indentation, and freedom from defects such as porosity, cracks, expulsion, or other gross defects. Once the proper schedule is obtained, the work can proceed, with periodic checks on spot quality to assure that nothing has changed. The data in Tables 5.11 and 5.12 are for only one alloy—AZ31B—but the parameters for all alloys are enough the same that this table can serve as the initial guide to the determination of a schedule for any magnesium alloy.

Mechanical Properties

Spot welds are strongest when loaded in shear, and joints should be designed with this in mind. Because the spots are discontinuous and are sources of stress concentration, the fatigue strength is not as good as with other joints. If this mode of failure is critical for the service of the part, some other joining method should be used.

Since the nugget formed by spot welding is rapidly frozen cast metal, the mechanical properties are characteristic of that state. The mechanical properties of a spot-welded joint are expressed as the shear force per spot to cause failure. Given a sound, defect-free nugget, the strength of a spot weld is related directly to the thickness of the work pieces and the spot diameter. If the work pieces are of different thicknesses, the thinner piece governs. While the relation between piece thickness and spot diameter is the major variable, strength is also affected by the exact procedures used to produce the spot, including type of machine, duration of pulse, number of pulses, electrical current, electrode force used, presence or absence of a forging force, a post-heat cycle, and others. It is not, therefore, possible to list the exact properties to be expected from spot welding; the properties obtained in each specific case must be measured.

However, because the principle variables governing strength are workpiece thickness and spot diameter, a guide to the strengths to be expected can be given as in Table 5.13. The numbers given were obtained by regression analysis of all the data found in the literature. They are close enough to what will be obtained in practice to permit preliminary design of a structure that uses spot welding for the joints. Final design will require the obtaining of data on the strength of welds produced in practice.

5.1.3 Miscellaneous Welding

Electron beam, laser beam, friction, explosive, stud, and ultrasonic welding can all be performed on magnesium alloys. There are no special precautions specific to magnesium except for electron beam welding. The nature of this technique is such that vapor is produced immediately under the beam. Molten metal from surrounding volumes then flows into the hole produced. The high vapor pressure of magnesium results in larger holes than is common for other metals, and porosity in the weld can result. In particular, alloys containing more than about 1% Zn should be avoided since the still higher vapor pressure of Zn exacerbates the situation. Practices such as beam oscillation and use of excess metal around the joint to minimize the problem must be adopted.

5.1.4 Brazing

While not a common joining method, brazing is possible for some magnesium alloys. The recommended method is dip brazing, although furnace or torch brazing are also possible. Table 5.14 [14] lists those magnesium alloys that are considered brazable, brazing alloys, and those properties of importance for the process.

Mechanical cleaning of the joint, sometimes followed by chemical cleaning, must be thorough for a good brazed joint. The joint is prepared, the brazing material placed in it, the assembly pre-heated, and then immersed in the flux bath for about 3 min. Following brazing, thorough cleaning of the joint for complete removal of all flux is a requirement. Failure to do so will result in excessive corrosion during use.

The flux used for the bath must melt at a temperature lower than that of the brazing alloy, promote flow of the molten brazing alloy, and protect the assembly from oxidation during the brazing step. It must not contain any oxygen-containing compounds such as nitrates since these can react violently with the magnesium.

If the brazing alloy being used is in the form of powder, it must be mixed with a volatile organic material to form a paste. The paste is placed in the joint, the assembly pre-heated, and then put in the brazing flux.

Because brazing is carried out at a high temperature, there is significant property degradation, as detailed for AZ31B-0 and ZE10-0 in Table 5.15 [14]. Properties of brazed joints for these two alloys are given in Figures 5.2 through 5.5 [14].

5.1.5 Soldering

Soldering is not recommended for magnesium for two reasons. First, the alloys used for solder, when in contact with magnesium, result in severe galvanic corrosion problems. Second, the alloying between the magnesium and the solder results in the formation of brittle compounds, making the joint weak. Solder has been used in special cases, such as repair of photoengraving plates, soldering for electrical contacts, and the filling of dents prior to painting, but should not be used to make a joint for structural purposes.

TABLE 5.1 Characteristics for Arc Welding Processes

Type	Current	Characteristics
Tungsten arc	Alternating	Excellent penetration. Most commonly used for all types of joints. There is less joint preparation than with direct current, electrode positive welding because there is better penetration of the arc.
Tungsten arc	Direct, electrode positive	Shallow penetration, with wide weld deposits; smooth weld bead. A higher heat input is required. There is good control of weld heat for thin material. An oversize tungsten electrode is required.
Tungsten arc	Direct, electrode negative	Deep penetration. Used primarily in mechanized welding where deep penetration is needed. Very clean base metal is required as the arc has no cleaning action.
Metal arc	Direct, electrode positive	Spray or short-arc metal transfer used, depending on the material thickness. Used for high-production welding, the rate being about 2 to 4 times faster than with tungsten-arc welding.

TABLE 5.2 Relative Arc Weldability of Alloys

Weldability rating	Casting alloys	Wrought alloys
Excellent	EZ33A, K1A, M1A	AZ10A, AZ31B, AZ31C, HK31A, HM21A, HM31A, ZE10A
Good	AM100A, AZ81, AZ91C, AZ92A, EK30A, EK41A, HK31A, HZ32A, QE22A, QH21A, WE54A, ZE41A, ZE63A	AZ61A, AZ80A, HZ11A, ZC11A, ZK21A, ZK30A
Fair	AZ63A, ZH62A	ZK40A
Not recommended	ZK51A, ZK61A	ZK60A

Note: High-pressure die castings made by the cold-chamber process are not weldable. Some hot-chamber die castings can be welded.

TABLE 5.3 Typical Joint Designs Used for Gas Shielded Arc Welding of Various Thicknesses of Magnesium Alloy Sheet and Plate

Type of Joint	Applicable range of work-metal thicknesses, in (a)					
	GTAW (b)			GMAW (c)		
	ac	DCEN	DCEP	arc	Pulsed arc	Spray arc
A (d)	0.025-0.25	0.025-0.5	0.025-0.1875	0.025-0.1875	0.090-0.25	0.1875-0.375
B (e)	0.25-0.375	0.25-0.375	0.1875-0.375	(f)	0.1875-0.25	0.25-0.5
C (g)	0.375 (h)	0.375 (h)	0.375 (h)	(f)	(f)	0.5 (h)
D (j)	0.04-0.25	0.04-0.25	0.04-0.25	0.0625-0.1875	0.0625-0.25	0.1875-0.5
E (k)	0.1875 (h)	0.1875 (h)	0.1875 (h)	(f)	0.125-0.25	0.25 (h)
F (m)	0.025-0.25	0.025-0.5	0.025-0.1563 (h)	0.0625-0.1563	0.090-0.1875	0.1563-0.375
G (n)	0.0625-0.1875	0.0625-0.375	0.0625-0.125	0.0625-0.1563	0.090-0.25	0.1563-0.75
H (p)	0.1875 (h)	0.375 (h)	0.125 (h)	(f)	0.25-0.375	0.375 (h)
J (q)	0.040 (h)	0.040 (h)	0.025 (h)	0.040-0.1563	0.090-0.25	0.1563 (h)

(a) Suggested minimum and maximum thickness limits.
(b) Using 300 A AC for DCEN or 125 A for DCEP.
(c) Using 400 A DCEP.
(d) Single-pass, full-penetration weld. Suitable for thin material.
(e) Full-penetration weld. Suitable for thick material. On material thicker than suggested maximum, use double-V-groove butt joint to minimize distortion.
(f) Not recommended because spray-arc welding is more practical or economical, or both.

(g) Full-penetration weld. Used on thick material. Minimizes distortion by equalizing shrinkage stress on both sides of joint.
(h) No maximum. Thickest material in commercial use could be welded in this type of joint.
(j) Single-pass full-penetration especially if a square corner is required.
(k) Single-pass or multiple-pass full-penetration weld. Used on thick material to minimize welding. Produces square joint corners.
(m) Single weld T-joint. Thickness limits are based on 40% joint penetration.
(n) Double weld T-joint. Suggested thickness limits based on 100% joint penetration.
(p) Double weld T-joint. Used on thick material requiring 100% joint penetration.
(q) Single or double weld joint. Strength depends on size of fillet. Maximum strength in tension on double weld joints is obtained when lap equals five times the thickness of the thinner member.

Conversion factor: in × 25.4 = mm.

Source: Ref. 3.

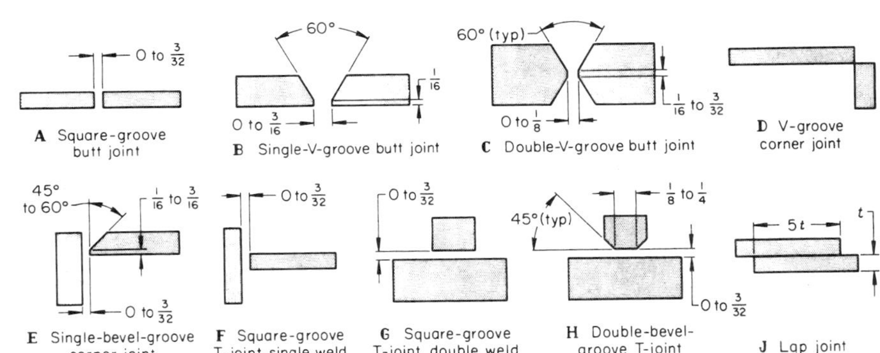

A Square-groove butt joint **B** Single-V-groove butt joint **C** Double-V-groove butt joint **D** V-groove corner joint **E** Single-bevel-groove corner joint **F** Square-groove T-joint, single weld **G** Square-groove T-joint, double weld **H** Double-bevel-groove T-joint **J** Lap joint

TABLE 5.4 Operating Conditions for Gas Metal Arc Butt Welding (a)

			Electrode					
Base-metal thickness, mm	Type of groove	No. of passes	Diameter mm	Feed rate, mm/sec	Consumption, kg/m	Current, amperes	Voltage, volts	Argon flow rate, m^3/h
Short-circuiting mode of transfer								
0.64	Square (b)	1	1.00	60	0.045	25	13	1.1-1.7
1.00	Square (b)	1	1.00	100	0.065	40	14	1.1-1.7
1.60	Square (b)	1	1.60	80	0.130	70	14	1.1-1.7
2.30	Square (b)	1	1.60	100	0.173	95	16	1.1-1.7
3.20	Square (c)	1	2.40	55	0.217	115	14	1.1-1.7
4.00	Square (c)	1	2.40	70	0.267	135	15	1.1-1.7
4.85	Square (c)	1	2.40	85	0.332	175	15	1.1-1.7
Pulsed-arc mode of metal transfer (d)								
1.60	Square (b)	1	1.00	150	0.101	50	21	1.1-1.7
3.20	Square (b)	1	1.60	120	0.202	110	24	1.1-1.7
4.85	Square (b)	1	1.60	200	0.339	175	25	1.1-1.7
6.35	Single-V (e)	1	2.40	120	0.469	210	29	1.1-1.7

Spray-arc mode of metal transfer (f)								
6.35	Single-V (e)	1	1.60	225	0.303	240	27	1.4-2.3
9.50	Single-V (e)	1	2.40	120-130	0.411	329-350	24-30	1.4-2.3
12.70	Single-V (e)	2	2.40	135-150	0.765	360-400	24-30	1.4-2.3
15.90	Double-V (g)	2	2.40	140-155	0.902	370-420	24-30	1.4-2.3
25.40	Double-V (g)	4	2.40	140-155	1.501	370-420	24-30	1.4-2.3

(a) Welding speed of 10 to 15 mm/sec.
(b) Zero root opening.
(c) 3.30 mm root opening.
(d) Pulse voltage of 55 V, except for metal 4.85 mm thick, which uses a pulse voltage of 52 V.
(e) 1.65 mm root face and zero root opening.
(f) Settings also apply to fillet welds in the same thickness of metal.
(g) 3.20 mm root face and zero root opening.
Conversion factors: mm × 0.0394 = in, mm/sec × 0.1969 = ft/min, kg/m × 0.6720 = lb/ft, m^3h × 35.3147 = ft^3/h.

TABLE 5.5 Operating Conditions for Gas Tungsten Arc Butt Joints

Base-metal thickness, mm	Joint design (a)	No. of passes	Electrode diameter, mm	AC current A (b)	Argon rate, m^3h	Filler metal	
						Diameter, mm	Consumption, kg/m
1.00	A	1	1.60	35	0.340	2.40	0.029
1.16	A	1	2.40	50	0.340	2.40	0.036
2.00	A	1	2.40	75	0.340	2.40	0.043
2.54	A	1	2.40	100	0.340	2.40	0.058
3.20	A	1	2.40	125	0.340	3.20	0.065
4.85	A	1	3.20	160	0.425	3.20	0.079
6.35	B	2	4.00	175	0.566	3.20	0.188
9.50	B	3	4.00	175	0.566	4.00	0.411
9.50	C	2	4.75	200	0.566	3.20	0.173
1.27	C	2	4.75	250	0.566	3.20	0.339

(a) A: Square groove butt joint, zero root opening
B: 60° bevel, single-V-groove butt joint; 1.60 mm root face, zero root opening
C: 60° bevel, double-V-groove butt joint; 2.40 mm root face, zero root opening

(b) Thorium-containing alloys will require about 20% higher current. With helium shielding, required welding current will be reduced by about 20 to 30 A.

Conversion factors: mm × 0.0394 = in, m^3/h × 35.3147 = ft^3/h, kg/m × 0.6720 = lb/ft.

TABLE 5.6 Groove Depths in Backing Bars, mm

Base metal thickness, mm	Gas tungsten	Gas metal	
		No root opening	Root opening
0.65	0.40	0.50	0.50
1.00	0.50	0.75	0.50
1.60	0.65	1.00	0.75
2.30	0.75	1.50	1.00
3.20	0.75	1.80	1.00
4.00	1.00	1.80	1.30
4.80	1.00	1.80	1.30
6.35	1.30	1.80	1.50
9.55	1.50	2.00	1.50

Conversion factor: mm × 0.0394 = in.

TABLE 5.7 Stress Relief Treatments After Welding

Alloy	Temperature, °C	Time, min
Sheet		
AZ31-O	260	15
AZ31-H24	150	60
HK31-H24	315	30
HM21-T81	400	30
ZE10-O	230	30
ZE10-H24	135	60
Extrusions		
AZ10-F	260	15
AZ31-F	260	15
AZ61-F	260	15
AZ80-F	260	15
AZ80-T5	200	60
HM31-T5	425	60
ZK60-F	260	15
ZK60-T5	150	60
Castings[a]		
AM100	260	60
AZ63	260	60
AZ81	260	60
AZ91	260	60
AZ92	260	60
EZ33	250	600
HZ32	350	120
ZE41	330	120
ZH62	330	120

[a]Since these are for stress relief only, maximum strength will not be produced. For maximum strength, see the recommendations of Table 5.8.

TABLE 5.8 Preheat and Postweld Treatments for Castings

Alloy	Temper		Max preheat temperature, °C	Postweld treatment	
	Before weld	After weld		Temperature, °C	Time, hours
AZ63A	T4	T4	380	385	0.5
	T4 or T6	T6	380	385	0.5
				+220	5
	T5	T5	260	220	5
AZ81A	T4	T4	400	415	0.5
AZ91C	T4	T4	400	415	0.5
	T4 or T6	T6	400	415	0.5
				+215	4
AZ92A	T4	T4	400	410	0.5
	T4 or T6	T6	400	410	0.5
				+260	4
AM100A	T6	T6	400	415	0.5
				+220	5
EK30A	T6	T6	260	200	16
EK41A	T4 or T6	T6	260	200	16
	T5	T5	260	200	16
EZ33A	F or T5	T5	260	345	2
				+215	5
HK31A	T4 or T6	T6	260	315	1
				+200	16

TABLE 5.8 (*Continued*)

Alloy	Temper		Max preheat temperature, °C	Postweld treatment	
	Before weld	After weld		Temperature, °C	Time, hours
HZ32A	F or T5	T5	260	315	16
K1A	F	F	None	None	
QE21A	T4 or T6	T6	450	510 +200	0.5 - 1 8
QE22A	T4 or T6	T6	260	530[a] +200	8 8
QH21A	T4 or T6	T6	450	510 +200	0.5 - 1 16
ZE41A	F or T5	T5	315	330 +175	2 16
ZH62A	F or T5	T5	315	330 +175	2 16
ZK51A	F or T5	T5	315	330 +175	2 16
ZK61A	F or T5 T4 or T6	T5 T6	315 315	150 500 +130	48 2 - 5 48

[a]Quench in water at 60 to 85°C before second heat treatment.

TABLE 5.9 Mechanical Properties of Weld Metal

	AZ61A					EZ33A				
		YS		TS			YS		TS	
Casting method	%E	MPa	ksi	MPa	ksi	%E	MPa	ksi	MPa	ksi
Sand	4.0	89	13	192	28	3.0	110	16	158	23
Weld bead	8.5	103	15	233	34	2.3	144	21	182	26
High-pressure die	8.0	130	19	219	32					

TABLE 5.10 Mechanical Properties of Arc-Welded Butt Joints

Alloy	Rod used	Condition	%E	YS MPa	YS ksi	TS MPa	TS ksi
Sheet							
AZ31B-H24	AZ61A	Unwelded	15	219	32	288	42
		Weld bead on	5	151	22	253	37
		Weld bead flush	5	130	19	247	36
AZ31B-O	AZ61A	Unwelded	21	151	22	253	37
		Weld bead on	11	130	19	247	36
		Weld bead flush	10	116	17	240	35
HK31A-H24	EZ33A	Unwelded	9	205	30	260	38
		Weld bead on	2	151	22	212	31
		Weld bead flush	4	137	20	219	32
HM21A-T8	EZ33A	Unwelded	10	171	25	233	34
		Weld bead on	2	137	20	192	28
		Weld bead flush	4	130	19	212	31
Extrusions							
AZ31B-F	AZ61A	Unwelded	15	199	29	260	38
		Weld bead on	7	151	22	253	37
		Weld bead flush	5	130	19	247	36
AZ61A-F	AZ61A	Unwelded	16	226	33	308	45
		Weld bead on	7	164	24	274	40
		Weld bead flush	6	144	21	260	38
AZ80A-F	AZ61A	Unwelded	11	247	36	336	49
		Weld bead on	5	178	26	274	40
		Weld bead flush	3	150	22	247	36
AZ80-T5	AZ61A	Unwelded	8	260	38	377	55
		Weld bead on	2	192	28	274	40
		Weld bead flush	2	164	24	233	34
HM31A-T5	EZ33A	Unwelded	8	260	38	301	44
		Weld bead on	1	150	22	178	26
		Weld bead flush	2	130	19	212	31
ZK21A-F	AZ61A	Unwelded	10	226	33	288	42
		Weld bead on	4	—		219	32
		Weld bead flush	5	116	17	233	34

TABLE 5.11 Guide to Spot Welding Procedures

Thickness, mm	Electrode		Electrode force, kgf		Forge delay cycles	Weld heat cycle		Welding current, amperes		Spot spacing, mm, minimum	Spot diameter mm
	Diameter, mm	Tip radius, mm	Weld	Forge		Pulse time cycles	No. pulses	Weld	Postheat		
Three-phase converter											
0.5	12.7	76.0	365			1	2	25400		6.5	4.8
1.0	16.0	76.0	545			1	2	28300		14.0	5.3
1.3	16.0	101.6	725			2	1	31000		11.0	4.8
1.3	16.0	101.6	635	1590	2	2	2	29000	10300	11.0	4.8
1.6	16.0	101.6	795			3	1	35200		12.5	5.6
1.6	16.0	101.6	545	870	3	3	1	43600	24800	12.5	7.4
3.2	22.0	152.5	2045			5	6	66900		24.0	11.7
Single-phase AC											
0.5	9.5	76.0	160			3		18000		6.5	3.6
1.0	12.7	76.0	225			5		26000		9.5	5.1
1.3	12.7	101.6	250			5		29000		11.0	5.8
1.6	12.7	101.6	270			6		31000		12.5	6.9
3.22	12.7	152.5	455			12		42000		24.0	9.7
Three-phase dry disc rectifier											
0.5	16.0	76.0	135	270	0.6	1	1	21000	14700	6.5	3.6
1.0	16.0	76.0	220	455	1.2	2	2	26000	18000	9.5	5.1
1.3	16.0	76.0	265	575	1.5	3	3	28500	20000	11.0	5.6
1.6	16.0	101.6	320	700	1.8	3	3	29300	20500	12.5	6.9
3.2	22.0	152.5	575	1265	7.7	10	6	48000	33400	24.0	9.7

Conversion factors: mm × 0.0394 = in, kgf × 2.2046 = lbf.

TABLE 5.12 Guide to Spot Welding Procedures, Condenser-Discharge Stored Energy Machines, Transformer Turns Ratio of 480:1

	Electrode					
Thickness, mm	Diameter mm	Tip radius, mm	Electrode force kgf	Charging V, KV	Minimum spot spacing, mm	Spot diameter, mm
0.5	12.5	76.0	295	1.4	6.5	3.6
1.0	12.5	76.0	330	2.2	9.5	5.1
1.3	16.0	101.6	380	2.2	11.0	5.8
1.6	16.0	101.6	490	2.2	12.5	6.9
2.5	19.0	152.4	820	2.2	22.0	8.6
3.2	22.0	152.4	1035	2.2	24.0	9.7

Conversion factors: mm × 0.0394 = in, kgf × 2.2046 = lbf.

TABLE 5.13 Typical Mechanical Strength of Spot Welds (kgf/spot)

Thickness, mm	Spot diameter, mm								
	2.5	4.0	5.0	6.5	7.5	9.0	10.0	11.5	12.5
Alloy AZ31B									
0.5	8	70	131						
0.8	29	90	152	213					
1.0	43	105	166	227	289				
1.3	61	122	183	245	306	367			
1.6	84	145	206	267	329	390	451		
2.0	113	175	236	297	359	420	481	543	
2.5	148	210	271	332	394	455	515	578	639
3.2	192	254	315	376	437	499	560	621	683
Alloy ZE10A									
0.5	61	121							
0.8	47	115	184						
1.0	41	110	178	247	315				
1.3	25	106	187	268	349	429			
1.6		103	199	295	392	489	585		
2.0		91	193	294	396	497	599	700	
2.5		81	197	312	427	543	658	774	889
3.2		68	199	330	461	592	723	854	985
Alloy HK31A									
0.5	48	98							
0.8	51	108	164						
1.0	57	113	168	224					
1.3	56	122	189	255	321				
1.6	55	135	216	296	376				
2.0	65	147	230	313	396	479	562	645	
2.5	70	164	258	352	446	540	634	728	822
3.2	78	184	290	396	502	608	714	821	927
Alloy HM21A									
0.5	67	111							
0.8	59	110	163						
1.0	53	107	161	215	269				
1.3	47	110	174	236	300	363			
1.6	40	115	190	265	340	415	491		
2.0	28	110	191	273	355	437	519	601	
2.5	15	109	203	298	392	487	581	676	770
3.2		107	216	325	434	543	652	761	870

Conversion factor: mm × 0.0394 = in, kgf/spot × 2.2046 = lbf/spot.

TABLE 5.14 Dip Brazable Alloys

Alloy	Melting range, °C	Brazing temperature range for filler, °C		
		AZ92A	AZ125A	GA432
AZ10A	630-645	605-615	580-595	495-505
AZ31B	565-625	*	580-595	495-505
AZ61A	510-615	*	*	495-505
K1A	650	605-615	580-595	*
M1A	640-650	605-615	580-595	495-505
ZE10A	595-645	*	580-595	495-505
ZK60A	520-635	*	*	495-505
Filler alloys				
AZ92A	445-600			
AZ125A	410-565			
GA432	330-360			

*Not a brazable combination.

TABLE 5.15 Properties of Base Alloy after Brazing

Time, min	Temperature, °C	AZ31B-O					ZE10A-O				
			TYS		TS			TYS		TS	
		%E	MPa	ksi	MPa	ksi	%E	MPa	ksi	MPa	ksi
0	—	23.5	140	20	250	36	22.5	140	20	225	33
1.5	510	20.0	135	20	245	36	21.5	65	9	200	29
0.5	565	17.5	95	14	235	34	16.5	70	10	185	27
1		18.0			230	33	15.0	75	11	185	27
2		18.0	110	16	230	33	14.0	70	10	180	26
3		16.0	105	15	225	33	12.5	60	9	175	25
5		15.0	105	15	225	33	11.5	60	9	170	25
0.5	580	19.0	120	17	235	34	16.5	75	11	190	28
1		17.0	115	17	235	34	14.5	75	11	190	28
2		17.5	115	17	235	34	12.0	70	10	185	27
3		15.5	110	16	230	33	10.0	70	10	185	27
5		13.0	110	16	225	33	10.5	60	9	175	25
0.5	595	17.0	115	17	235	34	15.0	80	12	190	28
1		16.0	110	16	230	33	14.0	75	11	185	27
2		14.5	105	15	225	33	13.0	75	11	185	27
3		15.0	100	15	225	33	12.0	75	11	185	27
5		14.0	100	15	215	31	10.5	70	10	175	25
0.5	605	14.0	110	16	225	33	16.0	70	10	190	28
1		12.0	85	12	215	31	15.5	75	11	185	27
2		9.0	75	11	200	29	16.5	75	11	180	26
3							15.5	65	9	180	26
							15.0	70	10	180	26

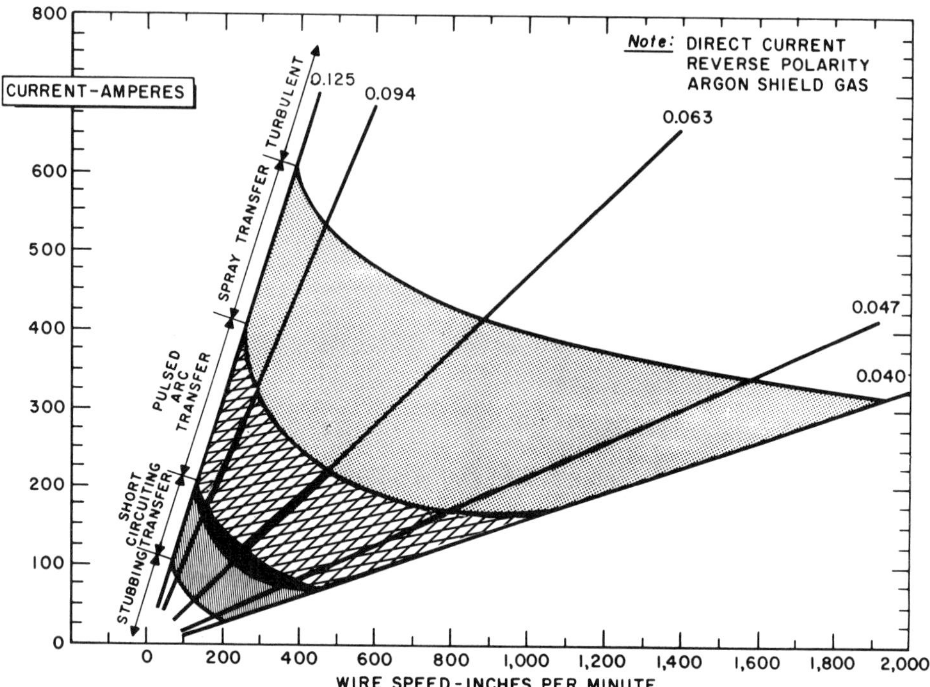

FIGURE 5.1 Relation between current and wire speed for various diameter filler wires and types of metal transfer. Note: direct current, reverse polarity, argon shield gas. Conversion factor: in × 25.4 = mm.

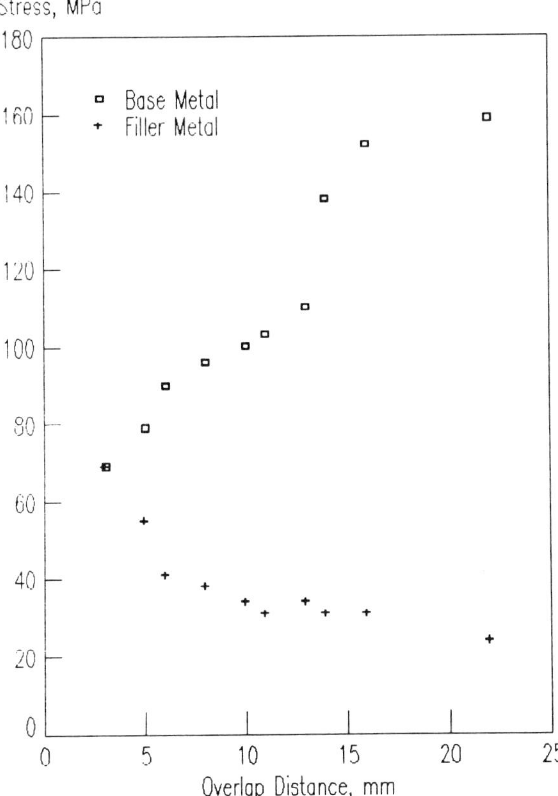

FIGURE 5.2 Average unit shear stress in the filler metal and average unit tensile stress in the base metal as a function of overlap distance. All failures were in the filler metal. □, Base metal (AZ31B); +, filler metal (AZ125A). Conversion factors: mm × 0.0394 = in, MPa × 0.1450 = ksi.

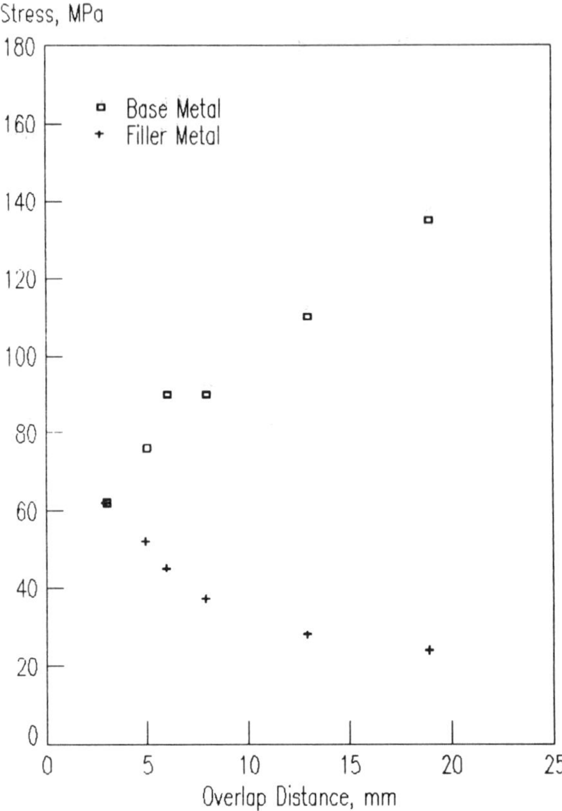

FIGURE 5.3 Average unit shear stress in the filler metal and average unit tensile stress in the base metal as a function of overlap distance. Failures were in the filler metal with overlap through 5 mm. Beyond 5 mm, failures were in the base metal. □, Base metal (ZE10A); +, filler metal (AZ125A). Conversion factors: mm × 0.0394 = in, MPa × 0.1450 = ksi.

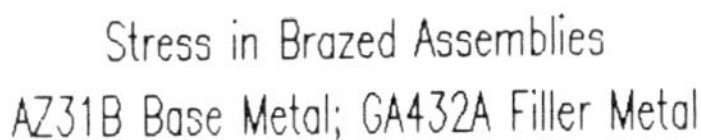

Stress in Brazed Assemblies
AZ31B Base Metal; GA432A Filler Metal

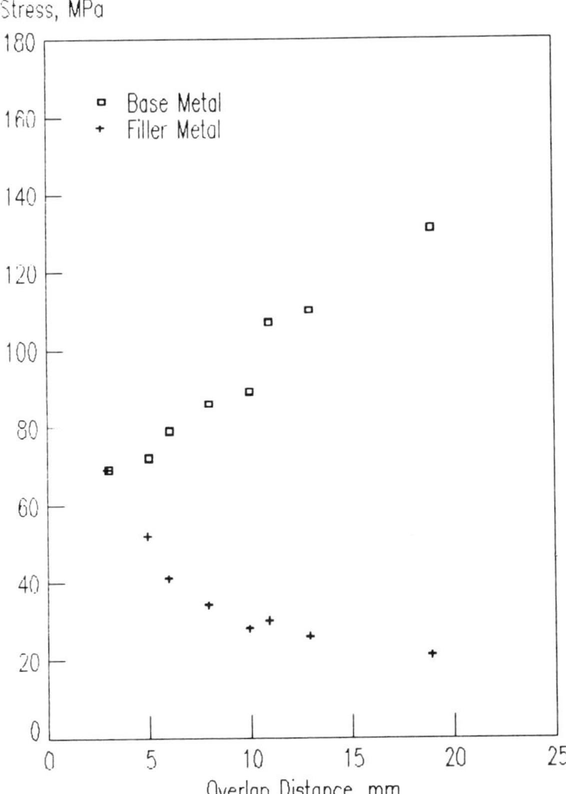

FIGURE 5.4 Average unit shear stress in the filler metal and average unit tensile stress in the base metal as a function of overlap distance. Failures were in the filler metal with overlap through 13 mm. Failure was in the base metal at 19 mm. □, Base metal (AZ31B); +, filler metal (GA432A). Conversion factors: mm × 0.0394 = in, MPa × 0.1450 = ksi.

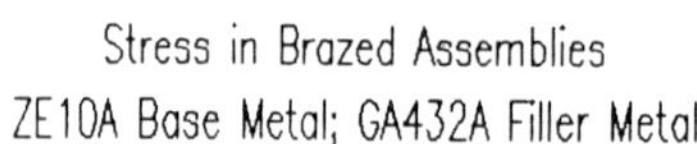

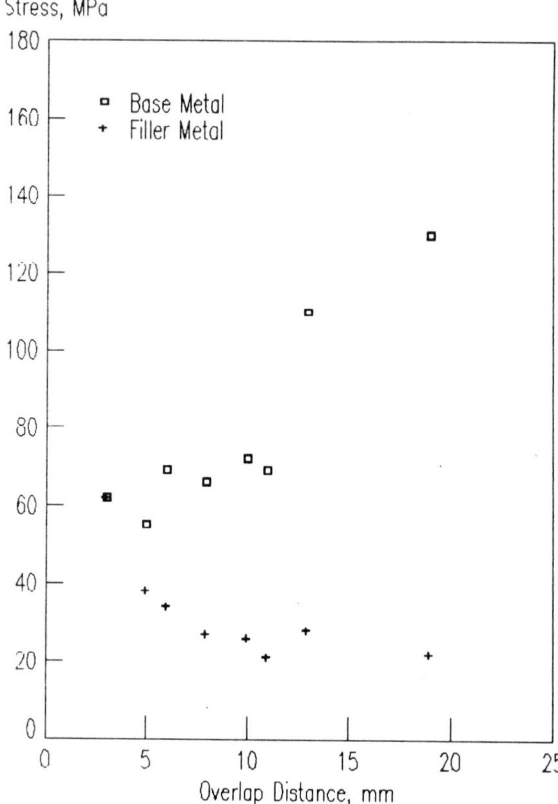

FIGURE 5.5 Average unit shear stress in the filler metal and average unit tensile stress in the base metal as a function of overlap distance. Failures were in the filler metal with overlap through 6 mm. Beyond 6 mm, failures were in the base metal. □, Base metal (ZE10A); +, filler metal (GA432A). Conversion factors: mm × 0.0394 = in, MPa × 0.1450 = ksi.

5.2 MECHANICAL JOINTS

Mechanical joints include rivets, screws, bolts, self-clinching devices, and interference fits. Since magnesium is not suitable for the joining material, such as a bolt or rivet, a common factor in all mechanical joints for magnesium is the use of a dissimilar metal. The possibility of galvanic corrosion thereby introduced must be guarded against in the preparation of all of these joints. The section on galvanic corrosion in Chapter 14 should be carefully reviewed for methods of protecting against galvanic corrosion. These methods must be used when using metallic fasteners.

5.2.1 Riveting

The riveting of magnesium is not greatly different from the riveting of other materials. Selection of the rivet material and protection of the joint are important. Care should be given to rivet spacing, both that between rivets and that from the edge. Proper hole preparation and hole size are important.

Rivets should be no closer together than three rivet diameters, and no closer to the edge than 2.5 rivet diameters. The edge should be smooth and free from any defects introduced by shearing. If this is not the case, then the edge distance must be increased to three times the rivet diameter.

It is important to keep stress concentrations minimized. The hole drilled for the rivet should be slightly larger than the diameter of the rivet being used, with a clearance of about 0.1 mm (0.004 in) usually being sufficient. The hole can be countersunk by machining, or the sheet can be dimpled. If the latter, the hole should first be drilled slightly smaller than the rivet diameter, the sheet dimpled, and then the final hole size drilled. Squeeze riveters rather than pneumatic hammers should be used to obtain better control of driving pressure, thus minimizing damage to the sheet from overdriving.

Dimpling is limited to sheet no thicker than 1.25 mm (0.05 in). The operation must be done hot, using tools that will heat the sheet locally for the short period of time needed to form the dimple. If the sheet is in a cold-rolled condition, the time and temperature must be such that excessive annealing does not occur. The time can usually be controlled to be no more than about 5 seconds, and, if this is the case, AZ31B-H24 can be heated to 300°C, HK31A-H24 to 440°C, and HM21A-T8 to 455°C [12]. If the entire sheet must be heated, see Chapter 6 on forming for allowable times and temperatures.

The design of the joint should follow the principle that the rivet, rather than the sheet, will fail first. Thus the size and the spacing of the rivets is selected on the basis of the bearing strength of the sheet as well as the strength needed for the joint itself. The design bearing ultimate strengths for sheet are as follows:

AZ31B-O: 410 MPa, 59 ksi
AZ31B-H24: 470MPa, 68 ksi
HK31A-O: 350 MPa, 51 ksi
HK31A-H24: 390 MPa, 57 ksi
HM21A-T8: 365 MPa, 53 ksi

See Chapters 8-12 on mechanical properties for more complete property data and for data for forms other than sheet.

5.2.2 Bolting

Bolts or studs can be threaded or cast directly into magnesium, in which case coarser, more rounded threads than are used for steel or aluminum, should be used. Aluminum alloy 6061, if sufficiently strong, results in the least galvanic corrosion. If steel must be used for strength, the bolts should be plated with aluminum, zinc, tin, or cadmium to minimize galvanic corrosion problems.

While threading directly in the magnesium is satisfactory for many purposes, a stronger joint is obtained by using steel bushings or inserts. This is especially true if the bolt will be threaded in and out during service. Pressed- or shrunk-in inserts can be used, but the various screwed-in types are preferable. One of the more popular inserts is the precision helical coil that is used between the tapped hole and the bolt. The insert should be at a depth of at least 1.5 times the bolt diameter, and preferably as much as 3 times.

Pressed- or shrunk-in inserts should have an interference at room temperature of as little as needed for adequate holding of the insert. A unit strain of 0.001 is usually adequate and results in a stress level that is low enough to avoid problems in most instances. Unit strains as low as 0.0003 have been used successfully for some applications. Unit strains as high as 0.003 have been used, but the stress level is then high enough that there is danger of stress corrosion if used with susceptible alloys, and increased potential for fatigue damage for all alloys. Magnesium's high coefficient of thermal expansion relative to other metals requires special consideration if interference inserts will be used in structures that then serve at elevated temperatures. It may be necessary to increase the room-temperature interference in order to retain adequate holding power at the elevated temperature.

The various kinds of locking and self-clinching nuts can all be used satisfactorily on magnesium, with no changes from the procedures used with other metals.

Washers at least three times the bolt diameter should be used under both bolt heads and nuts, and should preferably be of a compatible aluminum alloy such as 5052 or 6061 provided the torque is low enough that the aluminum will not deform. The washers serve two purposes: to prevent damage to the magnesium surface as the fastener is tightened; and to minimize galvanic attack, if aluminum is the washer material.

Since dissimilar metals are used for bolting, problems with galvanic corrosion must be recognized and prepared for. See the section on galvanic corrosion in Chapter 14 on corrosion for a full discussion of the problem and methods of handling. It is important that solutions for a specific application be detailed during design.

5.3 ADHESIVE BONDING

This is an extremely versatile method, being suitable for the joining of any form or size of magnesium to almost any other material. Because there is no disturbance of the metal, as by drilling or melting, there are no stress concentrations. Because the adhesive has a low modulus of elasticity, it will not transmit stress as readily as a "stiffer" joint. Thus, the method is especially suited for those applications that require a good fatigue strength. Because the adhesive is an insulator and prevents water from penetrating the joint, there is no danger of galvanic corrosion.

With the extensive development that has occurred in adhesive technology, a wide choice of adhesive types, strengths, and application methods is available to the designer. All types can be applied to magnesium, and the specific properties of the adhesive should be sought from the literature or the manufacturers. Table 5.16 lists a few that have been used for magnesium joining, as a preliminary guide. Techniques specific to magnesium relate to curing temperatures and times, and to surface preparation.

Most adhesives require a curing time at an elevated temperature. It is important to assure that the times and temperatures used do not decrease the properties of the magnesium alloys being bonded. The effects of time and temperature on the properties of alloys, which are detailed in the chapters dealing with mechanical properties, should be consulted.

Proper surface preparation prior to application of the adhesive is important. This can be simply cleaning off oil and grease followed by sanding. However, for best results, the surface should be prepared in the same way as for applying paint by use of chemical conversion coatings or anodic coatings. Some military specifications in the United States require an anodic coating plus a primer before applying the adhesive. This is good practice if severe corrosion conditions are expected in service, but the joint strength is decreased by the primer. The shear strength of the adhesive is, in general, greater than that of a conversion coating or a paint film. Therefore, the films applied during surface preparation should be thin, commonly about 0.003 mm (0.0001 in) or less for conversion or anodic coatings, and 0.008 mm (0.0003 in) or less for paint films. Service conditions can be categorized in the following four ways [13]:

1. Interior exposure with short life expectancy or low strength
2. Interior exposure with long life expectancy or high strength
3. Exterior exposure under commercial conditions
4. Millitary requirements; emphasis on corrosion protection

Surface preparation procedures appropriate for each of these service conditions are:

1. Degreasing plus mechanical cleaning
2. Degreasing plus chrome pickling
3. Degreasing followed by thin anodizing
4. Degreasing followed by both thin anodizing and priming with a thin film

See Chapter 14 on finishing for details of the coating procedures.

TABLE 5.16 Characteristics of Some Adhesives Used for Magnesium

Type of composition	Curing conditions			Shear strength at 20°C MPa
	Temperature, °C	Time, min	Pressure, MPa	
Phenolic rubber-base resin	160	32	0.4-3.5	15-18
Ethoxyline resin—two liquids	20	1440	contact	10-15
Epoxy resin paste plus liquid activator	95	60	contact	20
Rubber base	200	8	1.4	12-16
Vinyl phenolic	150	8 (preheat)		
	135-200	70 to 4	varied	7-12
Epoxy type resin	95	45	contact	8-12
	95	15	0.10	10-12
Modified epoxy resin	175	60	contact	19-30
Phenolic	150	15	0.2	16

Conversion factor: MPa × 0.1450 = ksi.

REFERENCES

1. Dow Chemical Company Bulletin Number 141-300-70, 1970.
2. Dow Chemical Company Bulletin, "Arc Welding Magnesium," 1965.
3. American Society for Metals, "Welding Brazing and Soldering," *Metals Handbook*, Vol. 6, 9th ed., p. 427, American Society for Metals, Metals Park, Ohio.
4. P. Klain, "Stress-Relaxation and Stress-Relief of Some Magnesium Alloys," *Welding Research Supplement*, September 1955.
5. Magnesium Elektron, Ltd., "Joining of Magnesium Alloys," Bulletin, June 1979.
6. R. W. Fenn, Jr. and L. F. Lockwood, "Low-Temperature Properties of Welded Magnesium Alloys," *Welding Journal Research Supplement*, August 1960.
7. R. J. Cross, "Design and Fabrication of Welded Magnesium Tanks for Aircraft," *Light Metal Age*, January 1947.
8. Dow Chemical Company Bulletin, "Magnesium in Design," 1967.
9. D. R. Apodaca and J. G. Louvier, "Static and Fatigue Properties of Repair-Welded Aluminum and Magnesium Premium Quality Castings," *Metals Engineering Quarterly*, *7*(1):23 (1967).
10. Dow Chemical Company Bulletin, "Resistance Welding Magnesium," 1970.
11. "Resistance Welding Equipment Standards," American National Standard C-88.2.
12. Dow Chemical Company Bulletin, "Riveting and Mechanical Fastening of Magnesium," 1964.
13. Dow Chemical Company, "Preparation of Magnesium Surfaces for Adhesive Bonding," Technical Service and Development Letter Enclosure, February 21, 1962.
14. Dow Chemical Company, "Dip Brazing Magnesium," TS&D Letter Enclosure, May 16, 1968.

BIBLIOGRAPHY

Lockwood, Lloyd, "Spot Welding of Wrought HK31A, HM21A, and ZE10A Magnesium Alloys," American Welding Society, Feburary 1960.

Lockwood, Lloyd, "Gas-Metal-Arc Spot Welding of Magnesium," *Welding Journal*, February 1963.

Lockwood, Lloyd, "Automatic Gas Tungsten-Arc Welding of Magnesium," *Welding Journal*, July 1964.

Lockwood, Lloyd, "Gas Shielded Stud Welding of Magnesium," *Welding Journal*, *46*(4):168S (1967).

Lockwood, Lloyd, "Repair Welding of Thin-Wall Magnesium Castings," *Transactions of AFS*, 1967.

6

Forming

6.1 GENERAL CONSIDERATIONS

The word forming is used to signify a large variety of techniques for changing the shape of a piece of metal through plastic deformation. These can be classified into two general types: those that involve free straining such as bending and drawing and those that involve constrained straining such as forging and coining.

When free-straining methods are used, the metal must be capable of distributing the strains uniformly so that local necking does not occur. Frequently, metal that has been stretched and thinned must then transfer the stress to other parts of the piece that are then, in turn, stretched and thinned. Thus, the first metal to be stretched must become capable of supporting higher unit stresses for the completion of the forming operation. The important material characteristics for good formability are therefore a capability for large, uniformly distributed strains and of hardening sufficiently to support an increasing unit stress without further strain.

When constrained-straining methods are used, the material must be capable of large strains, but any required uniformity of strain distribution is forced by the method. Hardening is not required for the operation itself.

The plastic deformation mechanisms available to magnesium have a direct bearing on the forming of magnesium parts. The principles and their consequences briefly reviewed below need to be taken into account when designing any part that will need to be formed.

Principle: The extent of plastic deformation that is possible increases greatly as the temperature is raised. Consequences: While limited forming is possible at room temperature, enough greater deformation is possible as the temperature is raised that it is common to form magnesium at elevated temperatures. Since, depending upon the alloy and temper, exposure to high temperature can result in a lowering of the room temperature mechanical properties of the metal being deformed, a compromise between degree of formability and lowering of properties is sometimes needed.

Principle: Some of the plastic deformation mechanisms are diffusion controlled and hence time-dependent; this effect increases with increasing temperature. Consequences: More severe forming is possible as the speed of forming is decreased. This effect can sometimes be used to arrive at a more desirable compromise between degree of forming required for a part and the lowering of the mechanical properties because of exposure to elevated temperatures.

Principle: Wrought products, such as sheet or extrusions, have a preferred orientation such that extensive twinning occurs when a compressive stress is applied parallel to the direction of working. Consequences: The neutral axis shifts 5 to 10% toward the tension side of a bend, resulting in a shortening of the length after bending.

Principle: Because of restricted plastic deformation modes at room temperature, there is sensitivity to stress concentrations. Consequencies: Parts must be designed with as generous radii as possible in order to minimize the effects of stress concentrations on the serviceability of the part. Tooling must be well polished to eliminate scratches and gouges.

6.2 FREE-STRAINING METHODS

All free-straining methods include one or more elements of bending, drawing, and stretch forming. An understanding of the requirements and limitations for magnesium for these three types of forming can be applied to methods such as spinning, joggling, dimpling, or any others that may be used. Two types of failure occur: extension by tension beyond the point of uniform strain, resulting in excess, localized thinning or tearing; and buckling upon the application of compressive stresses, resulting in wrinkling.

6.2.1 Bending

Bending is both the simplest and the most widely used of all forming methods. It is often the only operation performed in forming a piece, and is frequently a component of other methods, such as drawing, joggling, or dimpling. Simple bending is the wrapping of a piece of flat metal around a tool having a given radius of curvature. The angle made by the two parts of the sheet after bending can be anything desired. If the angle is 90 degrees, and the neutral axis of the bend is in the center of the sheet cross section, the percent tensile elongation on the outside of the bend is $100t/(2R + t)$, where t is the sheet thickness and R is the radius of curvature of the tool around which the sheet is bent, both in consistent units. As the bent sheet is held against the tool after bending, there is both a plastic and an elastic component to the elongation. When the tool is removed, any elastic elongation is recovered, and there is "springback." The proportion of the elongation that is elastic, and thus responsible for springback, depends upon the properties of the alloy being bent, decreasing for all alloys as the temperature is increased and the speed of bending is decreased.

Suggested minimum bend radii for various alloys and tempers in sheet form are given in Table 6.1 [1], and for extrusions in Table 6.2 [1]. The data suggest radii that are expected to result in successful bends 99% of the time. The effect of accepting a lower success rate, or higher scrap rate, is shown in Figure 6.1 for AZ31B-O [2]. If a scrap rate of only one in 10,000 is required, the minimum bend radius for this alloy is then 7t at room temperature. The data in Tables 6.1 and 6.2 are also for bends with a 90° angle. The minimum radius changes with the angle as shown in Figure 6.2 [3]. Little or no change occurs as the angle becomes larger than 90°, but the radius decreases markedly as the angle becomes smaller than 90°.

Suggested allowances to be made for springback after making a 90° bend are given in Table 6.3 for AZ31B. Since the allowance to be made in any given case is determined by the procedures used (for example, tooling, angle of bend used, speed of making the bend, any dwell time after the bend), the exact correction must be determined for each case. Since the amount of elastic elongation remains constant at a given temperature and speed of bend, the proportion of the elastic contribution to the total elongation increases as the total elongation decreases. Thus springback increases as the angle of bend decreases and as the radius of curvature increases. As the temperature is raised, the contribution of elastic deformation decreases markedly, resulting in much less springback.

Table 6.4 [4] gives recommended time and temperature exposure limits for alloys in sheet and extrusion form, which can be used as a general guide for limiting exposure during forming. However, the tables in Chapters 10 and 11 showing the effect of exposure at elevated temperatures on room-temperature properties should be consulted for arriving at the best compromise between formability and mechanical properties. While castings are not generally formed in the same sense as are wrought products, limited forming operations may be performed, as in straightening castings or in staking. The effect of exposure to elevated temperatures on the room temperature properties of castings as given in Chapter 8 and 9 should be consulted.

Residual stresses accumulate during forming operations, the amount depending upon the forming operation and the temperature at which it is performed. If the forming temperature for an alloy is lower than the minimum temperature shown in Table 6.4, stress relief may be required to prevent distortion either during further operations or during service. All of the AZ-type alloys containing more than 1.5% aluminum do require stress relief when forming is carried out below the minimum temperatures shown in Table 6.4 in order to prevent stress corrosion failure during service. Table 6.5 [1] gives suggested stress-relieving schedules that will result in substantial lowering of the stresses without greatly lowering the mechanical properties.

As pointed out above, the deformation by twinning that occurs on the compression side of the bend results in a net shortening of the piece. This needs to be accounted for so that the final piece will have the proper dimension. The nomograph given in Figure 6-3 [5] aids in this calculation. The effect is quite small for thin pieces, but can become significant as thickness increases.

6.2.2 Drawing

The act of forming a shell by using a punch to push a flat piece of metal through a die is known as drawing. What is produced can vary from a simple cylinder to a wide variety of shapes with flanges. Arbitrarily, a shallow draw is considered one where the depth of the cup is less than about one-half the diameter; anything deeper than this relative to the diameter is called a deep draw. If the draw required to produce a given shape is so deep that the metal fails during the draw, multiple draws, each of less depth than the total, are used. The metal may or may not be annealed between the draws required to reach the final shape.

Up to four parameters are used to define the severity of a draw: the percent reduction from the blank diameter to the cup diameter, $100(1-d/D)$, the die radius of curvature which is a measure of the severity of the bend required as the blank is pushed through the die, the radius of curvature between two side walls for box-like shells, and the radius of curvature of the punch, which forms the radius between the side wall and the bottom. These parameters must be kept within workable limits if the draw, or series of draws, is to be successful.

As a round blank is drawn into a cup, the metal that will be forced through the die must first be compressed from its radius to the radius of the die opening. Since the volume occupied by the metal does not change, the blank becomes thicker as it reaches the die, and remains thicker as it bends to go through the die. After being bent over the radius, and furing its extension into the wall of the cup, stretching and thinning occurs. The metal that is under the punch is essentially undeformed during the operation. Thus, during drawing, there is first compression and thickening followed by bending and then tension and thinning. The same principles apply when a rectangular shell is made, except that the compresison and thickening is confined to the vertical corner regions. The walls are simply bent and stretched. During the compression stage, there is a strong tendency for the metal to buckle, resulting in the formation of wrinkles. This is prevented by providing a blank holder to exert enough force to prevent the buckling. The thinner the sheet being drawn, the greater must be the blank holding force to prevent wrinkling.

Since, during the draw, the metal that has reached the wall is thinner than the metal that has just come around the die radius, the metal in the wall must support a higher unit stress than the metal at the radius in order to continue stretching the entire blank into the wall. This is possible, when the draw is done at room or low temperatures because the metal that has been stretched has been strengthened by work hardening. There is added flexibility when the draw is at elevated temperatures. The punch can be kept at a lower temperature than the die, resulting in higher strength in the cooler wall metal than in the warmer metal at the die.

Magnesium can be drawn at room temperature to a reduction of about 10-15%, resulting in a depth of less than one fifth of the cup diameter, a quite shallow draw. While multiple draws can be made at room temperature to arrive at a final deep draw, in practice all deep drawing of magnesium is done at elevated temperatures. The effect of temperature on lubrication practice and the modifications of tool design required to account for thermal expansion must therefore be considered.

At high temperatures, magnesium alloys can be deep drawn in one operation to a cylinder by a reduction of as much as 75%, resulting in a depth almost four times the cup diameter. For rectangular-shaped shells, the maximum percent reduction is less. The maximum percent reduction attainable in any specific case is influenced by the alloy, the temperature of the die, the temperature of the punch, the speed of the punch movement, and the tool design. In general, as the temperature is raised and the speed of the draw is reduced, the greater will be the amount of draw possible, as is illustrated in Table 6.6 [5]. The speed of drawing can vary from about 4 to as much as 2500 cm/min (1.5 in/min to 82 ft/min); the speed selected will depend upon the severity of the draw and any compromise required between using single and multiple draws, the latter requiring additional tooling.

The mechanical properties in the drawn cup depend upon the exposure of the metal to temperature and to the amount of working the metal receives. The temperature of the blank should be kept within the limits for maximum exposure given in Table 6.4 for maintenance of properties in those parts of the metal that will not be worked, such as the bottom of the can. The information in Chapter 11 should be consulted for deviations from Table 6.4 that can be made if some lowering of mechanical properties is permissible. Those parts of the metal that are deformed, such as the wall of a cup, can be considerably strengthened, as is illustrated for AZ31B-O and ZE10-H24 and -0 for a cup drawn to a reduction of 55% at 200°C:

	Before drawing					After drawing				
		TYS		TS			TYS		TS	
Alloy	%E	MPa	ksi	MPa	ksi	%E	MPa	ksi	MPa	ksi
ZE10-O	29	172	29	234	34	15	269	39	310	45
ZE10-H24	25	200	29	262	38	15	262	38	324	47
AZ31B-O	22	179	26	269	39	11	290	42	345	50

The limits for magnesium are shown in Table 6.7 [5] for the parameters defining the severity of the draw. In addition to taking into account the limits shown in Table 6.7, tool design must provide sufficient clearance to allow for the thickening that occurs as the blank is drawn to the die. A guide for the clearances to provide is [6]:

Blank thickness		
mm	in	Clearance
<0.40	<0.016	1.07-1.09t
0.40-1.3	0.016-0.050	1.08-1.10t
1.3-3.2	0.050-0.125	1.10-1.12t
>3.2	>0.125	1.12-1.14t

where t is the blank thickness.

Hydraulic presses are usually preferred for the deep drawing of magnesium alloys because they can operate at slower and more uniform speeds. Mechanical presses are used, however, for making draws that are not too severe. The force required for deep drawing magnesium to a cylinder can be estimated by use of the following empirical formula [4]:

$$L = \pi dtS(D/d - k)$$

where

L = Press load, in Newtons
d = Cup diameter in mm
D = Blank diameter in mm
t = Blank thickness in mm
S = Tensile strength at temperature in MPa
k = A constant, which is between 0.8 and 1.1 for magnesium
Conversion factor: Newtons × 0.225 = lbf

The values for the tensile strength of magnesium alloys at elevated temperatures shown in Chapter 11 can be used to estimate the force that will be needed to make any given draw. The hold-down force needed to prevent wrinkling depends upon blank thickness and temperature, but is between 1 and 10 percent of the drawing force.

Lubrication is required to prevent galling and build up on the tools and to minimize friction between the tools and the work piece. If the friction is too high, the stresses on the thinned metal in the wall will become high enough to result in tearing. Up to a temperature of 120°C, the oils, greases, and waxes normally used for deep drawing of other metals at room temperature can be used. Between 120 and 230°C, a soap lubricant is satisfactory. This is an aqueous solution and is applied by dipping or brushing and then drying in air. It is easily removed by washing in hot water. Above 230°C, colloidal graphite or molybdenum disulfide must be used. Since both materials are cathodic to magnesium, they must be completely removed in order to prevent the potential of severe galvanic corrosion in service. This can be done mechanically, such as by wire brushing, or by use of a chromic-nitrate pickle (see Chapter 14). It is helpful to the cleaning process if the blank is free of chemical conversion coatings, since they tend to increase the difficulty of removing the graphite or molybdenum disulfide.

The tools must be kept free of buildup of any kind. Rough tools result in gouging of the workpiece surface which is unsightly and can lead to stress concentrations that will promote failure in service. Some deposits transferred to the magnesium can lead to corrosion problems. The tools can be cleaned mechanically, or, in the case of magnesium buildup, with an inhibited hydrochloric acid solution.

6.2.3 Stretch or Compression Forming

If the piece to be formed is held in tension during the forming operation such that only tensile stresses are present even on the "compression" side of a bend, the operation is known as stretch forming. The particular feature of interest with this method is that the neutral axis is displaced outside of the piece being formed and no twinning takes place. The forming limits for magnesium are the same as in simple bending, taking into account the total elongation imposed on the piece by both the imposed tensile stress and the bending. Temperature considerations remain the same.

An effect opposite to that of stretch forming can be obtained by putting the piece into compression during the forming operation. The neutral axis is displaced outside of the piece, but on the opposite side as for stretch forming. There is no tensile elongation during forming, but the compressive strains induced can lead to buckling and wrinkling.

6.2.4 Other Methods

Many other methods of forming have been developed for attaining special effects, such as rubber forming, spinning, roll forming, dimpling, and joggling. All of these techniques combine the basic features of bending, drawing, and stretch or compression forming in various ways. The limitations and special precautions for magnesium that are applicable to bending or drawing must also be observed with all of the other types.

TABLE 6.1 Minimum Bend Radii[a] for Sheet (90° Bend; 99% Success Rate)

		Temperature, °C						
Alloy	Temper	20	100	150	200	260	315	370
AZ31B	0	5.5	5.5	4.0	3.0	2.0		
	H24	8.0	8.0	6.0				
HK31A	0	6.0	6.0	6.0	5.0	4.0	3.0	2.0
	H24	13.0	13.0	13.0	9.0	8.0	5.0	3.0
HM21A	T8	9.0	9.0	9.0	9.0	9.0	8.0	6.0
	T81	10.0	9.0	9.0	9.0	9.0	8.0	6.0
ZE10A	0	3.0	2.0	1.0				
	H24	8.0	8.0	6.0				

[a]Expressed as radius/sheet thickness.

TABLE 6.2 Minimum Bend Radii[a] for Extrusions (90° Bend; 99% Success Rate)

Alloy	Temper	R/t	Temperature, °C	R/t	Temperature, °C
AZ31B	F	2.4	20	1.5	290
AZ61A	F	1.9	20	1.0	290
AZ80A	F	2.4	20	0.7	290
	T5	8.3	20	1.7	230
HM31A	T5	11.0	20	6.0	425
ZK21A	F	15.0	20	5.0	315
ZK60A	F	12.0	20	2.0	290
	T5	12.0	20	6.6	260

[a]Expressed as radius/thickness.

TABLE 6.5 Suggested Springback Allowances for Right Angle Bends (0.4-1.6 mm (0.016-0.064 in) Thick Sheet)

Temperature, °C	Bend, R/t	AZ31B-O, degrees	AZ31B-H24 degrees
20	4	8	10
	5	11	13
	10	17	21
	15	25	29
100	3	4	5
	5	5	7
	10	8	12
	15	13	17
150	2	1	2
	5	3	4
	10	5	7
	15	8	11
230	2	0	0
	5	1	1
	10	2	2
	15	4	4
290	Up to 15	0	0

TABLE 6.4 Time and Temperature Exposure Limits

Alloy	Temper	Form	Temperature, °C					
				Minutes of exposure				
			Mimimum	1	3	10	30	60
AZ31B	-0	Sheet	120					290
	-H24	Sheet	120	225	200	180	175	150
	-H26	Sheet	120	225	200	180	175	150
	-F	Extrusion	120					290
AZ61A	-F	Extrusion	200					290
AZ80A	-F	Extrusion	140				290	
	-T5	Extrusion	140					195
HK31A	-0	Sheet	200					400
	-H24	Sheet	200					345
HM21A	-T8	Sheet	200			425		345
	-T81	Sheet	200					345
HM31A	-T5	Extrusion	290					425
ZK60A	-F	Extrusion	150					290
	-T5	Extrusion	150				200	

Note: If material is deformed below the minimum temperature shown, stress relief may be required.

TABLE 6.5 Times and Temperatures for Stress Relief

Alloy	Temper	Temperature, °C	Time, min
Sheet			
AZ31B	0	260	15
	H24	150	60
HK31A	H24	290	30
HM21A	T8	370	30
	T81	370	30
Extrusions			
AZ31B	F	260	15
AZ61A	F	260	15
AZ80A	F	260	15
	T5	200	60
HM31A	T5	425	60

TABLE 6.6 Drawability of AZ31B-O at Varying Temperatures and Speeds

	% Reduction		
Drawing temperature, °C	600 cm/min (235 in/min)	1500 cm/min 590 in/min)	2400 cm/min 945 in/min)
20	14.3	14.3	–
120	30.3	30.3	–
200	58.6	57.1	53.8
260	62.5	61.3	53.8

(a) 76 mm (3 in) diameter cylindrical cup.
(b) Sheet thickness = 1.65 mm (0.064 in).
(c) Draw-ring radius = 6t.
(d) Punch radius = 10t.

TABLE 6.7 Limits for Drawing AZ31B-O Sheet

Parameter	Cylindrical cups	Rectangular cups
Draw ring radii	4t to 7t	4t to 7t
Bottom corner radii	3t	3t
Vertical corner radii		1/20d to 1/12d
Percent reduction	70	

Note: t = sheet thickness, d = depth of draw.

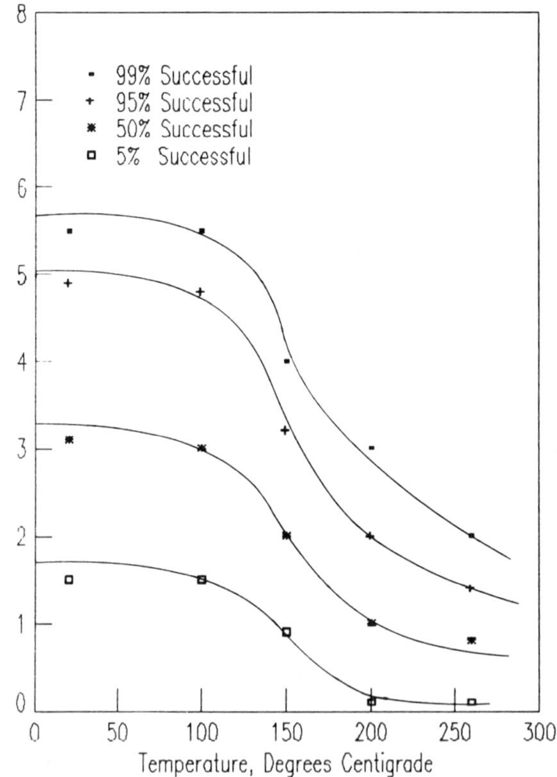

FIGURE 6.1 Press brake bending (AZ31B-O).

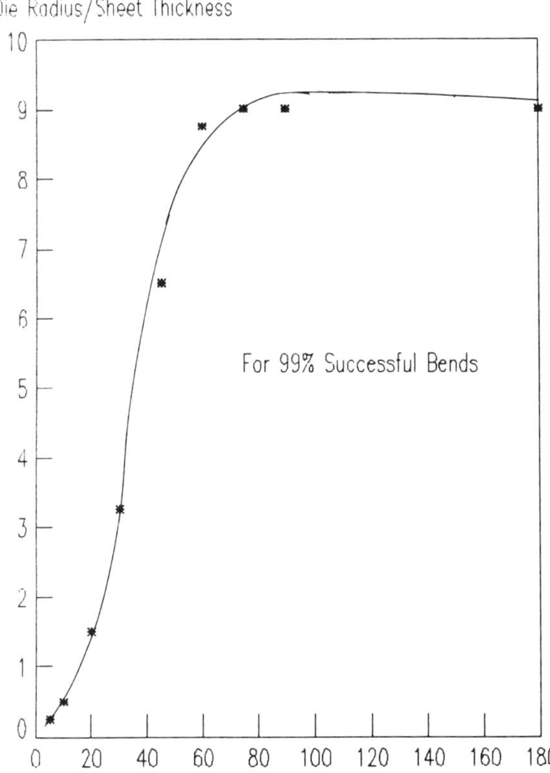

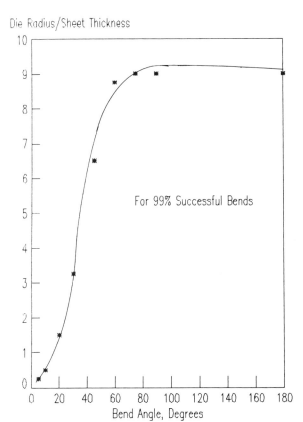

FIGURE 6.2 Effect of bend angle (HM21A-T8). For 99% successful bends.

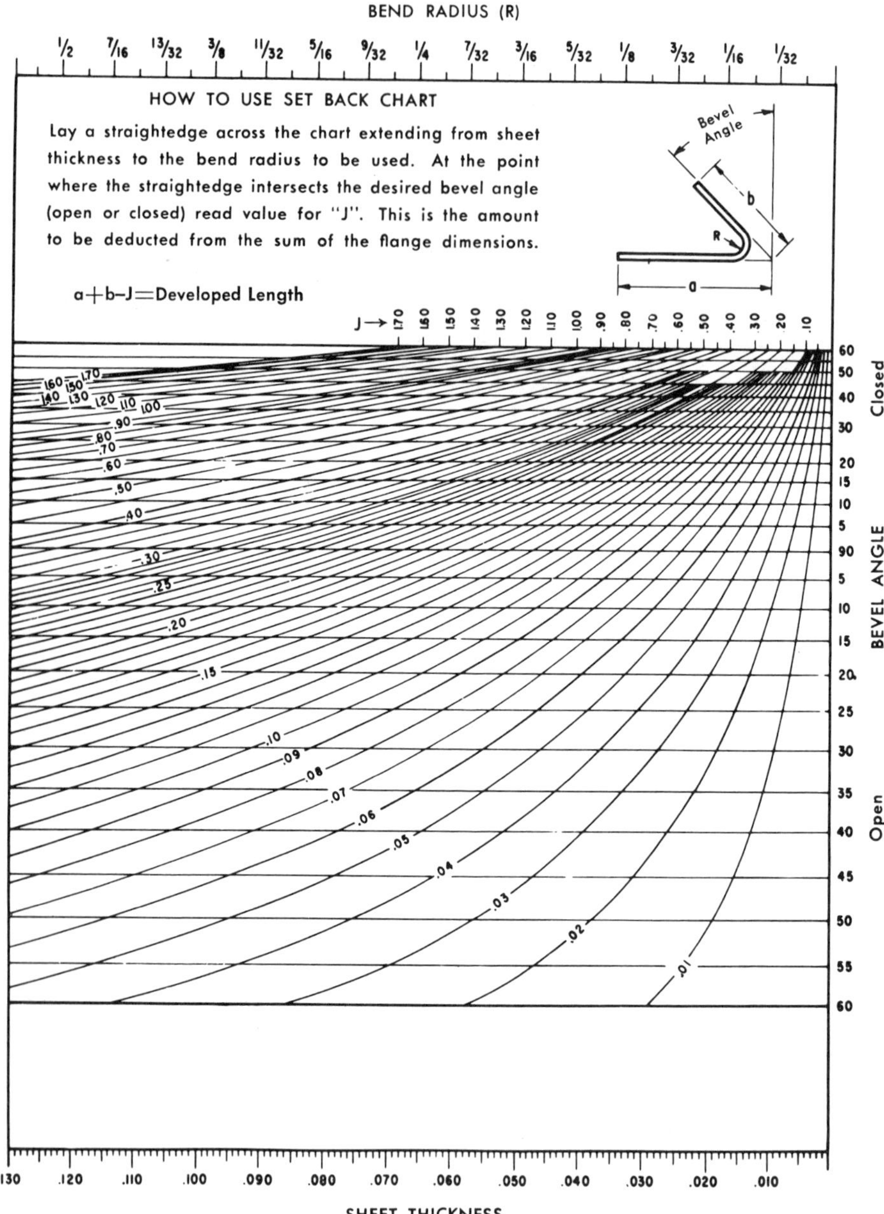

FIGURE 6.3 Nomograph for calculating length changes after bending.

6.3 CONSTRAINED-STRAINING METHODS

Any forming method that forces both the direction and the amount of strain throughout the piece is in the class of a constrained-strain method. The largest member of this class is forging, but it also includes coining, embossing, and impact extrusion.

6.3.1 Forging

Although metals are forged, as is done by the old-fashioned blacksmith, by pounding with a hammer from various directions until the desired contour is obtained, the usual method is to shape the piece by forcing the metal into die cavities. Because large strains are the rule, forging is performed at elevated temperatures for all metals.

As large strains are imposed, a metal develops a preferred orientation, both in the crystallographic sense, and in the sense that second-phase particles and the long axis of individual grains line up in the direction of the strain. Mechanical properties in the finished piece depend upon the direction of measurement with respect to the preferred orientation; in general properties measured parallel to the direction of strain are higher. An important part of forging design is to assure that metal flow during forging is directed so that maximum properties are developed where needed in the final part.

The mechanical properties developed in a part depend upon the alloy composition, the degree of work hardening present, the grain size, and the state of heat treatment. Forging, which is done at elevated temperatures, lends itself to control of all of these factors, with an accompanying high degree of control over mechanical properties. The forging schedule for a specific part should be chosen so that properties are maximized through optimization of the metallurgical variables.

A forging consists of webs, ribs, bosses, cavities, and holes. Ribs and bosses are stiffeners or bearing surfaces for such things as bolts. Cavities and holes serve to lighten the structure or to accommodate other parts. Webs hold all of the features together into one part. The important design features for a forging are the height, length, and width of ribs and bosses, the thickness of all sections, radii of curvature where two walls meet at an angle less than 180 degrees (fillets), and radii of curvature where two walls meet at an angle greater than 180 degrees (corners). Both fillets and corners can be horizontal or vertical with respect to the plane of forging.

Since the shape is formed in a die cavity, draft on the walls is usually required to allow extraction of the finished piece from the die. A draft angle of five degrees is normally used for magnesium, although it is possible to forge even with zero draft. Since greater care and closer tolerances are required for zero draft than for a forging with draft, the costs of a more expensive forging compared to those for machining the extra stock of the draft must be compared.

Because the design of the forging influences not only the service characteristics but also the metal flow during forging, there is a strong interaction between the design and the forging process resulting in turn in a strong influence of design on the mechanical properties developed in the finished piece. It is critically important that the designer consult with the forger at an early stage so that the design and the process will both be optimized to produce the best possible part for its projected service conditions.

Table 6.8 [4] shows section thickness recommendations for magnesium forgings. Minimum corner and fillet radii are given in Table 6.9 [4]. In the latter case in particular, the minima should be observed. In many cases, larger radii will be preferable, either because of serviceability or producibility. The advice of the forger should be sought before the final design is determined.

6.3.2 Impact Extrusion

If a slug of metal is placed in a die, and a punch then forced into the slug, the metal is forced up around the punch, between the punch and the die. Symmetrical, can-like shapes with very thin walls, close dimensional tolerances, and large depth-to-diameter ratios can be made with the process. With fully automated equipment, production rates can be as high as a hundred cans/minute.

Slugs can be made in any convenient way, either from wrought or from cast metal, but generally the least expensive method is by sawing slugs from an extruded rod. Slugs so prepared must be deburred and lubricated by coating with graphite or molybdenum disulfide before being impact extruded. The diametrical tolerance of the slug should be held to ±0.25 mm (0.010 in) of the diameter of the die at the extrusion temperature in order to assure symmetrical extrusion [7].

The temperature of extrusion varies from about 250 to 400°C depending upon the alloy and the shape being extruded. Pressures required vary from about 140 to 240 MPa (20 to 35 ksi), depending upon the alloy and temperature [8].

As long as the part is symmetrical about central vertical planes, a wide variety of shapes can be formed by this method, varying from simple cylindrical to straight-sided cans containing bosses, ribs, or multiple internal walls. Typical tolerances that can be held on a cylindrical can having a length-to-diameter ratio of 6 to 1 are [8]:

Wall thickness		± Tolerance	
mm	in	mm	in
0.50	0.02	0.050	0.002
0.75-1.00	0.030-0.040	0.075	0.003
1.15-1.40	0.045-0.055	0.100	0.004
1.50-2.55	0.060-0.100	0.125	0.005

On simple shapes, very sharp radii can be formed. On some more intricate parts, the metal flow pattern will sometimes dictate more generous radii. The extruder should be consulted where there is any doubt.

As with other forming methods requiring graphite or molybdenum disulfide for lubrication, the finished cans must be carefully cleaned before use. The chromic-nitrate pickle is suitable (see Chapter 14).

TABLE 6.8 Thickness Recommendations for Forgings

W × L -		W (maximum)		t_1 (minimum)	
mm^2	in^2	mm	in	mm	in
<650	1.0	180	7	3.0	0.12
650-1300	1.0-2.0	300	12	5.0	0.20
1300-2600	2.0-4.0	500	20	5.5	0.22
2600-5200	4.0-8.0	800	30	6.5	0.25

Note: t_2 varies from 3.0 mm (0.12 in) at D = 6.5 mm (0.25 in) to 6.5 mm (0.25 in) for D > 50 mm (2 in). The ratio of rib thickness to rib height varies from 1:2 for ribs less than 5.0 mm (0.20 in) in height to 1:5 for ribs over 40 mm (1.6 in) in height.

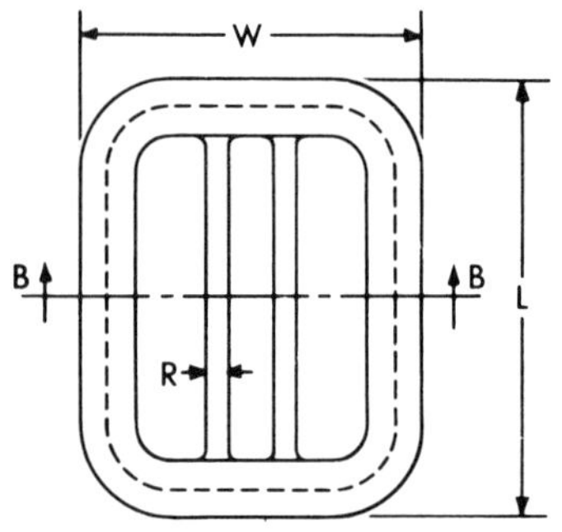

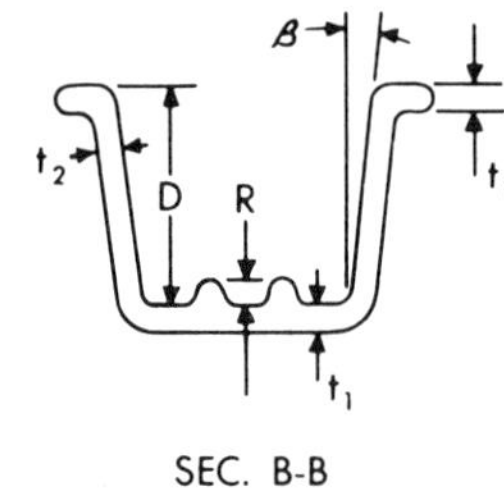

TABLE 6.9 Minimum Corner and Fillet Radii for Forgings

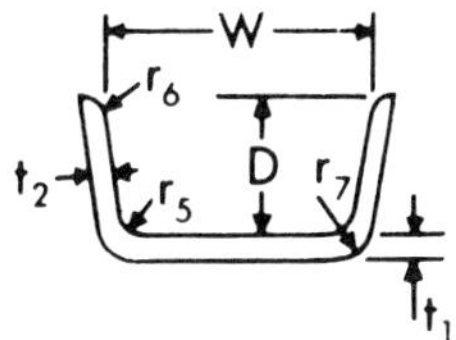

Channel sections W mm		<16	16-20	20-25	25-32	32-40	40-50	50-63	63-75	75-100	100-130
D mm	r_6 mm	r_5 mm									
<10	1.5	3.0	3.0	4.0	4.0	4.0	6.5	6.5	10.0	10.0	10.0
10-13	1.5		4.0	4.0	4.0	6.5	6.5	6.5	10.0	10.0	10.0
13-16	3.0			4.0	4.0	6.5	6.5	6.5	10.0	10.0	10.0
16-21	3.0				6.5	6.5	10.0	10.0	10.0	10.0	16.0
21-25	3.0					6.5	10.0	10.0	10.0	10.0	16.0
25-33	10.0						10.0	10.0	10.0	16.0	16.0
33-50	6.5								16.0	16.0	20.0

Note: $r_7 = r_5 + t_1$.

TABLE 6.9 (*Continued*)

mm				
I-sections			T-sections	
D	r_3	r_4	D	r_8
10.0	6.5	1.5	1.5	1.5
12.5	7.0	1.5	6.5	6.5
16.0	8.5	1.5	8.0	8.0
19.0	10.0	2.0	8.7	8.5
22.0	11.0	2.5	22.2	9.5
25.5	12.0	3.0	25.4	10.5
28.5	12.5	3.0	28.6	11.0
32.0	14.5	4.0	31.8	12.5
38.0	16.0	5.0	36.1	15.0
44.5	18.5	5.5	44.5	16.5
51.0	20.5	6.5	50.8	19.0
57.0	22.0	7.0	57.1	21.5
63.5	24.0	8.0	63.5	23.0
70.0	27.0	8.5	70.0	25.5
76.0	28.5	9.5	76.2	27.0

Conversion factor:
mm × 25.4 = in.

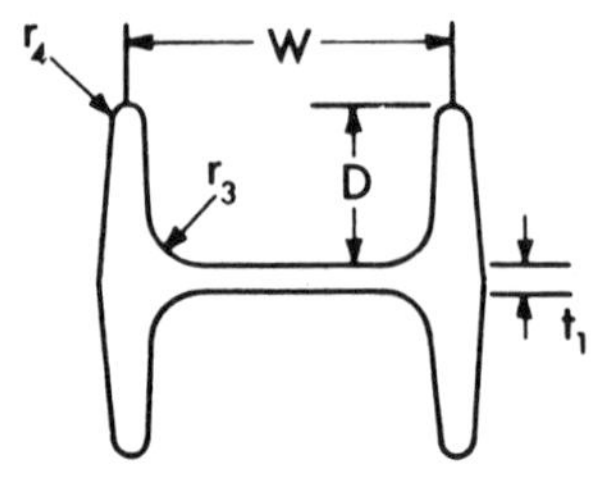

The values given for r_3 and r_4 apply when the W/D ratio is approximately 2 to 1. For W/D ratios greater than 2 to 1 the radii should be increased and for lower W/D ratios the radii may be decreased.

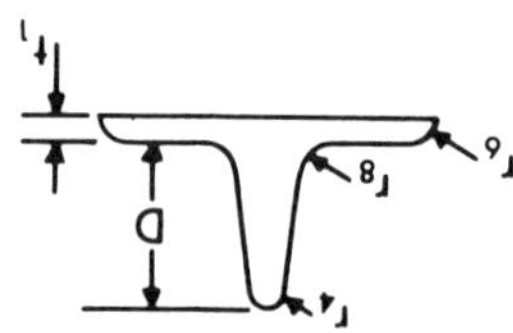

REFERENCES

1. American Society for Metals, "Forming of Magnesium Alloys," *Metals Handbook*, Vol. 4, 8th ed., p. 424, American Society for Metals, Metals Park, Ohio, 1969.
2. Dow Chemical Company, "Bend Radii of Magnesium-Alloy Sheet," Technical Memorandum No. 14,
3. Dow Chemical Company, "HM21A Magnesium Alloy Sheet and Plate," Technical Service and Development Letter Enclosure: Supplement to Bulletin 141-213. April 24, 1964.
4. Dow Chemical Company, "Magnesium in Design," Bulletin 141-213, 1967.
5. Dow Chemical Company, "Fabricating with Magnesium," Bulletin 141-477, 1982.
6. American Society for Metals, "Deep Drawing," *Metals Handbook*, Vol. 4, p. 162, American Society for Metals, Metals Park, Ohio, 1969.
7. T. L. Patton, "Magnesium Impact Extruded," *The Iron Age*, September 27, 1951.
8. T. L. Patton, "Designing Magnesium Impacts," *Machine Design*, February, 1954.

BIBLIOGRAPHY

1. P. B. Mellor, "Sheet-metal Forming," *International Metals Reviews*, Vol. 26, No. 1, 1981.
2. Amit K. Ghosh, "The Effect of Lateral Drawing-In on Stretch Formability," *Metals Engineering Quarterly*, Vol. 15, No. 3, August 1975.
3. I. L. Dillamore, R. J. Hazel, T. W. Watson, and P. Hadden, *Int. J. Mech. Sci.*, Vol. 13, p. 1094, 1971.
4. D. H. Avery, W. F. Hosford, and W. A. Backofen, *Trans. AIME*, Vol. 233, p. 71, 1965.
5. D. Lee and W. A. Backofen, *Trans. AIME*, Vol. 236, p. 1696, 1966.
6. C. A. Queener and Robert J. DeAngelis, "Elastic Springback and Residual Stresses in Sheet Metal Formed by Bending," *ASM Trans. Quart.*, Vol. 61, No. 4, p. 757, 1968.

Properties

Chapter 7 deals with the physical properties of all alloys and forms. Chapters 8-12 cover the mechanical properties of all alloys arranged according to the method of production. Thus, Chapter 8 deals with sand castings, Chapter 9 with permanent-mold castings, and Chapter 10 with extrusions.

Mechanical properties include tensile strength, bearing, shear, toughness, fatigue, stress corrosion, and damping. Each are covered in the several chapters for the various forms. However, there are some general comments that can be made about some of them.

Tensile strength properties are measured by applying a stress to a standard bar and measuring the strain that results. If the stress is low, the strain is proportional to the stress and is fully recovered after the stress is removed. The slope of this relation (stress/strain) is known as the elastic modulus. As the stress increases, the resultant strain increases more rapidly than the stress. Eventually, the bar breaks. The tensile strength is defined as the force required to break the bar divided by the original cross-sectional area. The yield strength is defined as the stress required to develop a strain that deviates from the elastic modulus by 0.2%. While other definitions for the yield strength are used—for example, the stress required to produce a strain that deviates 0.1% from the elastic modulus—the more common 0.2% deviation will be used throughout this book.

The most complete information on the tensile properties is given by the full stress-strain curve. However, a curve that shows the relation between stress and strain for significant amounts of strain, but not to the breaking point is almost as useful, and many such curves are given in the chapters of this section. The stress-strain curve can be used to determine reduced moduli, such as a secant modulus or a tangent modulus, both of which are useful when modifying strength-of-materials formulae developed for the elastic case so that they can be used when plastic deformation is occurring. Column curves and buckling can be so treated, for example.

Because some deformation mechanisms are related to diffusion, the strength of metals is sensitive to both the temperature and to the speed of testing. In general, the lower the temperature and the more rapid the testing, the stronger the metal. Details of these effects for magnesium alloys are given in the chapters of this section. As the temperature is raised significantly above room temperature, the effect of the speed of testing becomes more noticeable. In the extreme, placing a stress on a bar at an elevated temperature and holding it constant for a long period of time will result in a gradual extension of the bar. This effect, known as creep, can be important in some applications, and the details are given for magnesium alloys in the appropriate chapters. Creep data are usually given as the stress that will produce a given strain at a given temperature after a given number of hours (for example, the stress required to produce a strain of 0.2% at 250°C after maintaining the stress for 100 hours). The strain measured can be either the total strain from the beginning of load application, or the creep strain only. Creep strain is the extension that occurs over time after the initial loading is complete.

A variant of the creep curve is the isochronous stress-strain curve. This is obtained by measuring the strain at a given stress and temperature over a time period of up to ten hours. The data obtained are then rearranged to show stress-strain curves at a constant time of loading for different temperatures.

In a broad sense, toughness is the property of being able to take abuse. It has been measured in a variety of ways, and three are given for magnesium alloys in this book. The simplest is the ratio of the tensile strength in the presence of a notch to the tensile strength in the absence of a notch, which is an indication of the sensitivity of the alloy to stress concentrations in a structure.

The notched-bar Charpy or Izod tests measure the sensitivity of the alloy to both stress concentration and impact loading. The results are expressed as the energy required to break the bar.

If an extremely sharp notch, such as a fatigue crack, is present in a sufficiently large specimen, elastic theory can be used to calculate the stress intensity in the neighborhood of the crack tip. The stress intensity will depend upon the geometry of the piece and the orientation of the stress with respect to the crack. There is a critical stress intensity for any given material which will result in catastrophic failure. If this stress intensity is known, then the maximum crack size that can be tolerated for any given structure can be calculated.

Critical stress intensity factors for many of the magnesium alloys under conditions of plane strain are given in the chapters of this section.

Fatigue information is given for the alloys as measured on three different types of machines. The rotating-beam test uses a small cylindrical test bar that is rotated while being bent. The stress on an outer fiber varies from tension to compression as the bar rotates. The plate-bending test is performed on a flat plate suspended as a cantilever beam and bent. The outer fibers on each surface vary between tension and compression. The axial-load test is performed on a normal tensile specimen, with the load applied axially. In fatigue testing, it is important to specify not only the stress being applied and the nature of the stress variation but also what is known as the "R" value, the ratio of the minimum to the maximum stress. For the rotating-beam test, R is always equal to -1 since the outer fiber goes from the maximum load in tension to the same maximum load in compression. The R value for the other two tests can be controlled to whatever is desired.

Work at The Dow Chemical Company [1,2] has shown that, if the total life of the specimen is at least 10^5 cycles, the time to crack initiation occupies from 80 to 90% of the total, and the growth of the crack to failure the remaining 10 to 20%. There has been one study on the rate of crack growth [1]. A rotating-beam type bar was used with a diameter of 0.145 in (3.68 mm) having a notch with a depth of 0.080 in (2.03 mm) and root radius of 0.050 in (1.27 mm). The initiation of the crack was determined by microscopic examination on the surface; the progress of the crack was followed by measuring the electrical conductivity of the bar and using a correlation of that with the area of the crack. It is possible to use the data in the paper to arrive at crack growth rates for AZ63A-T4 at a stress level of 90 MPa (13 ksi) with the following result:

Crack length[a]		Growth rate	
mm	in	mm/cycle × 10^{-8}	in/cycle × 10^{-4}
2.243	0.008	1.6	6.3
2.322	0.091	3.3	13.0
2.362	0.093	6.9	27.2
2.476	0.097	10.4	40.9
2.583	0.102	18.3	72.0

[a]The length of the fatigue crack plus the notch depth.

It would be desirable to relate the growth rates to the range of stress intensity factors used in the test, but the data do not allow a calculation of stress intensity factors. The numbers obtained on growth rates are consistent with other values in the literature [3], and are probably typical of all magnesium alloys.

Stress corrosion is failure due to the simultaneous application of stress and corrosive conditions. The stress must be tensile. The corrosion is electrolytic since the application of a cathodic protection current completely eliminates failure. The corrosion need not be extreme or even obvious in order for the failure to occur. Stresses can be either design stresses occurring in service or residual stresses introduced by operations such as welding, machining, or quenching in water. Residual stresses can be relieved by proper annealing procedures.

The damping capacity of a material is a measure of its ability to convert the strain energy of vibrations into heat, thus decreasing the amplitude of free vibrations or requiring a constant input of energy to maintain a level of vibrations. It is usually expressed as specific damping capacity, which is the ratio of strain energy absorbed to total strain energy per vibration. It can be measured by determining the decrement of strain per vibration during free vibration, for which:

$$\%\ \text{Specific damping capacity} = 100(1-e^{-2d})$$

where

d = logarithmic decrement

REFERENCES

1. R. B. Clapper and J. A. Watz, "Determination of Fatigue-Crack Initiation and Propagation in a Magnesium Alloy," *Proceedings of the Second Pacific Area National Meeting*, American Society of Testing and Materials, Philadelphia, Pennsylvania, September 1956.
2. Dow Chemical Company, private communication.
3. S. Suresh and R. O. Ritchie, "Propagation of Short Fatigue Cracks," *International Metals Reviews*, *59*(6):445 (1984).

7

Physical Properties

Various physical properties of pure magnesium, arranged alphabetically, are given in Table 7.1. Those for alloys are in Tables 7.2 through 7.10. References are as noted in each table.

TABLE 7.1 Physical Properties of Pure Magnesium

Property	Temperature, °C	Value	References
Atomic number		12	
Atomic weight		24.310 (from distribution of isotopes)	1
Boiling point		1103 ±8°C	2-13
Coefficient of friction	20	0.36	14
Crystal properties			
Structure		Close-packed hexagonal	
a parameter	20	3.202 Angstroms	15
c parameter	20	5.199 Angstroms	15
Density	20	1.738 g/cm^3	15, 17
	100	1.724 "	16
	400	1.692 "	4
	500	1.676 "	4
	600	1.622 "	4
	650(s)	1.61 "	17
	650(1)	1.58 "	17
Electrical resistivity	0	4.10 ohm metre × 10^{-8} (microhm-cm)	18
	20	4.45	
	50	4.92	
	100	5.74	
	200	7.38	
	300	9.02	
	400	10.83	
	500	12.30	
	600	14.12	

	650(s)	15.35	
	650(1)	27.40	
	700	27.72	
	800	28.21	
	900	28.71	
Enthalpy	25	0 kJ/kg 0 Btu/lb	19
	150	131.8 " 56.7 "	
	260	253.1 " 108.9 "	
	650(s)	737.2 " 317.2 "	
	650(1)	1105.4 " 475.6 "	
	760	1247.7 " 536.8 "	
Hall constant	20	-1.06×10^{-16} ohm metre/(A/m)	20
Heat of combustion		25020 kJ/kg 10734 Btu/lb	14
Heat of fusion		368 kJ/kg 150 Btu/lb	19
Heat of sublimation		6109 kJ/kg 2628 Btu/lb	19
Heat of vaporization		5272 kJ/kg 2268 Btu/lb	19
Magnetic permeability		1.000012	26
Magnetic susceptibility		0.00627-0.00632 mks	26
Melting point	650		19
Skin effect (AC)		1 MHz 108 micro meters	27
		10 " 34.2 "	
		100 " 10.8 "	
		1000 " 3.42 "	
Specific heat	20	1025 J/(kg. K) 0.245 Btu/(lb. °F)[a]	19
	100	1034 " 0.247 "	

TABLE 7.1 *(Continued)*

Property	Temperature, °C	Value		References
	300	1150 J/(kg. K)	0.275 Btu/(lb. °F)	
	600	1327 "	0.317 "	
	650(s)	1360 "	0.325 "	
	650(1)	1322 "	0.316 "	
	700	1346 "	0.322 "	
Sublimation rate	25	3.4×10^{-11} mm/year	1.3×10^{-12} in/year	
	127	1.0×10^{-5} "	3.9×10^{-7} "	
	227	7.0×10^{-1} "	2.8×10^{-2} "	
	327	2.2×10^{2} "	8.7×10^{0} "	
	427	2.5×10^{3} "	9.8×10^{1} "	
	527	4.8×10^{5} "	1.9×10^{4} "	
	627	2.9×10^{6} "	1.1×10^{5} "	
Surface tension	700	55 Pa	550 dynes/cm^2	28
Thermal conductivity	-270	301 W/(m. K)	0.580 Btu in/(s.ft^2 °F)	21
	-265	1050 "	2.022 "	
	-250	1150 "	2.215 "	
	-225	417 "	0.803 "	
	-200	211 "	0.406 "	
	-173	169 "	0.325 "	1
	-150	160 "	0.289 "	21
	-100	158 "	0.304 "	
	-50	156 "	0.300 "	
	0	155 "	0.299 "	22
	20	154 "	0.297 "	
	50	149 "	0.287 "	23

	100	148 " 0.285 "	
	200	146 " 0.281 "	
	300	145 " 0.279 "	
	400	130 " 0.250 "	
Thermal diffusivity	20	0.86 cm^2/s	
	50	0.84 "	
	100	0.83 "	
	200	0.78 "	
	300	0.74 "	
	400	0.63 "	
Thermal expansion	-250	0.63 Unit strain × 10^{-6}/°C	24
	-225	5.4 "	
	-200	11.0 "	
	-150	17.9 "	
	-100	21.8 "	
	-50	23.8 "	
	0	25.0 "	18
	20	25.2 "	
	100	26.9 "	
	200	28.8 "	
	300	30.6 "	
	400	32.5 "	
	500	34.4 "	
	600	36.3 "	
Thermal neutron cross section		0.059 Barns/atom	25
Vapor pressure	25	1.5×10^{-15} Pa 1.48×10^{-20} Atm	13
	127	5.3×10^{-9} " 5.23×10^{-14} "	

TABLE 7.1 (*Continued*)

Property	Temperature, °C	Value		References
	227	4.0×10^{-5} Pa	3.95×10^{-10} Atm	
	327	1.4×10^{-2} "	1.38×10^{-7} "	
	427	9.0×10^{-1} "	8.88×10^{-6} "	
	527	2.0×10^{1} "	1.97×10^{-4} "	
	627	2.2×10^{2} "	2.17×10^{-3} "	
	650(S)	3.6×10^{2} "	3.55×10^{-3} "	
	650(1)	3.6×10^{2} "	3.55×10^{-3} "	
	727	1.4×10^{3} "	1.38×10^{-2} "	
	827	5.9×10^{3} "	5.82×10^{-2} "	
	927	1.9×10^{4} "	5.13×10^{-1} "	
	1027	5.2×10^{4} "	5.13×10^{-1} "	
	1103	1.0×10^{5} "	1.00×10^{0} "	
Viscosity	650(1)	1.25 mPa. s	0.0125 poise	8

[a] Also Cal/(g/°C).

TABLE 7.2 Densities of Alloys at 20°C

Alloy	g/cm^3	Alloy	g/cm^3
A8	1.81	HM21	1.78
AM60	1.78	HM31	1.80
AM100	1.81	HZ32	1.83
AS41	1.78	K1	1.74
AZ10	1.76	M1	1.76
AZ31	1.78	QE22	1.82
AZ61	1.80	QH21	1.82
AZ63	1.82	WE54	1.85
AZ80	1.80	AC61	1.83
AZ81	1.80	ZE41	1.84
AZ91	1.81	ZE63	1.87
AZ92	1.83	ZH11	1.76
EQ21	1.81	ZH62	1.87
EZ33	1.80	ZK10	1.78
HK31	1.79	ZK30	1.80
		ZK51	1.81
		ZK60	1.83
		ZK61	1.83
		ZM21	1.80

Source: Refs. 18, 32.

TABLE 7.3 Electrical Resistivity of Alloys

			Ohm meters × 10^{-8} at °C					
Alloy	Temper	Fabrication	20	50	100	150	200	250
A8	T4	Sand cast	13.4					
AM100	F	Sand cast	14.3					
	T4		17.2					
	T6		12.4	12.9	13.8	14.7	15.6	16.4
	T61		11.4					
AZ10	F	Extruded	6.4					
AZ31	Any	Any	9.2	9.3	10.4	11.2	12.0	12.7
AZ61	F	Extruded	12.5					
AZ63	F	Sand cast	12.2	12.3	13.5	14.2	14.9	15.6
	T5		11.0	11.1	12.3			
	T4		14.0					
	T6		11.8	11.9	13.0			
AZ80	F	Extruded	15.6	16.0	16.7	17.3	18.0	18.2
	T5		12.2	12.6	13.6	14.4	15.4	16.1
AZ855		Wrought	14.3					
AZ91	F	Pressure die	14.3	14.8	15.7	16.4	17.0	17.3
	F	Sand cast	13.6	14.1	15.0	15.7	16.4	16.7
	T4		16.2					
	T6		12.9	13.4	14.1			

AZ92	F	Sand cast	14.0					
	T5		12.4	12.9	13.7			
	T4		16.8					
	T6		12.4	12.9	13.7	14.6	15.5	16.2
EZ33	T5	Sand cast	7.0	7.5	8.3	9.2	10.1	11.0
HK31	T6	Sand cast	7.7	8.2	9.0	9.8	10.6	11.3
	H24	Rolled	6.1	6.6	7.5	8.4	9.3	10.1
	O		6.6	7.1	8.0	8.8	9.7	10.4
HM21	T8	Rolled	5.0	5.5	6.4	7.2	8.2	8.9
HM31	F	Extruded	6.6	7.1	8.0	8.8	9.7	10.4
	T5		5.7	6.2	7.1	8.0	8.9	9.4
HZ32	T5	Sand cast	6.5	7.0	7.8	8.6	9.5	10.2
K1	F	Sand cast	5.6	6.1	6.9	7.7	8.6	9.3
	F	Pressure die	5.3	5.8	6.7	7.6	8.6	9.4
M1	F	Extruded	5.4	5.8	6.7	7.5	8.4	9.1
QE22	F	Sand cast	7.3	7.8	8.6	9.4	10.3	11.0
	T6		6.8	7.3	8.1	8.9	9.8	10.5
QH21	T6	Sand cast	6.8					
WE 54	T6	Sand cast	7.0					
ZC61	T6	Extruded	7.0					
ZE41	T5	Sand cast	5.6	6.1	7.0	7.8	8.8	9.5
ZE63	T6	Sand cast	5.6					
ZH11	F	Extruded	6 .3					

TABLE 7.3 (*Continued*)

Alloy	Temper	Fabrication	Ohm meters × 10^{-8} at °C					
			20	50	100	150	200	250
ZH62	T5	Sand cast	6.4	6.9	7.8	7.6	9.6	10.3
ZK10	F	Extruded	5.3					
ZK30	F	Extruded	5.5					
ZK51	T5	Sand cast	6.4	6.9	7.7	8.4	9.3	10.0
ZK60	T5	Extruded	5.7	6.2	7.1	8.0	9.0	9.7

Source: Refs. 18, 25, 30, 32.

TABLE 7.4 Enthalpy of Alloys

	Enthalpy at temperature, °C in kJ/kg							
Alloy	25	150	200	300	500	600	650	750
AZ31	0		194	308	570		1118	1196
AZ80	0	131	187	295			1032	1110
AZ91	0	131	187	295			1030	1105
HK31	0		188	296	524		1082	1158
HM11	0	125	185	292	529	644	1065	1141
HM21	0		189	297	552	682	1102	1153
HM31	0		186	294	549	649	1090	1164
K1	0	128	189	295	533		1073	1151
ZK60	0		188	297			1072	1148

Conversion factor: kJ/kg × 0.4302 = Btu/lb.
Source: Ref. 18.

TABLE 7.5 Heat of Fusion of Alloys

	Heat of fusion	
Alloy	kJ/kg	Btu/lb
AZ31	339	146
AZ61	373	160
AZ80	280	120
AZ91	373	160
AZ92	373	160
EZ33	373	160
HK31	326	140
HM11	343	148
HM21	343	148
HM31	331	142
HZ32	373	160
K1	351	151
M1	373	160
QE22	373	160
ZH62	373	160
ZK60	318	137

Source: Refs. 18, 26.

TABLE 7.6 Melting Points of Alloys, °C

		Solidus	
Alloy	Liquidus	Equilibrium	Nonequilibrium
A8	600	470	460
AM60	615	640	504
AM100	595	465	427
AS41	620	565	
AZ10	645	630	620
AZ31	630	565	515
AZ61	615	510	450
AZ63	610	455	360
AZ80	610	490	454
AZ81	610	490	421
AZ91	595	470	425
AZ92	595	445	400
EQ21	640	540	
EZ33	640	545	
HK31	645	610	
HM21	645	605	
HM31	645	605	
HZ32	645	550	
K1	651	650	
M1	651	650	
QE22	640	550	
QH21	640	540	
WE54	640	550	
ZC61	620	455	
ZE41	640	510	
ZE63	630	515	
ZH11	645	625	
ZH62	630	500	
ZK11	645	600	
ZK30	635	600	
ZK51	640	550	
ZK60	635	520	
ZK61	635	520	

Source: Refs. 18,32.

TABLE 7.7 Specific Heat of Alloys

Alloy	J/(kg. K) at Temperature, °C					References
	20	100	300	600	650(1)	
A8	1000					25
AZ31	1040	1042	1148		1414	18,19,25
AZ61	1000					25
AZ80	975	1076	1155		1427	18
AZ91	979	1034	1155		1427	18
EQ21	1000					25
EZ33	1040					25
HK31	960		1088	1171	1427	18,25
HM21			1100	1180	1410	18
HM31			1084	1173	1339	18
HZ32	960					25
K1	1004	1045	1094	1494	1427	18
M2	1050					32
QE22	1000					25
QH21	1000					25
WE54	1000					30
ZC61	960					25
ZE41	960					25
ZE63	960					25
ZH11	960					32
ZH62	960					25
ZK10	1000					32
ZK30	960					25
ZK51	960					25
ZK60			1084	1182	1372	18

Conversion factor: J/(kg. K) × 2.39 × 10^{-4} = Btu/(lb. °F).

TABLE 7.8 Thermal Conductivity of Alloys

Alloy	Temper	Fabrication	W/(m. K) at °C					
			20	50	100	150	200	250
A8	T4	Sand cast	54.3					
AM100	F	Sand cast	51.2					
	T4		43.4					
	T6		58.3	62.0	67.3	72.1	76.2	80.8
	T61		63.0					
AM60	F	Pressure die	60.7					
AS41	F	Pressure die	66.9					
AZ10	F	Extruded	108.3					
AZ31	Any	Any	76.9	83.9	87.3	92.4	97.0	101.8
AZ61	F	Extruded	57.9					
AZ63	F	Sand cast	59.2	64.7	68.7	74.4	79.6	84.5
	T5		65.1	71.1	74.8			
	T4		52.2					
	T6		61.0	66.7	71.1			
AZ80	F	Extruded	47.3	51.0	56.7	62.3	67.3	73.7
	T5		59.2	63.3	68.2	73.4	77.3	82.1
AZ855		Wrought	51.2					
AZ91	F	Pressure die	51.2	54.7	59.9	65.4	70.8	77.1
	F	Sand cast	53.6	57.2	62.4	68.0	73.1	79.5

TABLE 7.8 *(Continued)*

Alloy	Temper	Fabrication	W/(m. K) at °C)					
			20	50	100	150	200	250
	T4		45.8					
	T6		56.2	59.9	66.0			
AZ92	F	Sand cast	52.2					
	T5		58.3	62.0	67.8			
	T4		44.3					
	T6		58.3	62.0	67.8	72.5	76.9	81.7
EZ33	T5	Sand cast	99.5	102.7	107.8	111.0	113.7	116.2
HK31	T6	Sand cast	90.9	94.4	99.9	104.6	108.7	113.3
	H24	Rolled	113.4	116.0	118.6	120.1	122.8	125.7
	O		105.2	108.2	111.6	115.7	118.1	122.4
HM21	T8	Rolled	137.3	138.1	137.9	139.8	138.2	141.5
HM31	F	Extruded	105.2	108.2	111.6	115.7	118.1	122.4
	T5		121.0	123.1	124.9	126.5	128.0	134.5
HZ32	T5	Sand cast	106.7	109.7	114.3	118.2	120.4	124.6

K1	F	Sand cast	123.1	125.0	128.4	131.2	132.2	135.8
	F	Pressure die	225.6	131.2	132.0	132.8	132.3	134.5
M1	F	Extruded	127.5	131.2	132.0	134.5	135.1	138.6
QE22	F	Sand cast	95.6	99.0	104.2	108.8	111.7	116.2
	T6		102.2	105.4	110.3	114.5	117.0	121.3
QH21	T6	Sand cast	102.2					
WE54	T6	Sand cast	107					
ZC61	T6	Extruded	99.5					
ZE41	T5	Sand cast	123.1	125.0	126.7	129.6	129.4	133.1
ZE63	T6	Sand cast	123.1					
ZH62	T5	Sand cast	108.3	111.2	114.3	116.9	119.2	123.5
ZK30	F	Extruded	125.5					
ZK51	T5	Sand cast	108.3	111.2	115.7	120.9	122.8	126.9
ZK60	T5	Extruded	121.0	123.1	124.9	126.6	126.7	130.6

Conversion factor: W/(m. K) × 1.93 × 10^{-3} = Btu in/(s. ft^2 °F).
Source: Ref. 30.

TABLE 7.9 Thermal Diffusivity of Alloys

Alloy	Temper	Fabrication	Thermal diffusivity (cm^2/s) at °C		
			20	100	200
A8	T4	Sand cast	0.300		
AZ31	Any	Any	0.415	0.474	0.506
AZ61	F	Extruded	0.322		
AZ80	F	Extruded	0.270	0.295	0.341
	T5		0.337	0.355	0.392
AZ91	F	Pressure die	0.289	0.323	0.363
	F	Sand cast	0.302	0.336	0.375
	T4		0.258		
	T6		0.317	0.356	
EZ33	T5	Sand cast	0.532		
HK31	T6	Sand cast	0.529		
	H24	Rolled	0.660		
	O		0.612		
HZ32	T5	Sand cast	0.607		
K1	F	Sand cast	0.705	0.712	0.722
	F	Pressure die	0.719	0.732	0.722
QE22	F	Sand cast	0.525		
	T6		0.562		
QH21	T6	Sand cast	0.562		
WE54	T6	Sand cast	0.578		
ZC61	T6	Extruded	0.566		
ZE41	T5	Sand cast	0.697		
ZE63	T6	Sand cast	0.686		
ZH62	T5	Sand cast	0.603		
ZK30	F	Extruded	0.726		
ZK51	T5	Sand cast	0.623		

TABLE 7.10 Thermal Expansion of Alloys

Mean coefficient (20-200°C), Unit strain/°C × 10^{-6}	Alloys
27.2-27.3	AZ61, AZ63, A8, AZ80, AZ855, AZ81, AZ91, AZ92
27.0-27.1	WE54, ZC61, ZE10, ZE41, ZE63, ZH62, ZK21, ZK30, ZK51, ZK60
26.8-26.9	AZ31, EK30, HK31, K1
26.5-26.7	EK41, EQ21, EZ33, HZ32, QE22, QH21

Source: Refs. 18,30.

7.1 DENSITY

Magnesium, with a value of 1.738 g/cm^3 at 20°C, has the lowest density of any structural metal. It is this property, combined with good mechanical properties, that is the basis for the majority of applications for the metal. Values for the alloys are shown in Table 7.2; even the densest of these has a low density of about 1.87 g/cm^3.

While not exact, as a first approximation the density of a magnesium alloy is proportional to the atomic percent of an element added multiplied by that element's density.

7.2 ELECTRICAL RESISTIVITY

Values are given for alloys in Table 7.3. The electrical resistivity of alloys is very sensitive to the amount of alloying element in solid solution and to the temperature at which the measurement is made, increasing both with increasing amounts of an element and with increasing temperature.

To a first approximation, the increase of electrical resistivity as an element is added to the magnesium lattice is proportional to the atomic percent of the addition. Thus, a given weight percent of aluminum added to magnesium raises the resistivity more than the same weight percent of zinc. When the added element is present as a separate phase, the resultant resistivity is that obtained by applying the law of mixtures to the multiphase system, using the resistivities of each of the phases. Thus an alloy has the highest resistivity in the T4 temper, the lowest in the T5 or T6 temper.

7.3 ENTHALPY

The enthalpy, or heat content, is calculated by setting the heat content arbitrarily at zero at 25°C and using specific heat and latent heat informaiton to arrive at the heat content for any temperature. Enthalpy information for some alloys are given in Table 7.4

7.4 HEAT OF FUSION

Data for alloys are given in Table 7.5.

7.5 MELTING POINT

Pure magnesium, as with any pure material, has a single melting point. Alloys, however, melt over a range of temperatures as shown in Table 7.6. The solidus is that point where liquid first appears as the temperature is raised; the liquidus is that point at which all solid disappears as the temperature is raised. Between the liquidus and solidus temperatures, solid and liquid coexist in equilibrium. The solidus and liquidus temperatures are for equilibrium conditions, but alloys are usually frozen too rapidly for equilibrium to be attained. The values for liquidus temperatures are close to those that are seen even under nonequilibrium conditions, but the values for the solidus can be quite different in practical cases. Values for the nonequilibrium solidus for magnesium-aluminum-zinc alloys as determined for metal frozen in a sand mold [31] are listed in Table 7.6. The nonequilibrium solidus is also sometimes known as an incipient melting point.

7.6 SPECIFIC HEAT

The energy required to raise a unit mass of material one degree Kelvin is known as the specific heat. The specific heat of a unit mass of magnesium is higher than that of other structural metals, as the following compilation shows [26]:

Metal	Specific Heat (J/(kg. K)
Mg	1025
Al	900
Ti	532
Ni	471

Metal	Specific Heat (J/(kg. K)
Fe	448
Cu	386
Zn	382
Mo	276
Ag	235
Ta	139

Thus, when magnesium is substituted for an equal mass of another metal, it acts as a better heat sink and stays cooler for a longer period of time. Figure 7.1 [28] is an illustration of the effect in an application involving transient heating.

By contrast, the specific heat of a unit volume of magnesium is lower than that of other metals:

Metal	Specific Heat (J/(cc. K)
Mg	178
Ta	231
Ti	240
Al	243
Ag	247
Zn	272
Mo	282
Cu	346
Fe	352
Ni	419

If an equal volume of magnesium is substituted for another metal, the magnesium will heat up more rapidly. If it is acting as a radiator, it will more readily dissipate heat from the part.

As Table 7.7 shows, differences in specific heat among magnesium alloys are small.

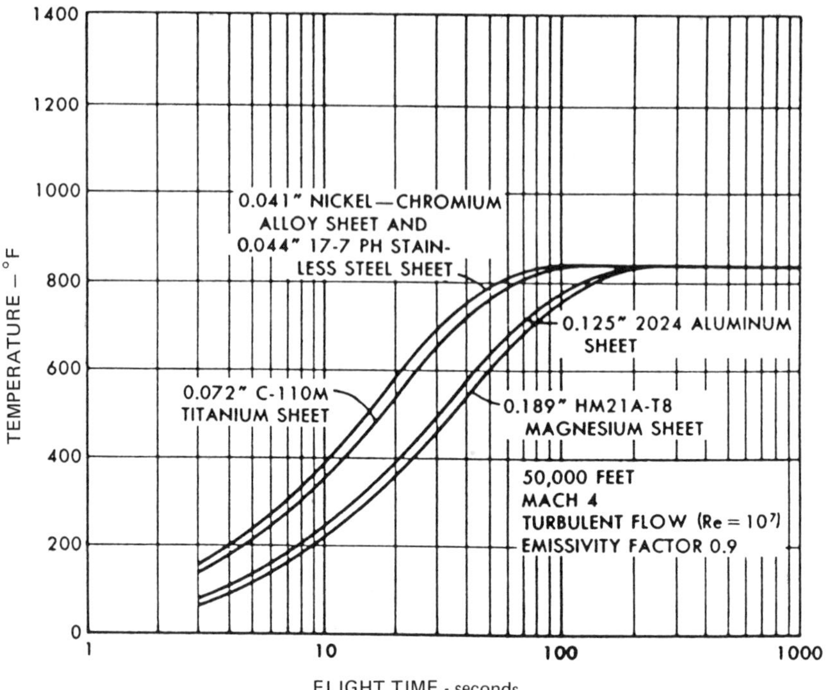

FIGURE 7.1 Effect of structure temperature and flight time on sheet alloys at mach 4 and 50,000 feet.

7.7 SUBLIMATION RATE

The sublimation rates shown in Table 7.1 for pure magnesium are calculated from the vapor pressure data given in the same table, using the Langmuir equation:

$$G = 79{,}464P(M/T)^{1/2}$$

where

G = sublimation rate in mm/yr
P - vapor pressure in Pascals
M = molecular weight
T = temperature in degrees Kelvin

The constant in the equation is specific to magnesium since it contains the specific gravity for magnesium in order to obtain the unit of mm/year for G.

The assumptions leading to the equation are that the surface is free of any films and that the external pressure is low enough that no leaving atoms return to the surface and replate. In practice, the pressure must be no higher than about 10^{-2} pascals (10^{-7} Atm.); since it is almost impossible to obtain a truly film-free magnesium surface, the observed rates of sublimation are lower than those calculated and shown in the table. The principal use for the table is for a conservative estimation of the life of magnesium structures sent into space. Even in this case, the normal procedure is to coat the magnesium in order to control absorption and emission of radiant energy and thus to control temperature. The life then depends upon the properties of the film rather than of the underlying magnesium.

7.8 THERMAL CONDUCTIVITY

The thermal conductivity data for pure magnesium at 0 and 20°C in Table 7.1 and those for alloys in Table 7.8 (except for WE54) are calculated from electrical resistivity data according to the relation published by Bungardt and Kallenback [21]:

$$\text{Thermal conductivity} = 418.4K(0.54 \times 10^{-10}/R + 0.4 \times 10^{-4})$$

where

Thermal conductivity = W/(m. K)
K = Degrees Kelvin
R = Resistivity in ohm meters

7.9 THERMAL DIFFUSIVITY

Thermal diffusivity, a measure of the rate at which heat will be transmitted through a metal, is defined as:

$$D = 10{,}000C/(C_p \times P)$$

where

D = Diffusivity in cm^2/sec
C = Thermal conductivity in W/(m. K)
C_p = Specific heat in J/(kg. K)
P = Density in kg/m^3

The factor of 10,000 is to convert from m^2/s to cm^2/s, which seems an easier dimension to visualize. Calculated values are given in Tables 7.1 and 7.9. Since the thermal diffusivity of magnesium is higher than that of other metals, parts used for high-temperature applications will run cooler when made of magnesium if there is a heat dissipation mechanism available such as fins or a circulating coolant. When this is the case, part design can be based on strength at a temperature 25-50°C lower than must be used for another metal.

7.10 THERMAL EXPANSION

Coefficients of expansion for alloys are in Table 7.10, while those for pure magnesium are in Table 7.1. There is only a very small effect of alloying on the values.

There is a small difference in the coefficient of expansion depending upon the crystallographic direction of measurement [29]. The 20°C value parallel to the basal plane is 27.1×10^{-6} unit strain/°C; perpendicular to the basal plane it is 24.3×10^{-6} unit strain/°C. This difference has a practical significance only if the magnesium being used has a very sharp preferred orientation. If this is the case, and the small difference is significant to the application, an estimate of the degree of orientation can be made and an adjustment to the coefficient calculated for different directions.

REFERENCES

1. R. C. Weast, *Handbook of Chemistry and Physics*, 57th Edition, CRC Press, 1976-1977.
2. M. E. Drits, "Super-light Construction Alloys," Foreign Technology Division, Wright-Patterson Air Force Base, Ohio, National Technical Information Service, U.S. Department of Commerce, November 8, 1973.
3. P. J. McGonigal, A. D. Kirshenbaum, and A. V. Grosse, "The Liquid Temperature Range, Density, and Critical Constants of Magnesium," *J. Physical Chemistry, 66*:737-740 (1962).
4. G. V. Raynor, *The Physical Metallurgy of Magnesium and its Alloys*, Pergamon Press, 1959.
5. C. Sheldon Roberts, *Magnesium and its Alloys*, John Wiley & Sons, Inc., New York, 1960.
6. D. R. Stull and R. A. McDonald, "The Enthalpy and Heat Capacity of Magnesium and of Type 430 Stainless Steel from 700 to 1000 K," Thermal Laboratory, Dow Chemical Company, 1955.
7. C. J. Smithells, *Metals Reference Book*, Vol. 1 and 2, 3rd ed., Butterworths, London, 1962.
8. E. F. Emley, *Principles of Magnesium Technology*, Pergamon Press, 1966.
9. *Comprehensive Inorganic Chemistry*, 1st ed., Pergamon Press, 1973.

10. Kirk-Othmar, *Encyclopedia of Chemical Technology*, Vol. 12, 2nd ed., John Wiley & Sons, Inc., New York, 1967.
11. C. J. Smithells, *Metals Reference Book*, 5th ed., Butterworths, London, 1976.
12. K. Raznjevic, *Handbook of Thermodynamic Tables and Charts*, McGraw Hill, New York, 1976.
13. University of California, "Selected Values for the Thermodynamic Properties of Metals and Alloys," Institute of Engineering Research, 1960.
14. J. W. Frederickson, "Pure Magnesium," *Metals Handbook*, Vol. 1, 8th ed., American Society for Metals, Metals Park, Ohio, 1961.
15. R. S. Busk, "The Lattice Parameters of Magnesium and Magnesium Alloys," *Transactions of the AIME 188*, 1460 (1952).
16. R. S. Busk, "Effect of Temperature on the Lattice Parameters of Mg Alloys," *Transactions of the AIME 194*, 207 (1952).
17. H. Grothe and C. Mangelsdorff, *Z. Metallkunde, 29*:352 (1937).
18. Dow Chemical Company, "Physical Properties of Magnesium and Magnesium Alloys," Technical Service & Development Letter Enclosure, April 10, 1967.
19. C. R. Stull and G. C. Sinke, "Thermodynamic Properties of the Elements," *Advances in Chemistry*, No. 18, American Chemical Society, 1956.
20. E. J. Salkovitz, A. J. Schindler, and F. W. Kammer, *Physical Review 105*, 887 (1957).
21. W. R. S. Kemp, A. K. Sreedhas, and G. K. White, *Proceedings of the Physical Society, 4*:317 (1950).
22. W. Bungardt and R. Kallenbeck, *Metallwissenshart und Technik, 4*:317, 365 (1950).
23. R. W. Powell, M. J. Hickman, and R. P. Tye, *Metallurgia, 70*(420):159-153 (1964).
24. R. J. Corruccini and J. J. Gniewid, "Thermal Expansion of Technical Solids at Low Temperature," *National Bureau of Standards Monograph 29*, U.S. Department of Commerce, 1961.
25. Magnesium Elektron Limited, "Elektron Magnesium Alloys," Bulletin 400, September 1983.
26. American Society for Metals, *Metals Handbook*, Vol. 2, 9th ed., American Society for Metals, Metals Park, Ohio, 1979.
27. Calculated from data on pp. 5-90, *American Institute of Physics Handbook*, McGraw Hill, New York, 1957.
28. Dow Chemical Company, "Magnesium in Design," Bulletin 141-213, 1967.
29. P. W. Bridgman, *Proceedings of the American Academy of Arts and Sciences 67*, 27 (1932).
30. Magnesium Elektron, Ltd., "WE54, A New Magnesium Casting Alloy for Use up to 300°C," Bulletin 466, 1985.

31. R. S. Busk, "Some Properties of Sand-cast Alloys in the Magnesium-rich Corner of the Magnesium-aluminum-zinc System," *Metals Technology*, June 1946.
32. Colin J. Smithells, *Metal Reference Book*, Butterworths, London and Boston, 1976.

8

Mechanical Properties of Sand Castings

8.1 GENERAL CHARACTERISTICS

A sand casting is made by pouring molten metal into a cavity made by extracting a pattern from sand that has been packed around it. While the shape of the cavity must be such that the pattern can be extracted, the use of cores, which are separate pieces of shaped sand placed in the primary cavity, minimizes this limitation. Shapes of great complexity and of large size can be produced, including hollow passageways of any shape or size inside external or internal walls of the casting. See, for example, Figures 2.12, 2.13, and 2.14. Sand casting is chosen as a method of producing a part when the production quantities required are small, or when great complexity or size is required together with high quality.

Since the properties obtained in a specific casting are dependent both upon the quality of the metal poured into the mold and upon the freezing conditions for that casting, quality control is an important aspect of assuring consistent castings. The properties that will be realized in a casting are dependent not only on the chemistry of the alloy and the heat treatments given to the casting, but upon the incidence of defects in the metal structure. The grain size in the casting depends both upon the molten metal treatment, and the pouring and freezing conditions. The solid metal is more dense than the liquid metal so that upon freezing there is a net shrinkage which can result in voids; the voids are known as sinks or shrinks if gross; if on a micro scale, they are known as microshrinkage or porosity. The latter lower mechanical properties to a degree dependent upon the amount of porosity and its orientation to the direction of the stress. The heat treatments specified for a given alloy assume a structure in the casting that will respond properly; if the structure is coarser than expected, for example, the heat treatments used will be improper.

Mechanical properties are checked by measuring the properties of separately-cast standard test bars, of special bars cast integrally with the casting, and of bars cut from random castings sacrificed for that purpose. Properties of the separately-cast test bars assure that the metal chemistry, the treatment of the molten metal, and the heat treatments used are all correct. Measurements of integrally-cast bars assure additionally that the metal actually poured into the casting mold is correct. The random check of properties in bars cut from castings assures that molding and pouring procedures are being followed with results as expected. The quality of each casting can be monitored by non-destructive testing with tools such as metallography, die-penetrant inspection, radiography, and visual inspection.

Because of the large variety of shapes produced by casting, the quality of the metal produced can vary widely from casting to casting and from place to place in a given casting. The skill of the foundryman is critical to the production of high-quality, consistent castings, and the designer should consult with the foundryman early in the design stage to assure that the optimum result will be obtained.

8.2 GENERAL DESIGN CONSIDERATIONS

Wall thickness should generally be at least 4 mm (0.160 in) although walls as thin as 2-3 mm (0.08-0.12 in) can sometimes be cast successfully. Abrupt changes in thickness should be avoided in order to facilitate casting. Radii between walls, and walls and ribs, should be generous, usually equal to the wall thickness, in order to facilitate casting and minimize stress concentrations in service. Figure 8.1 is a guide to good practice [1]. The foundryman should be consulted before the design is final.

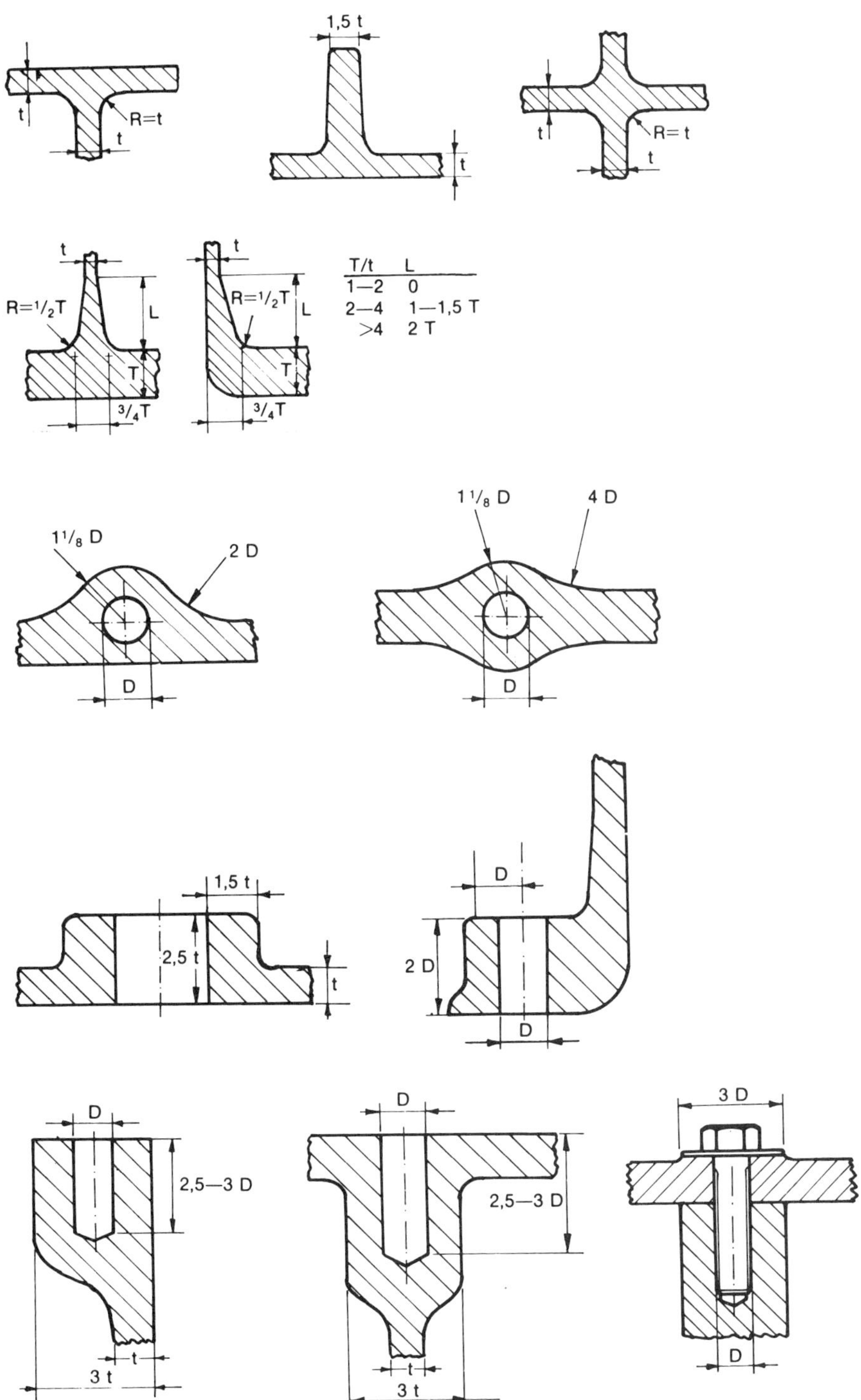

FIGURE 8.1 Recommendations for the design of sand castings. (From Ref. 1.)

8.3 ALLOYS USED FOR SAND CASTING

There are two major classes of alloys used for sand casting: those that are based on the addition of aluminum, and those that contain zirconium.

Those that have aluminum as the principle alloying element also usually contain some zinc for additional strength and improved salt-water corrosion resistance, and a small amount of manganese for improved salt-water corrosion resistance. Impurities such as iron, nickel, copper, and silicon may be controlled to low levels, for improved salt-water corrosion resistance. The alloys are characterized by low cost, ease of handling, and good resistance to atmospheric corrosion and staining.

Zirconium is added to magnesium because of the excellent grain refining that results. It is not added to those alloys that contain aluminum because the two elements combine to form an intermetallic compound that is insoluble in the molten metal, with only the element present in excess remaining in the alloy. There are several families of zirconium-containing alloys: those containing zinc as the primary alloying element, those containing rare-earth elements, those with thorium, those with silver, and, most recently, those with yttrium.

The addition of zinc to magnesium strengthens the metal greatly at room temperature. However, because of the fine grain size associated with the presence of zirconium, the mechanical properties decrease rapidly as the temperature is raised and the creep strength even at only warm temperatures is low. The castability is poor enough to make it difficult to obtain the best properties in complex castings or to produce castings that are pressure tight. These defects are minimized by the addition of rare-earth elements to alloys containing zinc. Thus 3% mischmetal is added to an alloy containing 3% zinc to produce EZ33A, an alloy with good room- and elevated-temperature properties and excellent pressure tightness. The addition of 1% mischmetal to an alloy containing 4% zinc (ZE41A), results in an alloy with very good room temperature properties, both in separately-cast bars and in bars cut from castings. ZE63A is an alloy with excellent castability and room-temperature properties. The requirement that ZE63A be heat treated in hydrogen in order to diffuse hydrogen into the alloy and react it with the rare-earth elements to form hydrides imposes a restriction on wall thickness.

Silver, like zinc, improves the room-temperature strength of magnesium. The effect is even more marked when rare-earth elements or thorium are also present, in which case not only the room-temperature but also the elevated-temperature properties are excellent. Thus EQ21A, QE22A, and QH21A all have good mechanical properties over a wide range of temperature and with the properties in casting sections equalling or even exceeding those characteristic of separately-cast bars.

Thorium, rare-earth elements, and yttrium all improve the elevated-temperature properties of magnesium markedly. In addition, yttrium additions result in high properties at room temperature. Thus WE54A has high strength over a wide range of temperatures.

Table 8.1 is a summary of the characteristics of the casting alloys. Alloys in the Mg-Al-Zn system are the least expensive from the point of view of alloying elements and have been used for the longest period of time. AZ81-AZ91 are the most generally used and can be considered as the work-horse alloys for sand casting. However, in recent years, ZE41A has been substituted more and more for AZ91 because of greater ease of handling in the foundry and better properties in the casting itself. The presence of Zr assures fine grain and the presence of mischmetal assures a minimum of microporosity. Alloys containing thorium have been popular in the past for high-temperature service, but are used less now because of the disposal problems associated with scrap containing thorium. For those applications that can afford the addition of silver, QE22 is a good overall choice. The recently developed alloy WE54A has the best overall properties of any magnesium alloy over the widest range of temperatures. Although yttrium is expensive, the alloy will undoubtedly gain wide acceptance.

TABLE 8.1 General Characteristics of Cast Alloys

Alloy	Temper	General characteristics
AM100A	T4, T6	Permanent mold alloy, pressure tight, weldable, good atmospheric stability.
AZ63A	T4, T6	Good salt-water corrosion resistance even with a high iron level, good toughness, difficult to cast. Very seldom used today.
AZ91C	T6	General purpose alloy, good strength at room temperature, useful properties up to 175°C, good atmospheric stability. The most commonly used of the alloys in the Mg-Al-Zn family.
AZ92C	T6	General purpose alloy, excellent strength at room temperature, useful properties up to 175°C, good atmospheric stability.
EQ21A	T6	Heat-treated alloy with a high yield strength up to 250°C, pressure tight, weldable.
EZ33A	T5	Creep-resistant up to 250°C, excellent castability, pressure tight and weldable.
HK31A	T6	Creep-resistant up to 345°C for short-time applications, pressure tight, weldable.
HZ32A	T5	Creep-resistant up to 345°C, pressure tight, weldable.
QE22A	T6	Heat-treated alloy with high yield strength up to 250°C, pressure tight, weldable.
QH21A	T6	Good creep resistance, high yield strength up to 300°C, pressure tight, weldable.
WE54A	T6	The first of a new family of alloys containing yttrium. Exceptional strength at both room and elevated temperatures.
ZE41A	T5	Easily cast, weldable, pressure tight, useful strength at elevated temperatures.
ZE63A	T6	Excellent castability, pressure tight and weldable, high developed properties in thin-wall castings.
ZH62A	T5	Stronger than, but as castable as ZE41A, weldable, pressure tight.
ZK51A	T5	Good strength at room temperature.
ZK61A	T6	Excellent strength at room temperature, only fair castability, but capable of developing excellent properties in castings.

8.4 TENSILE PROPERTIES

8.4.1 Room Temperature

In the tables, the following apply unless otherwise stated: %E is measured on a gauge length of 50.8 mm (2 in), tensile yield strength (TYS) is taken at a 0.2% deviation from the elastic modulus line, the strain rate to the yield strength is 0.005/min, the strain ratc between the yield strength and the tensile strength is 0.5/min.

Typical and minimum properties for separately-cast test bars and minimum average properties for bars cut from castings are given in Table 8.2. Typical properties are taken from Reference 3, the minima from Reference 2. While the ASTM values are reasonably representative, the designer should consult the current specification for the country of application. See the list of world-wide specifications in Chapter 3.

The minima quoted for bars cut from castings are an average of at least three coupons cut from the casting. One is taken from the thinnest, one from a thicker, and one from the thickest section; the table therefore shows what the minimum average quality of a casting can be expected to be. Since, however, the foundryman has considerable control over the quality level of metal in specific areas of the casting, the designer can specify places that are to have higher minima than those given in Table 8.2. As an example, one set of specifications for premium-quality castings with provision for designating such areas are given in Table 8.3. It is wise to consult with the foundryman before the design is final to assure that the optimum in quality is obtained in those places where it is needed for the application.

8.4.2 The Effect of Temperature

Tensile properties for a range of temperatures are given in Table 8.4. The beneficial effects of the rare-earth elements, of thorium, and of yttrium are apparent. The data were obtained by raising the test bar to temperature, holding ten minutes, and then testing at normal speeds.

A series of typical stress-strain curves for cast alloys are given in Figures 8.2 through 8.9.

Figures 8.10 and 8.11 compare the tensile properties for the four alloy types of AZ, QE, ZE and WE over the temperature interval from room temperature to 315°C. Based on tensile properties alone, WE54 is the strongest, but is likely to be the most expensive. AZ91 is quite acceptable, and is the least expensive.

8.4.3 The Effect of Exposure to Elevated Temperatures

The microstructures developed in an alloy by casting and heat treatment variables are rarely, if ever, the equilibrium structures, and a change with time is to be expected. The higher the temperature to which the metal is exposed, the shorter will be the time needed for changes to become apparent. Table 8.5 gives information on the effect of time at elevated temperatures on the tensile properties.

The density of an alloy depends not only on the density of the elements composing the alloy, but also on the state of the alloy. For example, when aluminum is dissolved in the magnesium lattice, the lattice dimensions become smaller and the solid solution is denser than would be calculated by a simple law of mixtures. Changing the state of heat treatment of a given alloy can therefore change the density. Table 8.6 shows linear dimensional changes to be expected for alloys upon exposure to various temperatures for various time intervals. If especially close tolerances are required for an application, this effect may need to be considered.

8.4.4 The Effect of Speed of Testing

Some of the important deformation mechanisms for magnesium involve diffusion, making the time of stress application an important consideration. The more rapidly the stress is applied, the higher will be the stress required for yielding to be apparent; the more slowly the stress is applied, the lower will be the stress required for yielding and deformation to become significant. The time required for strain to become apparent at a low stress becomes shorter as the temperature becomes higher.

8.4.4.1 High Strain Rates

The effect of increasing the strain rate at a number of temperatures is given in Table 8.7 for some of the cast alloys. The marked effect of temperature is apparent as is the fact that alloys containing rare-earth elements or thorium are less sensitive to speed at any temperature than are the magnesium-aluminum-zinc alloys.

8.4.4.2 Isochronous Stress-strain Curves

Isochronous stress-strain curves have been developed for a number of cast alloys, with the results shown in Figures 8.12 through 8.23. They are a variant of creep curves for short times.

8.4.4.3 Creep

Creep data for cast alloys are given in Tables 8.8 and 8.9. Total extension is given in Table 8.8, which includes the elongation experienced upon the initial application of stress as well as the creep extension itself. Any elastic extension that occurs during the initial loading is recovered upon removal of the stress. Table 8.9 gives the stresses required for creep extension only, ignoring any initial extension as the load is first applied.

Since creep is diffusion dependent, the stress for producing strain is not sensitive to gross defects such as porosity. While it is sensitive to grain size, the effect is that the larger the grain size the higher will be the stress needed to produce a given strain level. In many cases, the grain size of the metal in a casting will be larger than the grain size of separately-cast test bars. The net effect is that the creep properties of metal in castings is equal to, or sometimes higher than, the creep properties of separately-cast bars. Use of data from test bars is conservative.

TABLE 8.2 Room Temperature Tensile Properties (SI Units)

		Typical			Minima			Bars from castings		
			MPa			MPa			MPa	
Alloy	Temper	%E	TYS	TS	%E	TYS	TS	%E	TYS	TS
AM100A	F	2	85	150	—	70	140			
	T4	10	90	275	6	70	235			
	T6	4	110	275	2	105	235			
	T61	1	150	275	—	115	235			
AZ63A	F	6	95	200	4	75	180			
	T4	12	90	275	7	75	235	—	70	175
	T5	4	97	205	2	85	180			
	T6	5	130	275	3	110	235	—	100	175
AZ81A	T4	15	85	276	7	75	235			
AZ91C	F	2	95	165	—	75	160			
	T4	14	85	275	7	75	235	—	70	175
	T5				2	85	160			
	T6	5	130	275	3	110	235	—	100	175
AZ92A	F	2	95	165	—	75	160			
	T4	9	95	275	7	76	235	—	70	175
	T5	2	110	180	—	85	160			
	T6	2	145	275	1	125	235	—	110	175

EQ21A	T6	4	195	261	2	175	240	7.4	188	272
EZ33A	T5	3	105	160	2	95	140	—	85	105
HK31A	T6	8	105	220	4	90	185	—	80	160
HZ32A	T5	7	95	205	4	90	185	—	80	160
K1A	F		51	185	14	40	165			
QE22A	T6	4	205	275	2	170	240	—	160	220
QH21A	T6				2	185	240	2	170	230
WE54A	T6	4	200	275	2	185	260	5	200	275
ZE41A	T5	5	140	205	2.5	135	200	2.5	135	195
ZE63A	T6	7	190	295	5	185	275	—	165	255
ZH62A	T5	6	170	275	5	150	240	—	120	215
ZK51A	T5	8	165	275	5	140	235	—	115	210
ZK61A	T6				5	180	275	—	165	235

Note: The values for bars from castings are a minimum average of specimens, except for alloys WE54A and EQ21A. The values for WE54A are an average of 106 bars from 7 castings [24]; those for EQ21A are from 6 bars [27].

Source: Refs. 2, 3, 23, 24, 26, 27.

TABLE 8.2a Room Temperature Tensile Properties (English Units)

		Typical			Minima			Bars from castings		
			ksi			ksi			ksi	
Alloy	Temper	%E	TYS	TS	%E	TYS	TS	%E	TYS	TS
AM100A	F	2	12	22	—	10	20			
	T4	10	13	40	6	10	34			
	T6	4	16	40	2	15	34			
	T61	1	22	40	—	17	40			
AZ63A	F	6	14	29	4	11	26			
	T4	12	13	40	7	11	34	—	10	25
	T5	4	14	30	2	12	26			
	T6	5	19	40	3	16	34	—	14	25
AZ81A	T4	15	12	40	7	11	34			
AZ91C	F	2	14	24	—	11	23			
	T4	14	12	40	7	11	34	-	10	25
	T5				2	12	23			
	T6	5	19	40	3	16	34	-	14	25
AZ92A	F	2	14	24	—	11	23			
	T4	9	14	40	7	11	34	—	10	25
	T5	2	16	26	—	12	23			
	T6	2	21	40	1	18	34	—	16	34

EQ21A	T6	4	28	38	2	25	35	7.4	27	39
EZ33A	T5	3	15	23	2	14	20	–	12	15
HK31A	T6	8	15	32	4	13	27	–	12	23
HZ32A	T5	7	14	30	4	13	27	–	12	23
K1A	F		7.5	27	14	6	24			
QE22A	T6	4	30	40	2	25	35	–	23	32
QH21A	T6				2	27	35	2	25	33
WE54A	T6	4	29	40	2	27	38	5	29	40
ZE41A	T5	5	20	30	2.5	20	29	2.5	20	28
ZE63A	T6	7	28	43	5	27	40		24	37
ZH62A	T5	6	25	40	5	22	35		17	31
ZK51A	T5	8	24	40	5	20	34		17	30
ZK61	T6				5	26	40		24	34

Note: The values for bars from castings are a minimum average of specimens for all alloys except WE54A and EQ21A. The values for WE54A are an average of 106 bars from 7 castings [24]; those for EQ21A are from 6 bars [27].

Source: Refs. 2, 3, 23, 24, 26, 27.

TABLE 8.3 Minimum Tensile Properties from Designated Areas of Castings

Alloy	Temper	Class 1			Class 2			Class 3		
		%E	TYS	TS	%E	TYS	TS	%E	TYS	TS
SI Units			MPa			MPa			MPa	
AM100A	T6	3	140	260	1.5	125	240	1	110	205
AZ91C	T6	4	125	240	3	110	200	2	95	185
AZ92A	T6	3	170	275	1	140	235	0.75	125	205
HK31A	T6	6	110	230	3	95	200	1	85	170
QE22A	T6	4	195	275	2	180	255	2	160	230
ZE63A	T6	6	195	290	5	180	275	4	165	255
ZH62A	T5	5	160	260	3	145	235	2	130	215
ZK51A	T5	6	145	250	4	130	220	3	115	200
ZK61A	T6	6	200	290	4	180	255	2	160	235

English Units			ksi			ksi			ksi	
AM100A	T6	3	20	38	1.5	18	35	1	16	30
AZ91C	T6	4	18	35	3	16	29	2	14	27
AZ92A	T6	3	25	40	1	20	34	0.75	18	30
HK31A	T6	6	16	33	3	14	29	1	12	25
QE22A	T6	4	28	40	2	26	37	2	23	33
ZE63A	T6	6	28	42	5	26	40	4	24	37
ZH62A	T5	5	23	38	3	21	34	2	19	31
ZK61A	T5	6	21	36	4	19	32	3	17	29
ZK51A	T6	6	29	42	4	26	37	2	23	34

Source: Ref. 4.

TABLE 8.4 Tensile Properties of Separately-Cast Bars at Elevated Temperature

Alloy	Temper	Temperature, °C	%E	MPa TYS	MPa TS	ksi TYS	ksi TS
AZ81A	T4	20	15	85	275	12	40
		95	22	90	255	13	37
		150	32	85	195	12	28
		200	30	75	140	11	20
		260	25	70	95	10	14
AZ91C	T4	20	14	85	275	12	40
		95	26	95	235	14	34
		150	30	95	195	14	28
		200	30	90	140	13	20
	T6	20	5	130	275	19	40
		95	24	130	255	19	37
		150	31	115	185	17	27
		200	33	95	140	14	20
AZ92A	T5	20	2	110	180	16	26
		95	2	110	165	16	24
		150	4	95	160	14	23
		200	15	74	140	11	20
		260	32	55	110	8	16
		315	61	30	60	4	9
AZ92A	T6	20	2	145	275	21	40
		95	25	145	255	21	37
		150	35	115	195	17	28
		200	36	85	115	12	17
		260	33	55	75	8	11
		315	49	35	55	5	8
EQ21A	T6	20	4	195	261	28	38
		100	10	189	230	27	33
		150	16	180	211	26	31
		200	16	170	191	25	28
		250	15	152	169	22	25
		300	10	117	132	17	19
		325	9	92	105	13	15
		350	14	60	75	9	11
EZ33A	T5	20	3	105	160	15	23
		95	5	105	160	15	23
		150	10	95	150	14	22

TABLE 8.4 (*Continued*)

Alloy	Temper	Temperature, °C	%E	MPa TYS	MPa TS	ksi TYS	ksi TS
		200	20	85	145	12	21
		260	31	70	125	10	18
		315	50	55	85	8	12
HK31A	T6	20	8	105	220	15	32
		95	8	110	200	16	29
		150	12	105	185	15	27
		200	17	95	165	14	24
		260	19	90	160	13	23
		315	22	85	130	12	19
		370	26	55	90	8	13
HZ32A	T5	20	7	95	205	14	30
		95	15	95	120	14	17
		150	23	85	150	12	22
		200	33	70	115	10	17
		260	39	60	95	9	12
		315	38	55	85	8	10
		370	29	50	70	7	10
QE22A	T6	20	4	205	275	30	40
		95		195	235	28	34
		150		185	205	27	30
		200		165	185	24	27
		260		110	140	16	20
		315		60	85	9	12
QH21A	T6	20		210	270	31	39
		100		200	240	29	35
		150		190	220	28	32
		200		185	210	27	30
		150		165	185	24	27
WE54A	T6	20	5	200	275	29	40
		250	7.5	170	225	25	33
ZE41A	T5	20	5	140	205	20	30
		95	8	130	185	19	27
		150	15	115	165	17	24
		200	29	95	130	14	19
		260	40	70	95	10	14
		315	43	55	75	8	11

TABLE 8.4 (*Continued*)

Alloy	Temper	Temperature, °C	%E	MPa		ksi	
				TYS	TS	TYS	TS
ZE63A	T6	20		175	290	25	42
		100		130	235	19	34
		150		110	185	16	27
		200		95	130	14	19
ZH62A	T5	20	6	170	275	25	40
		95	20	160	230	23	33
		150	24	140	180	20	26
		200	28	105	130	15	19
		260	30	70	95	10	14
ZK51A	T5	20	8	165	275	24	40
		95	12	145	205	21	30
		150	14	115	160	17	23
		200	17	90	115	13	17
		260	16	60	85	9	12
		315	16	40	55	6	8

Source: Ref. 3, except for QE22A, ref. 5; QH22A and ZE63A, ref. 7; WE54A, ref. 24; EQ21A, ref. 27.

TABLE 8.5 Effect of Temperature Exposure on Tensile Properties: Separately-Cast Bars

		Exposure				MPa		ksi	
Alloy	Temper	°C	Hours	Test temperature, °C	%E	TYS	TS	TYS	TS
AZ92A	T6	95	0	20	2	145	275	21	40
			25		2	170	275	25	40
			100		2	160	265	23	38
			500		2	160	280	23	41
			1000		3	150	280	22	41
			2500		4	160	285	23	41
			0	95	25	145	255	21	40
			25		7	135	280	20	41
			100		15	135	275	20	40
			1000		9	130	270	19	39
			2500		10	145	275	21	40
		150	0	20	2	145	275	21	40
			25		1	175	260	25	38
			100		2	155	270	22	39
			500		2	170	270	25	39
			1000		4	145	270	21	39
			2500		2	165	270	24	39
			5000		2	165	265	24	38
			0	150	35	115	195	17	28
			25		42	110	190	16	28
			100		40	110	180	16	26
			1000		46	110	190	16	28
			2500		35	100	170	15	25
			5000		42	100	185	15	27

TABLE 8.5 (*Continued*)

Alloy	Temper	Exposure °C	Exposure Hours	Test temperature, °C	%E	MPa TYS	MPa TS	ksi TYS	ksi TS
		200	0	20	5	145	275	21	40
			25		1	175	275	25	40
			100		2	165	275	24	40
			500		2	155	275	22	40
			1000		2	155	270	22	39
			2500		3	135	245	20	36
			5000		3	135	245	20	36
			0	200	33	85	115	12	17
			25		37	80	115	12	17
			100		31	80	120	12	17
			1000		37	80	120	12	17
			2500		48	80	120	12	17
			5000		47	70	120	10	17
EZ33A	T5	200	0	20	3	105	160	15	23
			25		3	130	170	19	25
			100		2	135	170	20	25
			500		2	140	170	20	25
			1000		2	140	170	20	25
			5000		1	145	170	21	25
EZ33A	T5	200	0	200	20	85	145	12	21
			25		18	85	140	12	20
			100		17	85	140	12	20
			500		17	90	135	13	20

			1000		16	90	135	13	20
			5000		16	90	130	13	19
			0	315	50	55	83	8	12
			25		51	55	85	8	12
			100		53	55	85	8	12
			500		56	55	85	8	12
			1000		58	55	85	8	12
		260	0	20	3	105	160	15	23
			25		4	125	165	18	24
			100		3	125	165	18	24
			500		3	130	165	19	24
			1000		3	130	165	19	24
			5000		2	135	165	20	24
		260	0	260	31	70	125	10	18
			25		42	70	110	10	16
			100		43	70	105	10	15
			500		43	65	95	9	14
			1000		45	65	95	9	14
			5000		45	65	95	9	14
			0	315	50	55	85	8	12
			25		44	50	80	7	12
			100		50	50	80	7	12
			500		54	50	80	7	12
			1000		56	45	80	7	12
EZ33A	T5	315	0	20	3	105	160	15	23
			25		4	125	170	18	25
			100		3	125	170	18	25
			500		2	130	175	19	25

TABLE 8.5 (*Continued*)

		Exposure				MPa		LSO	
Alloy	Temper	°C	Hours	Test temperature, °C	%E	TYS	TS	TYS	TS
			1000		2	130	175	19	25
			5000		2	130	175	19	25
			25	315	55	55	80	8	12
			100		55	55	80	8	12
			500		55	55	80	8	12
			1000		55	55	80	8	12
			5000		54	55	80	8	12
HK31A	T6	200	0	20	8	105	220	15	32
			25		8	120	235	17	34
			100		8	125	240	18	34
			500		8	125	250	18	36
			1000		8	130	250	19	36
			5000		8	130	255	19	37
HK31A	T6	200	0	200	17	95	165	14	24
			25		18	100	165	14	24
			100		18	100	170	14	25
			500		18	100	170	14	25
			1000		18	100	175	14	25
			5000		17	100	180	14	26
			0	315	22	85	140	12	20
			25		21	85	140	12	20
			100		22	90	140	13	20
			500		23	90	140	13	20

			1000		24	90	140	13	20
			5000		26	90	140	13	20
		260	0	20	8	105	220	15	32
			25		6	110	230	16	33
			100		5	115	230	17	33
			500		5	115	230	17	33
			1000		4	115	230	17	33
			5000		4	115	230	17	33
HK31A	T6	260	0	260	19	90	160	13	23
			25		22	95	165	14	24
			100		22	90	160	13	23
			500		21	85	150	12	22
			1000		21	75	140	11	20
			5000		29	50	100	7	14
		260	0	315	22	85	140	12	20
			25		22	85	130	12	19
			100		22	85	130	12	19
			500		22	85	125	12	18
			1000		22	70	120	10	17
			5000		22	50	95	7	14
		315	0	20	8	105	220	15	32
			25		8	105	215	15	30
			100		8	105	205	15	30
			500		9	90	195	13	28
			1000		9	85	185	12	27
			5000		9	60	165	9	24
			0	315	22	85	140	12	20
			25		28	70	115	10	17
			100		30	55	105	8	15

TABLE 8.5 (*Continued*)

Alloy	Temper	Exposure °C	Exposure Hours	Test temperature, °C	%E	MPa TYS	MPa TS	ksi TYS	ksi TS
			500		37	40	90	6	13
			1000		40	35	75	5	11
			5000		55	30	40	4	6
		345	25	20	8	85	185	12	27
			100		10	80	175	12	25
			500		11	75	170	11	25
			1000		13	75	170	11	25
			1	345	25	75	105	11	15
			10		27	50	90	7	13
			25		40	50	85	7	12
			100		50	35	65	5	9
			500		62	20	40	2	6
			1000		70	15	30	2	4
		370	1	345	22	70	115	10	17
			5		27	55	100	8	14
QH21A	T6	None		20	4	205	275	30	40
		200	500	20	8	205	285	30	41
			1000	20	8	200	280	29	41
WE54A	T6	200	0.25	200	6.5	177	244	26	35
			1000		3	193	235	28	34
		250	0.25	250	10	167	233	24	34
			1000		11	165	213	24	31

Source: Ref. 6, except for QH21A, ref. 7; WE54A, ref. 24.

TABLE 8.6 Dimensional Changes on Temperature Exposure

Alloy	Temper	Hours	Percent change in length at temperature, °C								
			100	120	135	150	175	200	260	315	371
AZ91C	T4	10	-0.002	0.001		-0.005					
		100	-0.002	0.010		0.032					
		1000	-0.008	0.044		0.068					
		10000	0.045	0.077		0.078					
	T6	10	0.000	0.000		0.003					
		100	0.001	0.004		0.020					
		1000	0.006	0.026		0.048					
		10000	0.027	0.056		0.059					
EZ33A	T5	10						0.011	0.013	0.012	0.010
		100						0.013	0.016	0.015	0.012
		1000						0.013	0.018	0.017	0.013
		5000						0.013	0.019	0.018	0.014
HK31A	T6	10						0.003	0.003	0.002	0.006
		100						0.003	0.005	0.005	0.011
		1000						0.003	0.007	0.013	0.011
		5000						0.003	0.007	0.013	0.011
HZ32A	T5	10						0.011	0.008	0.008	0.006
		100						0.013	0.010	0.010	0.008
		1000						0.014	0.011	0.011	0.009
		5000						0.014	0.012	0.012	0.010

Source: Ref. 3.

TABLE 8.7 Effect of Strain Rate and Temperature on the Strength of Cast Alloys (SI Units)

Alloy	Test temperature, °C	TYS (MPa) at strain/min				TS (MPa) at strain/min			
		0.005	0.050	0.50	5.0	0.005	0.050	0.50	5.0
AZ91C-T6	25	135	135	135	135	280	280	280	280
	95	120	130	130	130	265	265	270	270
	150	100	110	115	125	195	200	220	245
	200	80	90	100	110	125	130	150	180
	260	60	75	85	100	90	95	110	140
	315	45	55	70	85	65	70	85	105
	370	25	40	50	65	40	45	55	85
	425	10	15	20	20	15	20	20	20
AZ92A-T6	25	160	160	160	160	270	270	270	270
	95	140	150	155	160	255	255	255	255
	150	110	125	140	150	200	210	230	245
	200	75	95	115	135	125	135	150	195
	260	55	70	90	110	85	90	105	140
	315	40	55	70	85	55	60	75	105
	370	20	30	50	70	30	35	50	75

EZ33A-T5	25	125	125	125	125	190	190	190	190
	95	100	105	110	110	165	165	165	165
	150	85	95	100	100	160	160	160	160
	200	75	75	85	90	155	155	155	155
	260	70	70	70	80	135	135	140	140
	315	65	65	65	70	100	110	120	125
	370	40	50	55	65	50	60	80	100
	425	15	25	35	50	25	30	40	70
	480	5	10	15	35	10	15	20	45
HK31A-T6	25	110	110	115	115	235	235	235	240
	95	105	105	110	110	185	185	195	205
	150	100	100	100	105	175	175	180	185
	200	95	95	95	100	160	160	160	160
	260	90	90	90	90	140	140	140	140
	315	85	85	85	85	130	130	130	130
	370	50	50	55	60	85	90	95	100
	425	30	35	35	40	45	50	60	75
	480	15	20	25	35	20	25	30	55
HZ32A-T5	25	100	100	100	100	220	220	220	220
	95	85	90	90	95	165	170	175	185

TABLE 8.7 (*Continued*)

Alloy	Test temperature, °C	TYS (MPa) at strain/min				TS (MPa) at strain/min			
		0.005	0.050	0.50	5.0	0.005	0.050	0.50	5.0
	150	70	80	85	90	135	140	145	160
	200	60	70	75	80	110	115	120	135
	260	50	55	65	70	85	90	95	105
	315	45	45	50	60	75	75	80	85
	370	40	40	45	55	70	70	75	75
	425	35	35	40	40	50	50	60	65
	480	15	20	25	30	25	30	35	45
ZH62A-T5	25	190	190	190	190	285	285	285	285
	95	160	170	175	185	230	235	240	255
	150	135	145	160	170	175	185	200	220
	200	100	115	135	150	125	130	150	180
	260	65	80	95	120	90	95	105	140
	315	50	60	70	90	65	70	80	110
	370	40	50	60	75	50	55	65	85
	425	25	35	40	60	35	40	45	60
	480	10	15	20	35	15	20	20	35

Source: Ref. 6.

TABLE 8.7a Effect of Strain Rate and Temperature on the Strength of Cast Alloys (English Units)

Alloy	Test temperature, °C	TYS (ksi) at strain/min				TS (ksi) at strain/min			
		0.005	0.050	0.50	5.0	0.005	0.050	0.50	5.0
AZ91C-T6	25	20	20	20	20	41	41	41	41
	95	17	19	19	19	38	38	39	39
	150	14	16	17	18	28	29	32	36
	200	12	13	14	16	18	19	22	26
	260	9	11	12	14	13	14	16	20
	315	7	8	10	12	9	10	12	15
	370	4	6	7	9	6	7	8	12
	425	1	2	3	3	2	3	3	3
AZ92A-T6	25	23	23	23	23	39	39	39	39
	95	20	22	22	23	37	37	37	37
	150	16	18	20	22	29	30	33	36
	200	11	14	17	20	18	20	22	28
	260	8	10	13	16	12	13	15	20
	315	6	8	10	12	8	9	11	15
	370	3	4	6	10	4	5	7	11

TABLE 8.7a (*Continued*)

Alloy	Test temperature, °C	TYS (ksi) at strain/min				TS (ksi) at strain/min			
		0.005	0.050	0.50	5.0	0.005	0.050	0.50	5.0
EZ33A	25	18	18	18	18	28	28	28	28
	95	14	15	16	16	24	24	24	24
	150	12	14	14	14	23	23	23	23
	200	11	11	12	13	22	22	22	22
	260	10	10	10	12	20	20	20	20
	315	9	9	9	10	14	16	17	19
	370	6	7	8	9	7	9	12	14
	425	2	4	5	7	4	4	6	10
	480	1	1	2	5	1	2	3	7
HK31A-T6	25	16	16	17	17	34	34	34	35
	95	15	15	16	16	27	27	28	30
	150	14	14	14	15	25	25	26	27
	200	13	13	13	14	23	23	23	23
	260	13	13	13	13	20	20	20	20
	315	12	12	12	12	19	19	19	19
	370	7	7	8	9	12	13	14	14
	425	4	5	5	6	7	7	9	11

	480	2	3	4	5	3	4	4	8
HZ32A-T5	25	14	14	14	14	32	32	32	32
	95	12	13	13	14	24	25	25	27
	150	10	12	12	13	20	20	21	23
	200	9	10	11	12	16	17	17	20
	260	7	8	9	10	12	13	14	15
	315	7	7	7	9	11	11	12	12
	370	6	6	7	8	10	10	11	11
	425	5	5	6	6	7	7	9	9
	480	2	3	4	4	4	4	5	7
ZH62A-T5	25	28	28	28	28	41	41	41	41
	95	23	25	25	27	33	34	35	37
	150	20	21	23	25	25	27	29	32
	200	14	17	20	22	18	19	22	26
	260	9	12	14	17	13	14	15	20
	315	7	9	10	13	9	10	12	16
	370	6	7	9	11	7	8	9	12
	425	4	5	6	9	5	6	7	9
	480	1	2	3	5	2	3	3	5

Source: Ref. 6.

TABLE 8.8 Creep Strength of Cast Alloys: Separately-cast Test Bars (Stress in MPa)

Alloy	Temperature, °C	Hours for 0.1% TE				Hours for 0.2% TE				Hours for 0.5% TE				Hours for 1.0% TE			
		1	10	100	1000	1	10	100	1000	1	10	100	1000	1	10	100	1000
AZ92A	95	50	40	35	30	80	70	55	45	130	110	90	70	160	140	115	100
T6	150	30	25	15	10	55	40	25	20	85	70	50	35	105	85	60	45
	200	15	10	5		30	20	10		50	35	15	5	60	40	25	10
	260	10				15	5			25	15			35	20		
EQ21A	150		149	138	123			155	134				152				
	200		109	78				95	62			116	76				
	250		46	29				36	19			42	24				
EZ33A	200	40	40	35	30	65	60	55	45	75	75	70	55	90	85	80	65
T5	260	30	20	15	10	50	35	20	15	65	45	30	20	70	50	35	20
	315	20	10	5		25	15	10	5	30	15	10	5	35	20	15	10

HZ32A	200	40	40	35	30	60	55	55	50	70	70	70	70	75	75	75	70
T5	260	40	35	30	25	55	50	40	35	65	60	55	45	70	70	65	50
	315	30	25	15	10	40	35	25	15	50	45	35	20	60	50	40	20
QE22A	150			140	105			165	125				150				
T6	200		105	75				85	55			105	70				80
	250		40	25	10			30	15			40	20				25
WE54A	250						56	41	26			69	45				
T6																	
ZH62A	95	45	40	35	30	85	80	75	60	140	125	110	95	170	155	140	115
T5	150	40	35	30	20	70	60	50	40	110	100	85	70	130	120	110	90
	200	30	25	20	10	50	35	25	15	75	55	40	25	95	70	50	30
	260	20	15	10	5	25	15	10	10	35	20	15	10	45	25	15	10
ZK51A	95	40	35	30	25	75	65	55	45	125	110	100	85	145	130	120	100
T5	150	30	25	20	15	55	45	35	25	90	75	65	50	110	95	80	60
	200	25	20	15	10	40	30	20	10	60	50	40	25	75	65	50	35

Note: Values for EQ21A and QE22A are for creep extension only. The initial extension is not included.
Source: Ref. 3, except for QE22A, ref. 7; WE54A, ref. 24; EQ21A, ref. 27.

TABLE 8.8a Creep Strength of Cast Alloys: Separately-cast Test Bars (Stress in ksi)

Alloy	Temperature, °C	Hours for 0.1% TE				Hours for 0.2% TE				Hours for 0.5% TE				Hours for 1.0% TE			
		1	10	100	1000	1	10	100	1000	1	10	100	1000	1	10	100	1000
AZ92A	95	7	6	5	4	12	10	8	7	19	16	13	10	23	20	17	14
	150	4	4	2	1	8	6	4	3	12	10	7	5	15	12	9	7
	200	2	1	1		4	3	1		7	5	2	1	9	6	3	1
	260	1				2	1			4	2			5	3		
EQ21A	150		22	20	18			22	19				22				
	200		16	11				14	9			17	11				
	250		7	4				5	3			6	3				
EZ33A T5	200	6	6	5	4	9	9	8	7	11	11	10	8	13	12	12	9
	260	4	3	2	1	7	5	3	2	9	7	4	3	10	7	5	3
	315	3	1	1		4	2	1	1	4	2	1	1	5	3	2	1

HZ32A	200	6	6	5	4	9	8	8	7	10	10	10	11	11	11	10	10
T5	260	6	5	4	4	8	7	6	5	9	9	8	7	10	10	9	7
	315	4	4	2	1	6	5	4	2	7	7	5	3	9	7	6	3
QE22A	150			20	15			24	18				22				
T6	200		15	11				12	8			15	10				12
	250		6	4	1			4	2			6	3				4
WE54A	250						8	6	4			10	6				
T6																	
ZH62A	95	7	6	5	4	12	12	11	9	20	18	16	14	25	22	20	17
T5	150	6	5	4	3	10	9	7	6	16	14	12	10	19	17	16	13
	200	4	4	3	1	7	5	4	2	11	8	6	4	14	10	7	4
	260	3	2	1	1	4	2	1	1	5	3	2	1	7	4	2	1
ZK51A	95	6	5	4	4	11	9	8	7	18	16	14	12	21	19	17	14
T5	150	4	4	3	2	8	7	5	4	13	11	9	7	16	14	12	9
	200	4	3	2	1	6	4	3	1	9	7	6	4	11	9	7	5

Note: Values for EQ21A and QE22A are for creep extension only. The initial extension is not included.
Source: Ref. 3, except for QE22A, ref. 7; WE54A, ref. 24; EQ21A, ref. 27.

TABLE 8.9 Creep Strength of Cast Alloys: Separately-Cast Test Bars (Stress in MPa)

Alloy	Temperature, °C	Time		Stresses to produce specified creep strain, %							
		hr	sec	0.05	0.1	0.2	0.5	1.0	2	5	10
EZ32A-T5	200		30						98	118	130
			60						97	117	128
			600						96	116	125
		100		52	66	71					
		500		41	54	65					
		1000		36	47	58					
	250		30				76	84	92	111	123
			60				74	83	91	110	120
			600				73	82	89	108	114
		100		23	28	32	36				
		500		11	19	24	30	34			
		1000			14	20	26	30			
	315		30				52	59	73	80	85
			60				51	58	69	76	83
			600				42	49	56	62	68
		100		5.6	7.4	8					
		500			5.2	6.5					
		1000			4.3	5.6					
ZE41A-T5	100		30				100	107	116	127	
			60				99	105	114	124	
			600				86	99	103	114	
		100			97	111	117				
		500				106	117				
		1000				103	116				

	150		30				86	90	94	99	
			60				83	88	91	96	
			600				71	76	81	86	
		100		77	86	97	101	107			
		500			75	88	96	100			
		1000			70	83	91	97			
	200		30				62	69	76	69	83
			60				59	66	73	76	79
			600				49	53	59	64	67
		100		29	43	52	67	73			
		500		22	28	37	52	64			
		1000		20	23	31	43	53			
	250	100		6	12	19	32	29			
		500		4	6	9	15	19			
		1000		4	5	7	12	15			
HK31A-T6	250		30				96	103	119	138	
			60				95	103	118	137	
			600				94	102	117	137	
	315		30				80	88	103	117	
			60				78	86	102	116	
			600				74	82	96	107	
HK32A-T5	250		30				58	65	71	84	102
			60				57	64	69	81	99
			600				56	63	68	74	86
		100		42	50	56	63	66			
		500		35	43	51	58	63			
		1000		31	39	49	56	61			

TABLE 8.9 (*Continued*)

Alloy	Temperature, °C	Time		Stresses to produce specified creep strain, %							
		hr	sec	0.05	0.1	0.2	0.5	1.0	2	5	10
	300	100		23	28	35	46	52			
		500		19	21	25	36	41			
		1000		27	19	21	32	36			
	315		30				55	60	64	73	76
			60				53	59	63	72	76
			600				50	59	61	71	74
	325	100		14	19	24	29	36			
		500		12	13	16	21	25			
		1000		10	12	13	15	20			
	350	100		10	12	18	21	23			
		500			9	10	12	14			
		1000			8	8	9	10			
	375	100			8	11	12	13			
		500					12	9			
		1000						8			

QE22A-T6	150	100			140	165					
		1000			105	125	150				
	200	10			105						
		100			75	85	105				
		1000				55	70	80			
	250	10			40						
		100			25	30	40				
		1000			10	15	20	25			
ZH62A-T5	150	100		51	66	82	96	102			
		500		36	56	69	85	94			
		1000		26	51	63	80	90			
	200		30				96	113	120	128	137
			60				93	109	117	124	133
			600				83	90	102	110	114
		100		26	32	45	56	62			
		500		15	22	26	40	49			
		1000		11	17	20	31	40			
	250		30				70	77	85	96	99
			60				65	74	80	90	94
			600				56	60	66	74	77
	315		30				54	59	64	70	74
			60				52	57	62	66	70
			600				44	49	53	56	58

Source: Refs. 7, 26.

TABLE 8.9a Creep Strength of Cast Alloys: Separately-Cast Test Bars (Stress in ksi)

Alloy	Temperature, °C	Time		Stresses to produce specified creep strain, %							
		hr	sec	0.05	0.1	0.2	0.5	1.0	2	5	10
EZ32A-T5	200		30						14	17	19
			60						14	17	19
			600						14	17	18
		100		7.5	9.6	10.3					
		500		5.9	7.8	9.4					
		1000		5.2	6.8	8.4					
	250		30				11	12	13	16	18
			60				11	12	13	16	17
			600				11	12	13	16	17
		100		3.3	4.1	4.6	5.2				
		500		1.6	2.8	3.5	4.4	4.9			
		1000			2.0	2.9	3.8	4.4			
	315		30				7.5	8.6	10.6	12	12
			60				7.4	8.4	10	11	12
			600				6.1	7.1	8.1	9.0	9.9
		100		0.8	1.1	1.2					
		500			0.8	0.9					
		1000			0.6	0.8					
ZE41A-T5	100		30				14	16	17	18	
			60				14	15	17	18	
			600				12	14	15	17	

Alloy	Temp	Hours	Min								
		100			14	16	17				
		500				15	17				
		1000				15	17				
	150		30				12	13	14	14	
			60				12	13	13	14	
			600				10	11	12	12	
		100		11	12	14	15	16			
		500			11	13	14	15			
		1000			10	12	13	14			
	200		30				9.0	10	11	10	12
			60				8.6	9.6	11	11	11
			600				7.1	7.7	8.6	9.3	9.7
		100		4.2	6.2	7.5	9.7	11			
		500		3.2	4.1	5.4	7	7.5	9.3		
		1000		2.9	3.3	4.5	6.2	7.7			
	250	100		0.9	1.7	2.8	4.6	5.7			
		500		0.6	0.9	1.3	2.8				
		1000		0.6	0.7	1.0	1.7	2.2			
HK31A-T6	250		30				14	15	17	20	
			60				14	15	17	20	
			600				14	15	17	20	
	315		30				12	13	15	17	
			60				11	12	15	17	
			600				11	12	14	16	
HZ32A-T5	250		30				8.4	9.4	10	12	15
			60				8.3	9.3	10	12	14
			600				8.1	9.1	9.9	11	12

TABLE 8.9a (*Continued*)

Alloy	Temperature, °C	Time		Stresses to produce specified creep strain, %							
		hr	sec	0.05	0.1	0.2	0.5	1.0	2	5	10
		100		6.1	7.3	8.1	9.1	9.6			
		500		5.1	6.2	7.4	8.4	9.1			
		1000		4.5	5.7	7.1	8.1	8.8			
	300	100		3.3	4.1	5.1	6.7	7.5			
		500		2.8	3.0	3.6	5.2	5.9			
		1000		2.5	2.8	3.0	4.6	5.2			
	315		30				8.0	8.7	9.3	11	11
			60				7.7	8.6	9.1	10	11
			600				7.3	8.6	8.8	10	11
	325	100		2.0	2.8	3.5	4.2	5.2			
		500		1.7	1.9	2.3	3.0	3.6			
		1000		1.5	1.7	1.9	2.2	2.9			
	350	100		1.5	1.7	2.6	3.0	3.3			
		500			1.3	1.5	1.7	2.0			
		1000			1.2	1.2	1.3	1.5			
	375	100			1.2	1.6	1.7	1.9			
		500					1.7	1.3			
		1000						1.2			

QE22A-T6	150	100			20	24					
		1000			15	18	22				
	200	10			15						
		100			11	12	15				
		1000				8.0	10	12			
	250	10				5.8					
		100				3.6	4.4	5.8			
		1000				1.5	2.2	2.9	3.6		
ZH62A-T5	150	100		7.4	9.6	12	14	15			
		500		5.2	8.1	10	12	14			
		1000		3.8	7.4	9.1	12	13			
	200		30				14	16	17	19	20
			60				13	16	17	18	19
			600				12	13	15	16	17
		100		3.8	4.6	6.5	8.1	9.0			
		500		2.2	3.2	3.8	5.8	7.1			
		1000		1.6	2.5	2.9	4.5	5.8			
	250		30				10	11	12	14	14
			60				9.4	11	12	13	14
			600				8.1	8.7	9.6	11	11
	315		30				7.8	8.6	9.3	10	11
			60				7.5	8.3	9.0	9.6	10
			600				6.4	7.1	7.7	8.1	8.4

Source: Refs. 7, 26.

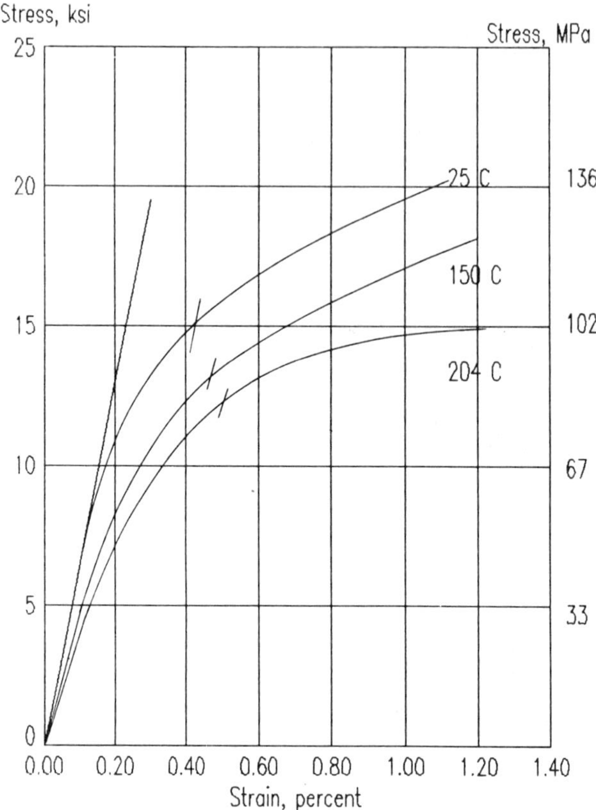

FIGURE 8.2 Stress-strain curves (AZ91C-T4).

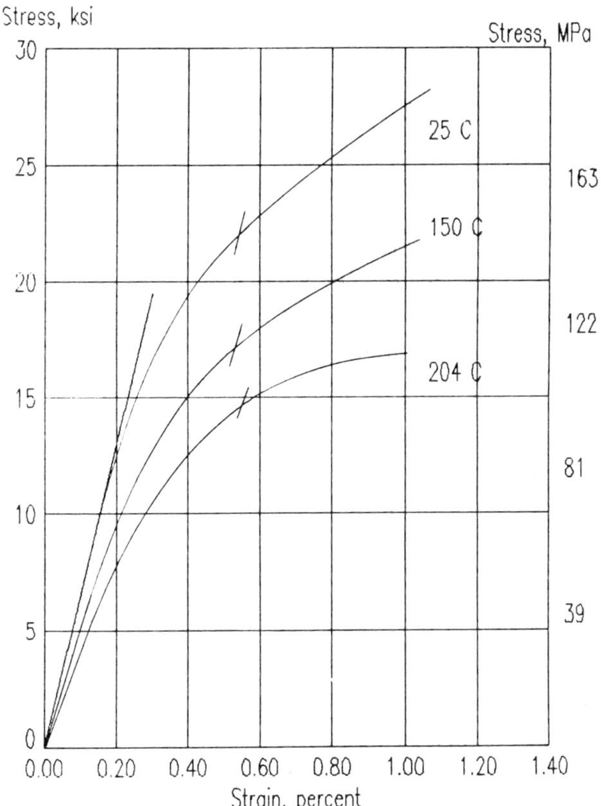

FIGURE 8.3 Stress-Strain Curves (AZ91C-T6).

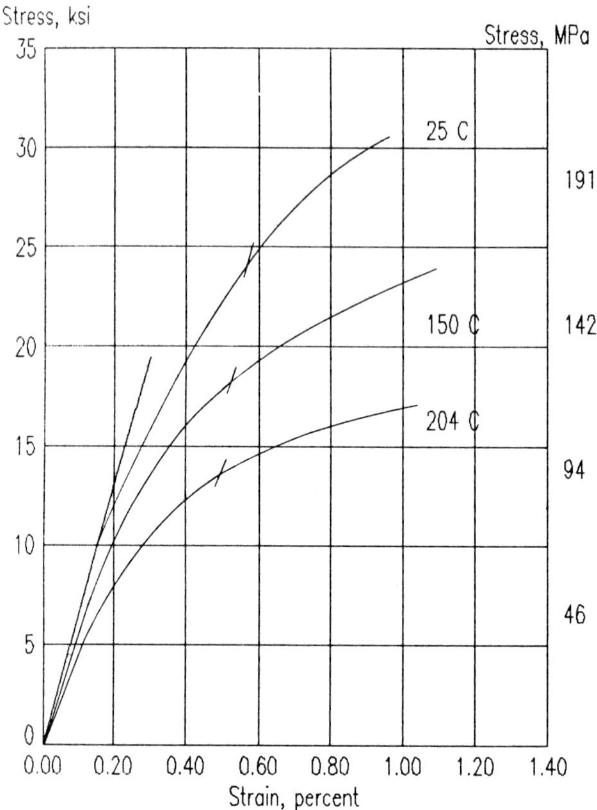

FIGURE 8.4 Stress-strain curves (AZ92A-T6).

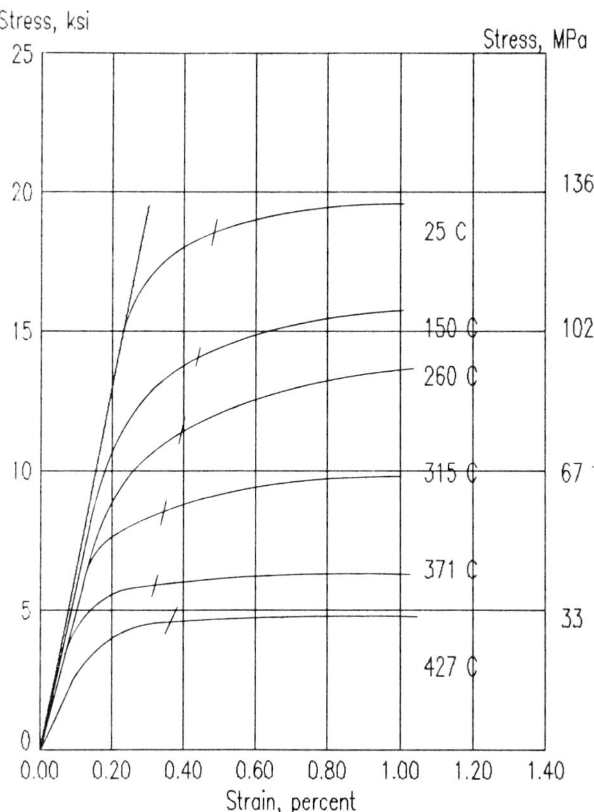

FIGURE 8.5 Stress-strain curves (EZ33A-T5).

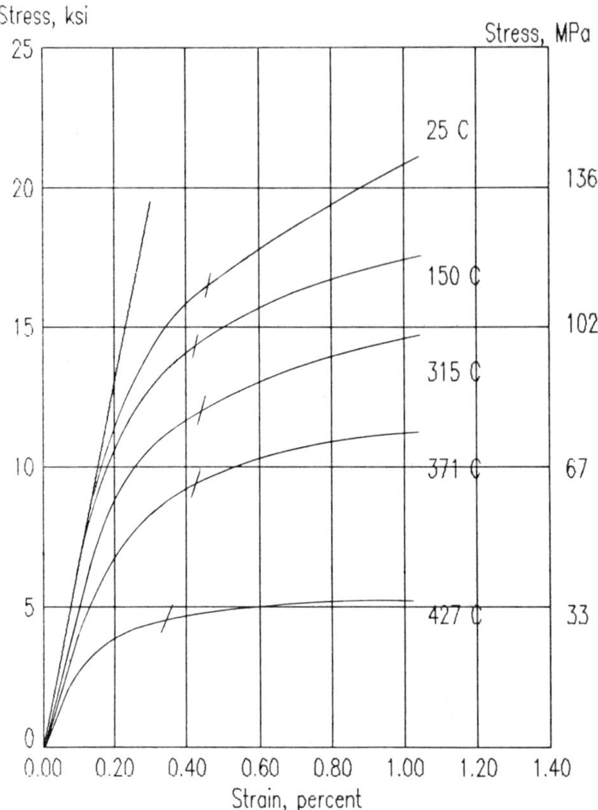

FIGURE 8.6 Stress-strain curves (HK31A-T6).

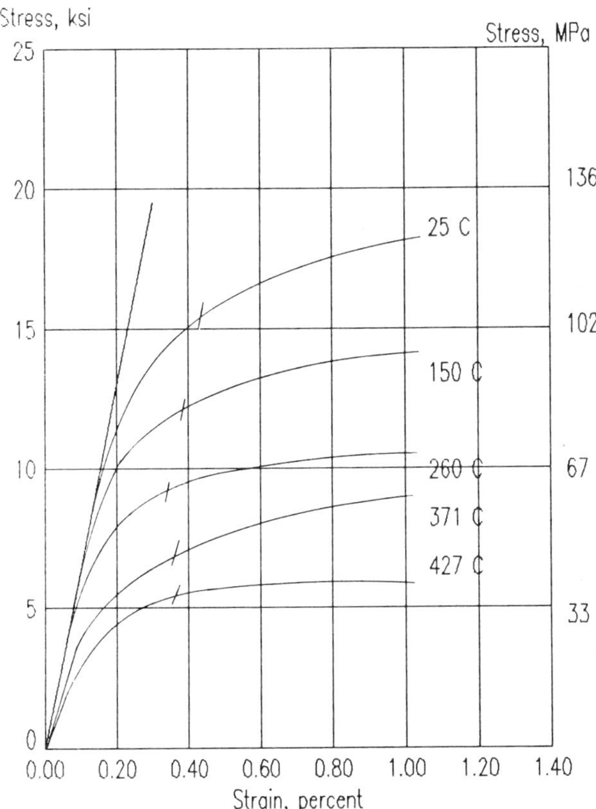

FIGURE 8.7 Stress-strain curves (HZ32A-T5).

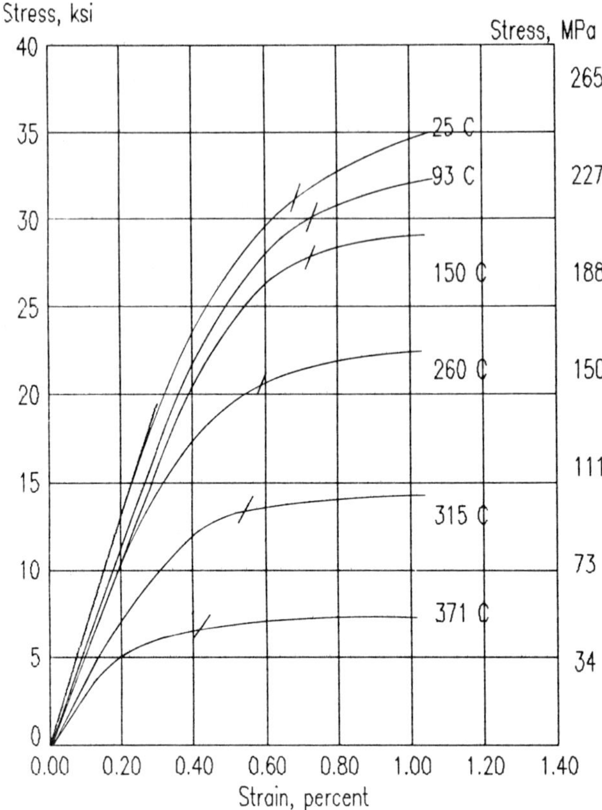

FIGURE 8.8 Stress-strain curves (QE22A-T6).

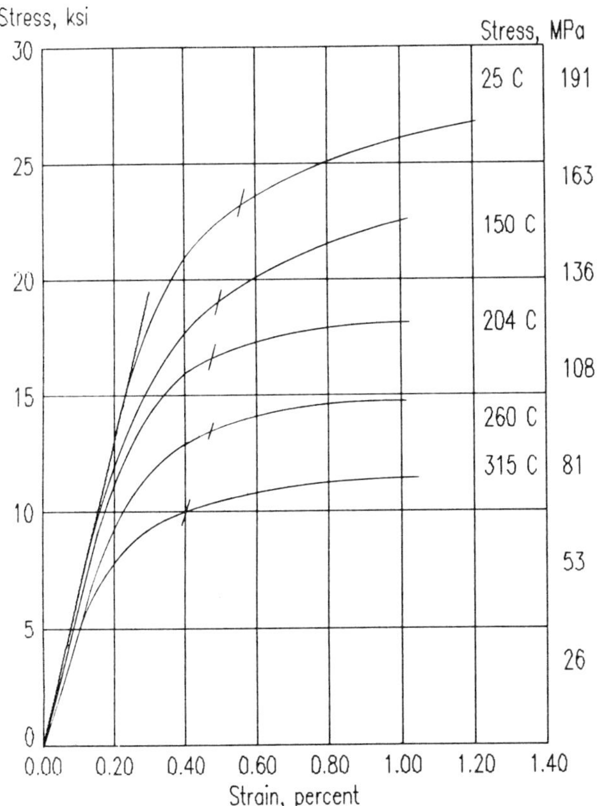

FIGURE 8.9 Stress-strain curves (ZE41A-T5).

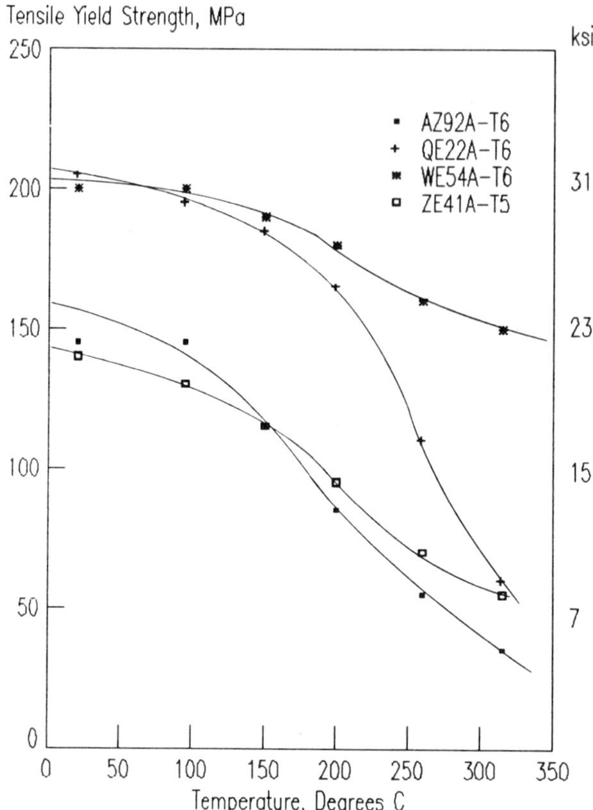

FIGURE 8.10 Effect of temperature on TYS. Comparison of four alloy types.

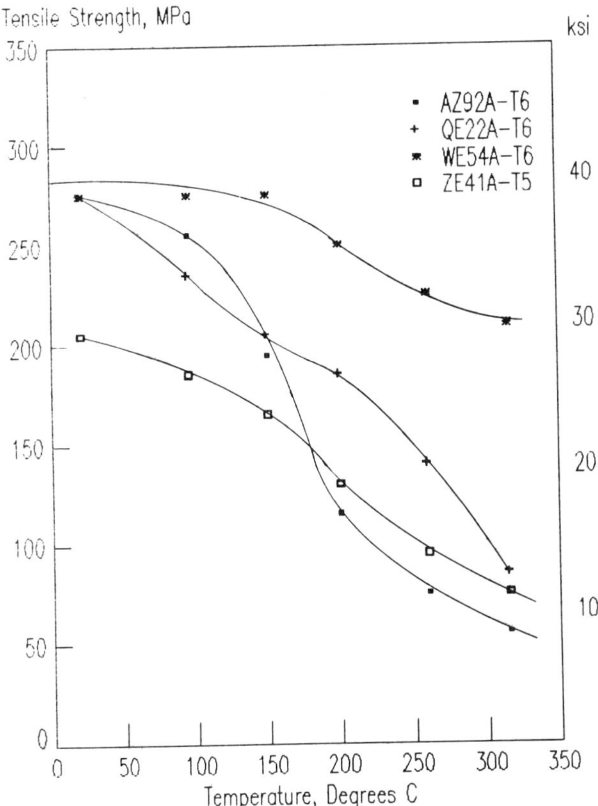

FIGURE 8.11 Effect of temperature on TS. Comparison of four alloy types.

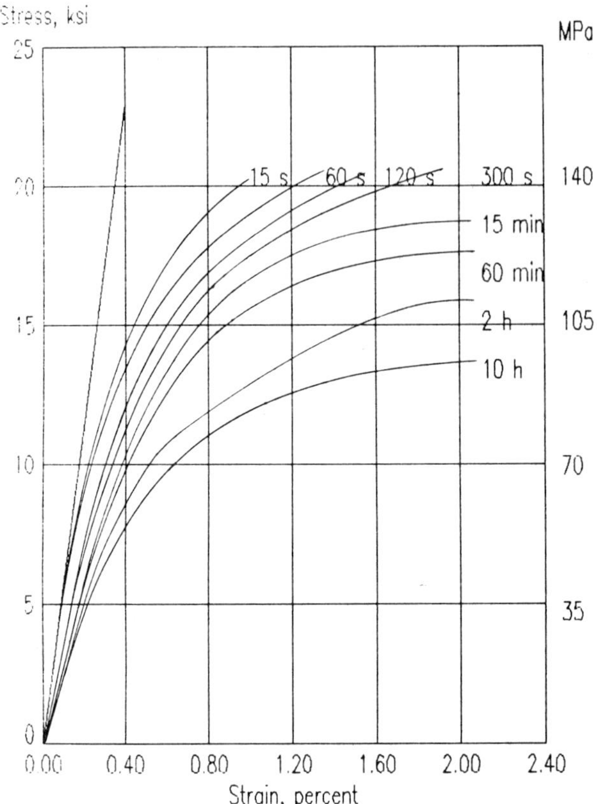

FIGURE 8.12 Sand cast AZ91C-T6 isochronous stress-strain curves (150°C).

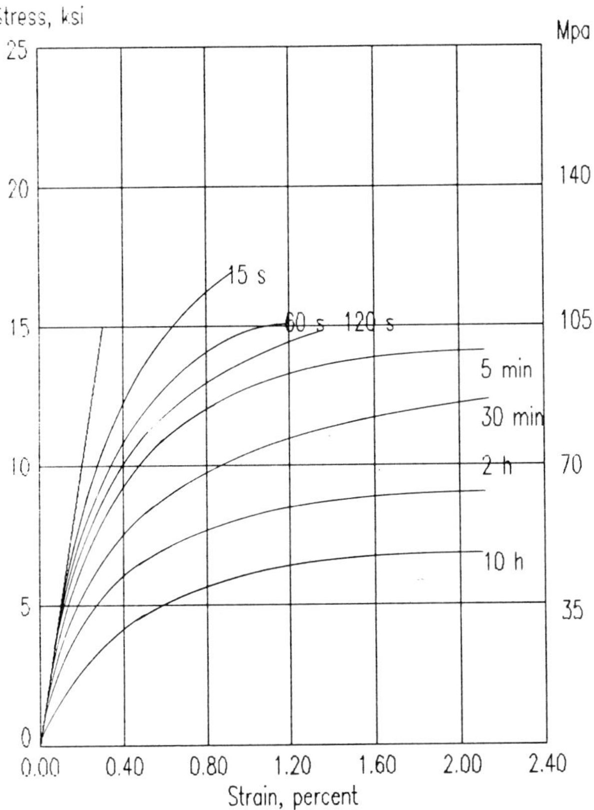

FIGURE 8.13 Sand cast AZ91C-T6 isochronous stress-strain curves (204°C).

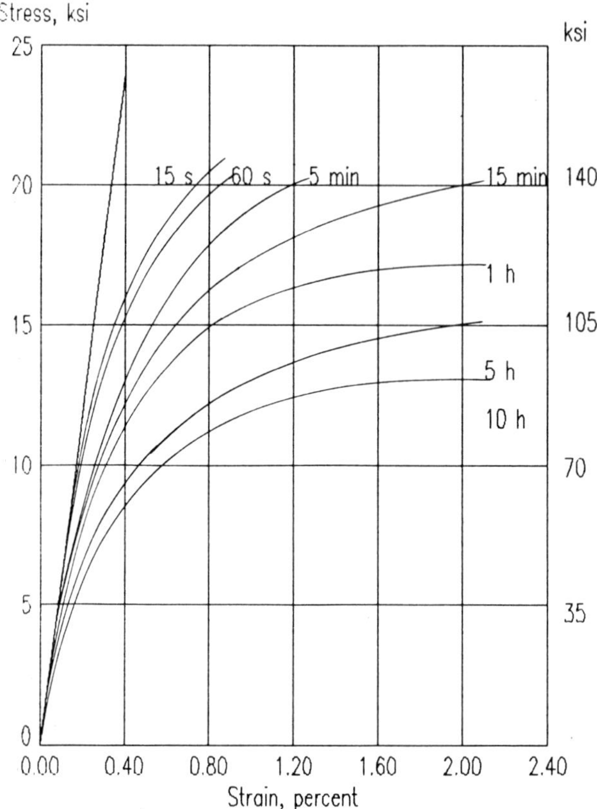

FIGURE 8.14 Sand cast AZ92A-T6 isochronous stress-strain curves (150°C).

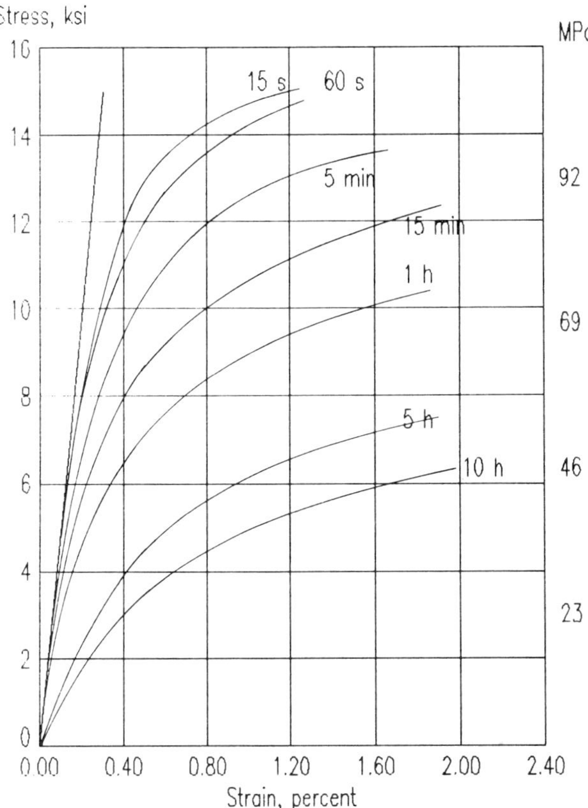

FIGURE 8.15 Sand cast AZ92A-T6 isochronous stress-strain curves (204°C).

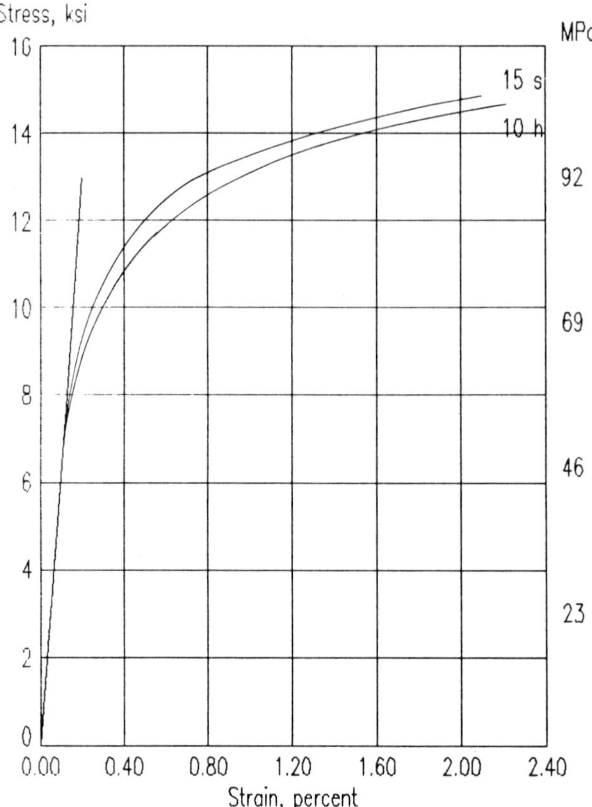

FIGURE 8.16 Sand cast EZ33A-T5 isochronous stress-strain curves (204°C).

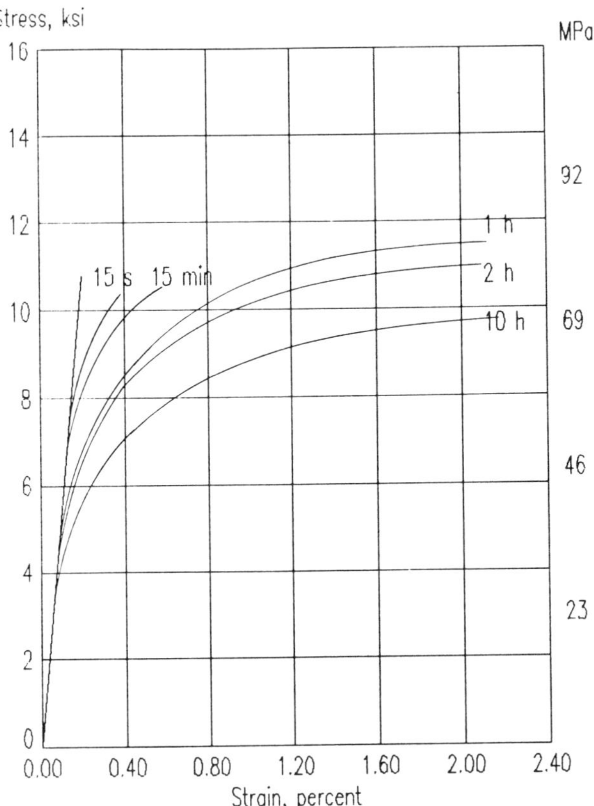

FIGURE 8.17 Sand cast EZ33A-T5 isochronous stress-strain curves (260°C).

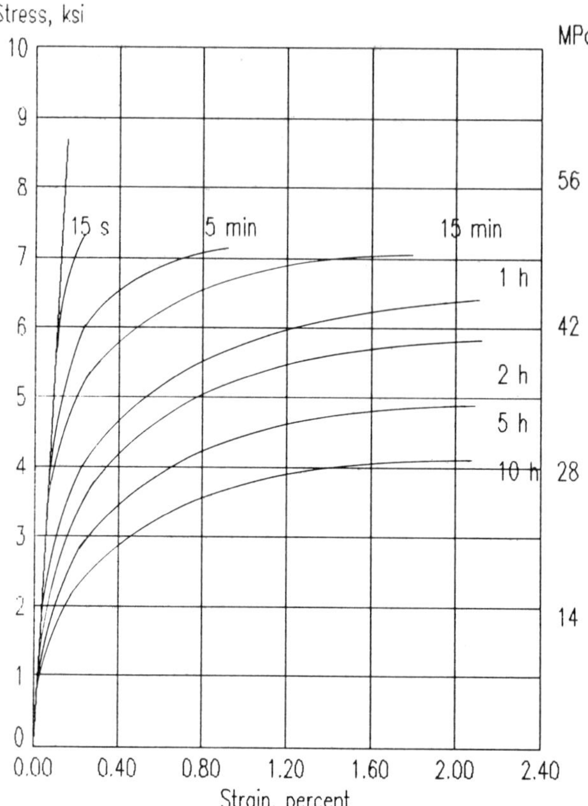

FIGURE 8.18 Sand cast EZ33A-T5 isochronous stress-strain curves (315°C).

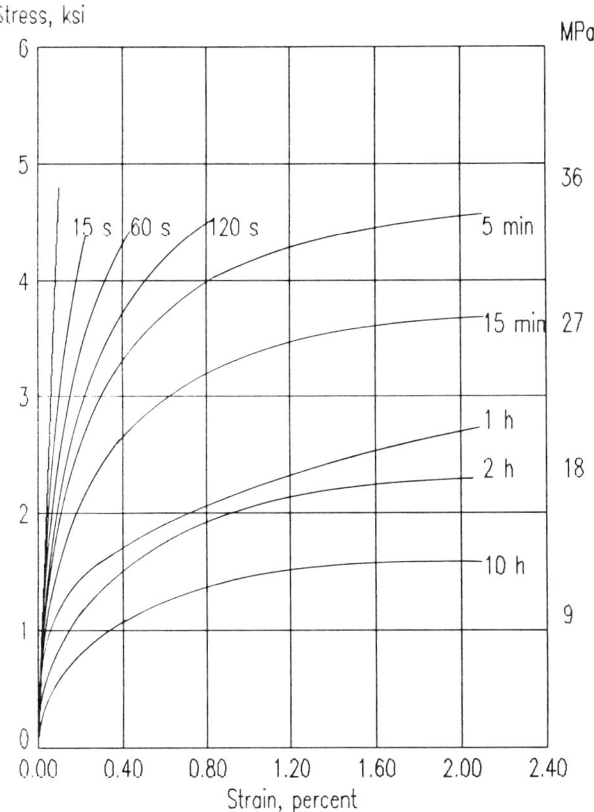

FIGURE 8.19 Sand cast EZ33A-T5 isochronous stress-strain curves (370°C).

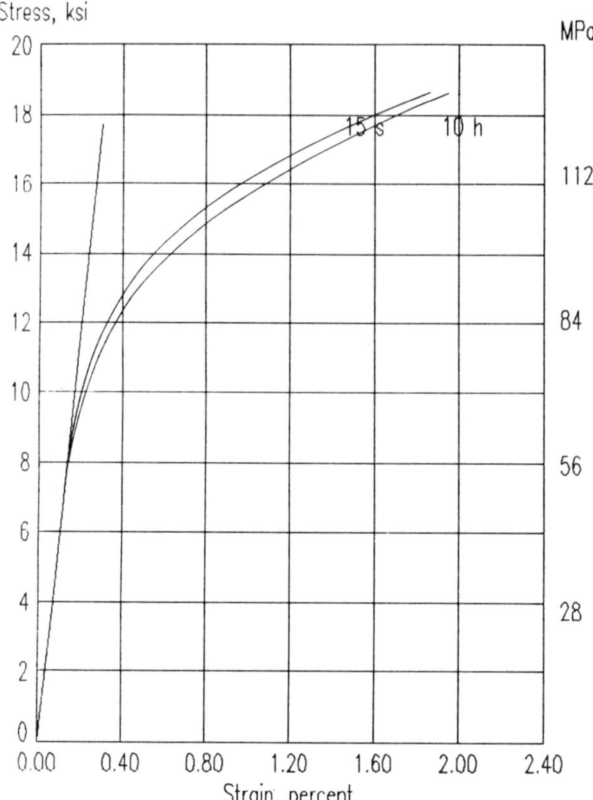

FIGURE 8.20 Sand cast HK31A-T6 isochronous stress-strain curves (204°C).

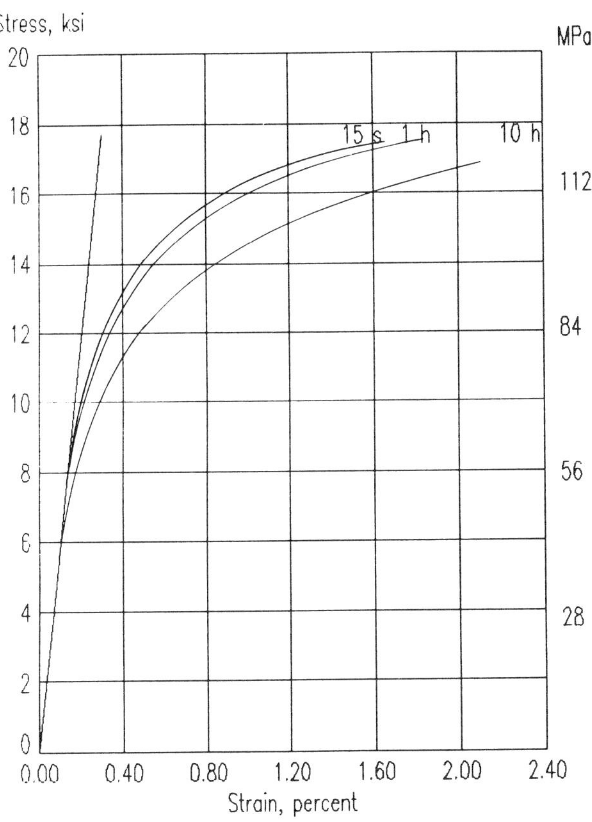

FIGURE 8.21 Sand cast HK31A-T6 isochronous stress-strain curves (260°C).

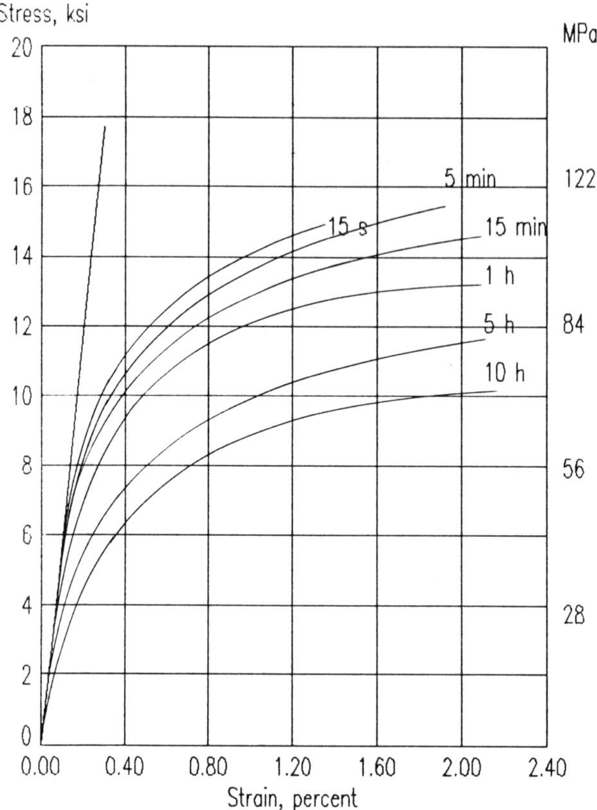

FIGURE 8.22 Sand cast HK31A-T6 isochronous stress-strain curves (315°C).

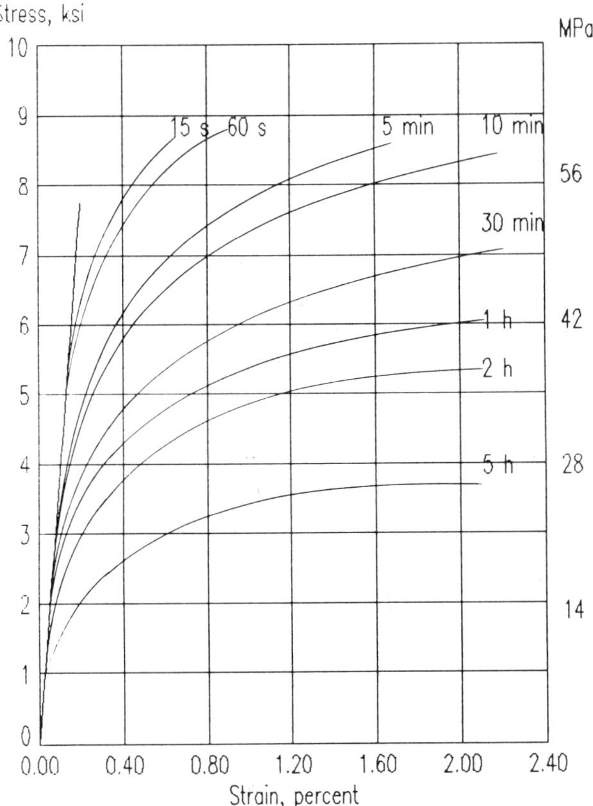

FIGURE 8.23 Sand cast HK31A-T6 isochronous stress-strain curves (370°C).

8.5 ELASTIC MODULUS

The elastic modulus of magnesium and its alloys at room temperature is 44.8 GPa (6500 ksi) at room temperature. The true elastic modulus is little affected by temperature. However, when the modulus is determined by applying a stress and measuring the resultant strain, the result is affected by any relaxation that may occur during the test. Since relaxation becomes more noticeable at elevated temperatures, the effect is more pronounced as the temperature is raised and the apparent modulus becomes lower. If the modulus is measured by determing the speed of sound, relaxation effects become minor, and the measured modulus is not as sensitive to temperature. These effects are evident in the following data for commercial electrolytic magnesium [25]:

	Modulus (stress-strain)		Modulus (dynamic)	
Temperature, °C	GPa	ksi	GPa	ksi
26	43.0	6240	45.0	6530
93	36.7	5330		
149	32.1	4650		
204	26.7	3870		
260	22.1	3210		
316	16.5	2400	39.9	5790

Since the measurements were made on pure magnesium, the effect of relaxation is seen even at room temperature.

The modulus of elasticity as a function of temperature can be obtained from Figures 8.2 through 8.23, using the 15-second data from the isochronous curves. As shown in Table 8.10, those alloys that have the highest creep strength (least relaxation) retain the highest apparent modulus as the temperature is raised. The slope of the isochronous stress-strain curves for times longer than 15 seconds show that the apparent modulus is dependent on the time of testing as well as the temperature. This is to be expected if relaxation is affecting the result.

For design purposes, the relaxed modulus as determined by static methods is a better choice than the dynamic modulus since a structure must withstand applied stresses that will invoke any relaxation that may occur.

TABLE 8.10 Elastic Moduli as a Function of Temperature

	Temperature, °C						
Alloy	25	150	204	260	315	370	427
Stress in GPa							
AZ91C-T4	44.8	31.7	28.3				
AZ91C-T6	44.8	37.9	31.7				
AZ92A-T6	44.8	40.0	33.8				
EZ33A-T5	44.8	44.0	44.8	35.9	37.2	33.8	
HK31A-T6	44.8	44.8	40.7	40.7	37.2	27.6	
HZ32A-T5	44.8	44.8		44.8		30.3	24.8
QE22A-T6	44.8	35.9		34.5	26.2	20.7	
ZE41A-T5	44.8	44.8	39.3	28.3			
Stress in ksi							
AZ91C-T4	6500	4600	4100				
AZ91C-T6	6500	5500	4600				
AZ92A-T6	6500	5800	49000				
EZ33A-T5	6500	5800	6500	5200	5400	4900	
HK31A-T6	6500	6500	5900	5900	5400	4000	
HZ32A-T5	6500	6500		6500		4400	3600
QE22A-T6	6500	5200		5000	3800	3000	
ZE41A-T5	6500	6500	5700	4100	4100		

Note: Data derived from the stress-strain curves of Figures 8.2 through 8.9 and from the 15-second isochronous stress-strain curves of Figures 8.12 through 8.23.

8.6 TOUGHNESS

Data on the toughness of cast alloys are given in Table 8.11.

The ratio of the tensile strength in the presence of a notch to the tensile strength in the absence of a notch is an indication of the sensitivity of the alloy to stress concentrations in a structure. The data in Table 8.11 were obtained on smooth cylindrical test bars compared with cylindrical bars with a circular notch. The diameter of the bar was 9.52 mm (0.375 in) and the root diameter of the notch, 6.73 mm (0.265 in).

The notched or Izod charpy tests measure the energy required to break the bar, and are a measure of the ability of the alloy to absorb impact without fracture.

The critical stress intensity factor, K_{Ic}, a material constant, is the largest stress intensity the material will support, under conditions of plane strain, without failing catastrophically. If K_{Ic} is known for the material, and the geometry and stress known for the part, the largest crack that can be tolerated can be calculated. The larger the critical stress intensity factor, the larger the flaw size that can be tolerated.

TABLE 8.11 Notch Sensitivity and Toughness of Cast Alloys

Alloy	Temper	Temperature, °C	Notched/unnotched TS Notch radius (mm) 0.229	0.025	0.008	Charpy V-notch (J)	K_{Ic} ksi $(in)^{0.5}$
AZ81A	T4	25				6.1	
AZ91C	F	25				0.79	
	T4	25	0.97	0.96	0.90	4.1	
		-78	0.90	0.81	0.81		
		-196	0.89	0.83	0.81		
	T6	25	1.06	1.02	0.86	1.4	10.4
		-78	1.00	0.78	0.74		
		-196	0.93	0.76	0.68		
AZ92A	F	25				0.7	
	T4	25				4.1	
	T6	25				1.4	
EQ21A	T6	20					14.9
EZ32A	T5	20				1.5[a]	
HK31A	T6	25	0.84	0.88	0.85		
		-78	0.85	0.86	0.82		
		-196	0.95	0.76	0.79		
HZ32A	T5	20				2.2[a]	

TABLE 8.11 (*Continued*)

Alloy	Temper	Temperature, °C	Notched/unnotched TS Notch radius (mm)			Charpy V-notch (J)	K_{Ic} Ksi $(in)^{0.5}$
			0.229	0.025	0.008		
QE22A	T6	25	1.24	1.06	1.06	2.0	12.0
		-78	1.08	0.82	0.85		
		-196	1.02	0.86	0.72		
WE54A	T6	20					10.4
QH21A	T6	25					17.0
ZE41A	T5	25				1.4	14.1
ZE63A	T6	25				0.07	19.1
ZH62A	T5	25				3.4[a]	
ZK51A	T5	20				3.5[a]	

Conversion factors: mm × 0.03937 = in; J × 0.7376 = ft-lbf.
[a]Notched Izod
Source: Notch/unnotched tensile strength; ref. 8; Charpy V-notch, refs. 7, 26; K_{Ic}, refs. 9, 24, 27.

8.7 BEARING AND SHEAR

Bearing and shear data obtained on separately-cast test bars for a number of alloys over a range of temperatures are given in Table 8.12. Comparison with earlier tables shows that the shear strength is 0.69 ± 0.14 of the tensile strength. This ratio can be used to estimate roughly the shear strength for alloys and tempers not shown in Table 8.12.

TABLE 8.12 Shearing and Bearing Strengths of Cast Alloys

Alloy	Temper	Temperature, °C	MPa			ksi		
			Bearing			Bearing		
			Yield	Ultimate	Shear	Yield	Ultimate	Shear
AM100A	F	25			125			18
	T4	25	310	475	140	45	69	20
	T61	25	470	560	145	68	81	21
AZ63A	F	25	275	415	125	40	60	18
	T4	25	270	410	150	39	59	22
		95	265	440	160	38	64	23
		200	210	315	100	30	46	14
	T5	25	275	455	130	40	66	19
	T6	25	355	475	165	51	69	24
		95	315	485	165	46	70	24
		200	215	305	90	31	44	13
AZ81A	T4	25	245	400	150	36	58	22
		95	245	425	155	36	62	22
		200	220	375	110	32	54	16
AZ91C	F	25	275	415		40	60	
	T4	25	255	415	150	37	60	22
		95	265	425	155	38	62	22
		200	220	360	105	32	52	15
	T6	25	360	460	165	52	67	24
		95	330	500	170	48	73	25
		200	230	325	105	33	47	15

AZ92A	F	25	315	345	125	46	50	18
	T4	25	315	470	140	46	68	20
	T5	25	317	345	140	46	50	20
	T6	25	460	540	180	67	78	26
		95	420	565	190	61	82	28
		200	230	310	100	33	45	14
EZ33A	T5	25	275	310	135	40	45	20
		200	180	310	115	26	45	17
		315	145	210	65	21	30	9
HK31A	T6	25	275	415	150	40	60	22
		200	210	320	110	30	46	16
		315	180	285	90	26	41	13
HZ32A	T5	25	250	410	140	36	59	20
		200	185	310	100	27	45	14
		315	130	210	70	19	30	10
K1A	F	25	125	315	55	18	46	8
ZE41A	T5	25	355	485	150	51	70	22
		200	255	360	115	37	52	17
		315	165	215	60	24	31	9
ZH62A	T5	25	340	500	160	49	72	23
		200	235	340	105	34	49	15
		315	140	185	55	20	27	8
ZK51A	T5	25	350	485	150	51	70	22

Note: Bearing strength determined with a 2d edge distance and an 8d width. Shear strength determined with a 3.175 mm (0.125 in) diameter pin.
Source: Refs. 7, 10, 11.

8.8 FATIGUE

8.8.1 Unnotched Specimens

Tables 8.13 and 8.14 give data on the fatigue performance of magnesium cast alloys using three different testing modes.

There are a small amount of data on the effect of temperature on the fatigue life of some of the alloys designed for use at elevated temperatures.

The values in the table are given as single numbers, but there is always scatter in fatigue data. As a general rule, the scatter in stress values will be within ±15 MPa (2 ksi) of the mean value that is given in the tables. The scatter in life at a given stress will depend upon the slope of the stress vs. life curve and is larger than the scatter in stress. Some studies have shown that the scatter in life tends to follow a geometrical normal distribution with the standard deviation varying from case to case from about 2 to about 3.5.

8.8.2 Notched Specimens

The effect of notches with two different stress concentrations is shown in Table 8.14. The stress concentration of each notch is the concentration calculated for elastic conditions for the geometry. Since the condition of pure elastic strain is not met during the fatigue test, the effective, or apparent, stress concentration is different. The plastic deformation that occurs at the root of the notch redistributes the stress at the base of the notch resulting in less concentration than is calculated on the basis of elastic theory. Table 8.15 shows the apparent stress concentration for the two notches obtained by dividing the unnotched by the notched values. The values are independent of the number of cycles. These numbers can be used to estimate the effect of stress raisers met in practice on the fatigue life of a structure.

The basic fatigue data are obtained by using smooth, polished bars prepared with every precaution to remove any source of stress concentration. In practice, this is seldom achieved, and it is necessary to know the effect of surface conditions that will be met in practice on the fatigue strength.

Table 8.15 gives some data on the effect of cast surfaces and of anodic finishes on the fatigue results. In general, cast surfaces act as though they are notched, with an elastic stress concentration factor of 2 (an apparent concentration of 1.5). The anodic coatings are less severe.

TABLE 8.13 Plate Bending and Axial-Loaded Fatigue: Machined and Polished Surfaces

			MPa					
			Krouse plate bending (R = -1)			Axial (R = 0.25)		
Alloy	Temper	Temperature, °C	10^5	10^6	10^7	10^5	10^6	10^7
AZ91C	F	25	85	75	60			
	T4	25	105	85	75			
	T6	25	105	85	65			
AZ92A	F	25	90	75	65	165	160	150
	T4	25	105	85	60	235	215	205
	T6	25	125	100	90	210	195	190
EZ33A	T5	25	105	90	75			
		200	95	60	50			
HK31A	T6	25	125	75	60			
		200	100	55	50			
		315	60	45	35			
HZ32A	T5	25	105	70	52			
		200	105	60	45			
ZH62A	T5	25	125	85	75			
ZK51A	T5	25	110	70	60			
ZK61A	T6	25	85	60	50			

Source: Ref. 12.

TABLE 8.13a Plate Bending and Axial-Loaded Fatigue: Machined and Polished Surfaces

			ksi					
			Krouse plate bending (R = -1)			Axial (R = 0.25)		
Alloy	Temper	Temperature, °C	10^5	10^6	10^7	10^5	10^6	10^7
AZ91C	F	25	12	11	9			
	T4	25	15	12	11			
	T6	25	15	12	9			
AZ92A	F	25	13	11	9	24	23	22
	T4	25	15	12	9	34	31	30
	T6	25	18	14	13	30	28	28
EZ33A	T5	25	15	13	11			
		200	14	9	7			
HK31A	T6	25	18	11	9			
		200	14	8	7			
		315	9	7	5			
HZ32A	T5	25	15	10	8			
		200	15	9	7			
ZH62A	T5	25	18	12	11			
ZK51A	T5	25	16	10	9			
ZK61A	T6	25	12	9	7			

Source: Ref. 12.

TABLE 8.14 Fatigue Strength in MPa at Indicated Number of Cycles (Rotating Beam (R = -1), Room Temperature)

		Smooth, machined, and polished						Notched				
								SCF[a] = 2			SCF[b] = 5	
Alloy	Temper	10^5	10^6	10^7	5×10^7	10^8	5×10^8	10^6	5×10^7	10^8	10^6	10^8
A8	F	108	90	88	86			73[b]	63[b]			
	T4	104	97	90	90			82[b]	69[b]			
AZ91C	F	110	105	95		85		70		55		
	T4	135	115	105		95		75		60		
	T6	125	105	90		80		75		55		
AZ92A	F	115	105	95		90	85	80		55		
	T4	125	110	105		95	90	100		75	55	40
	T6	130	115	100		95	85	80		60	40	30
EQ21A	T6	113	100	98	96			58	54			
EZ33A	T5	100	85	75	70	70		50	50	45	40	30
HK31A	T6	115	90	75		60		60		45		

TABLE 8.14 (*Continued*)

		Smooth, machined, and polished						Notched				
								SCF[a] = 2			SCF[b] = 5	
Alloy	Temper	10^5	10^6	10^7	5×10^7	10^8	5×10^8	10^6	5×10^7	10^8	10^6	10^8
HZ32A	T5	105	90	75	70	70		60	60	50	40	30
QE22A	T6	150	125	115	105	105		85	65	60		
QH21A	T6	135	110	105	105			65	65			
WE54A	T6		102	99	97							
ZE41A	T5	124	97	96	94			91[b]	83[b]			
ZE63A	T6				120				70			
ZH62A	T5	120	85	83	82			80[b]	76[b]			
ZK51A	T5	105	75	70		60		70		55	60	50

[a]SCF = Stress Concentration Factor.
[b]SCF = 1.8.
Source: Refs. 7, 12, 13, 26, 27.

TABLE 8.14a Fatigue Strength in ksi at Indicated Number of Cycles (Rotating Beam (R = -1), Room Temperature)

		Smooth, machined, and polished						Notched				
								SCF[a] = 2			SCF[a] = 5	
Alloy	Temper	10^5	10^6	10^7	5×10^7	10^8	5×10^8	10^6	5×10^7	10^8	10^6	10^8
A8	F	16	13	13	12			11[b]	9[b]			
	T4	15	14	13	13			12[b]	10[b]			
AZ91C	F	16	15	14		13		10		8		
	T4	20	17	15		14		11		9		
	T6	18	15	13		12		11		8		
AZ92A	F	17	15	14		12	13	12		8		
	T4	18	16	15		14	13	14		11	8	6
	T6	19	17	14		14	12	12		9	6	4
EQ21A	T6	16	15	14	14			8	8			
EZ33A	T5	14	13	11	10	10		7	7	7	6	4
HK31A	T6	17	13	11		9		9		6		
HZ32A	T5	15	13	11	10	10		9	9	7	6	4

TABLE 8.14a (*Continued*)

		Smooth, machined, and polished						Notched				
								SCF[a] = 2			SCF[a] = 5	
Alloy	Temper	10^5	10^6	10^7	5×10^7	10^8	5×10^8	10^6	5×10^7	10^8	10^6	10^8
QE22A	T6	22	18	17	15	15		13	9	9		
QH21A	T6	20	16	15	15			9	9			
WE54A	T6		15		14	14						
ZE41A	T5	18	14	14	14			13[b]	12[b]			
ZE63A	T6				17				10			
ZH62A	T5	17	12	12	12			12[b]	11[b]			
ZK51A	T5	15	11	10		9		10		8	9	7

[a]SCF = Stress Concentration Factor.
[b]SCF = 1.8.
Source: Refs. 7, 12, 13, 26, 27.

TABLE 8.15 Apparent Stress Concentrations at Notches. Fatigue in R. R. Moore Testing at Room Temperature

		Apparent stress concentration				
					Dow 17 anodic coat[a]	
Alloy	Temper	SCF = 2	SCF = 5	As-cast surface[a]	Thin	Thick
A8	F	1.3				
	T4	1.2				
AZ91C	F	1.5				
	T4	1.6				
	T6	1.4			1.1	1.1
AZ92A	F	1.5				
	T4	1.2	2.2			
	T6	1.5	3.0	1.5		
EQ21A	T6	1.7				
EZ33A	T5	1.6	2.2	1.7	1.1	1.1
HK31A	T6	1.6		1.3		
HZ32A	T5	1.4	2.3	1.4		
QE22A	T6	1.6				
QH21A	T6	1.7				
ZE41A	T5	1.2				
ZE63	T6	1.7				
ZH62A	T5	1.0		1.4		
ZK51A	T5	1.1		1.3		

[a]Krouse plate bending with R = -1.
Source: Based on the data in Table 8.11 and Refs. 12, 17.

8.9 STRESS CORROSION

Since the corrosion part of stress corrosion is electrolytic, the effect is sensitive to alloy composition, being more severe in the presence of cathodic phases in the microstructure. While all magnesium alloys will stress corrode to some extent, the most susceptible are those containing aluminum as an alloying element. The broad class of alloys containing zirconium are sufficiently insensitive that stress corrosion is not a problem for castings. See the chapter on wrought alloys for information on the stress corrosion of these alloys. Paint, or other good coatings eliminate the stress corrosion as long as the integrity of the coating is maintained.

Residual stresses after casting, welding, or machining can be adequately relieved by either the T5 or the T4 treatments. If quenching is a part of the T4 treatment, stress relief is adequate during the heating for the T6 treatment.

Stress corrosion data for magnesium casting alloys exposed to a rural atmosphere while loaded in constant tension are given in Figures 8.24 through 8.26 [8]. While many short-term tests for measuring stress corrosion have been used, simple exposure over a long time most nearly represents service life expectations. Failure can occur with the aluminum-containing alloys at stresses that might be encountered in service or residually. On the other hand, although the alloy will stress corrode in a technical sense, QE22A-T6 will not fail even at stresses that are higher than would normally be encountered in service. This behavior is true for all of the non-aluminum containing alloys, in cast form.

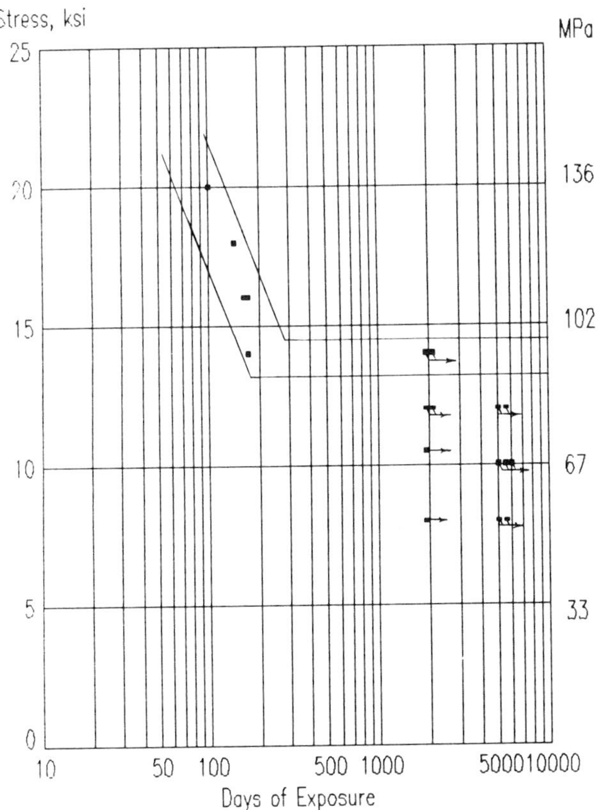

FIGURE 8.24 Stress corrosion of AZ91C; rural exposure; tension; -T4 and -T6.

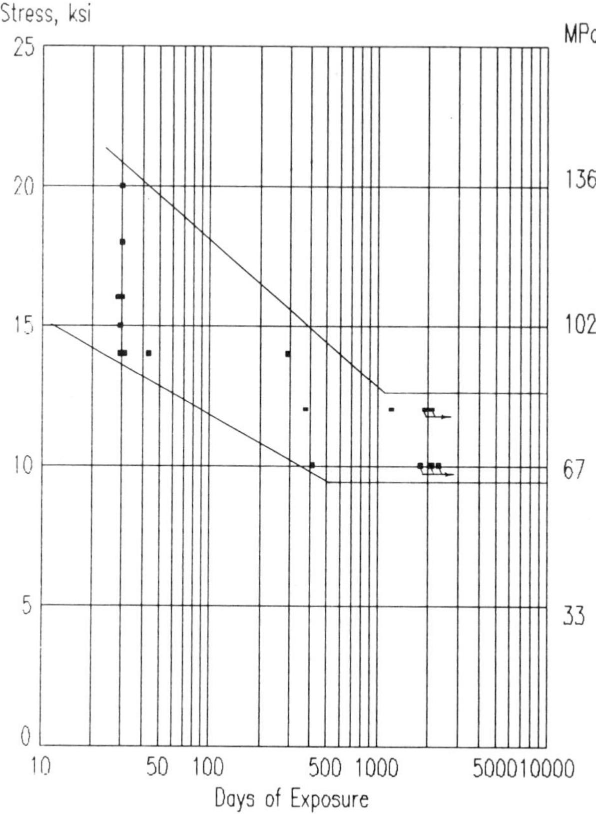

FIGURE 8.25 Stress corrosion of AZ92A; rural exposure; tension; -T4 and -T6

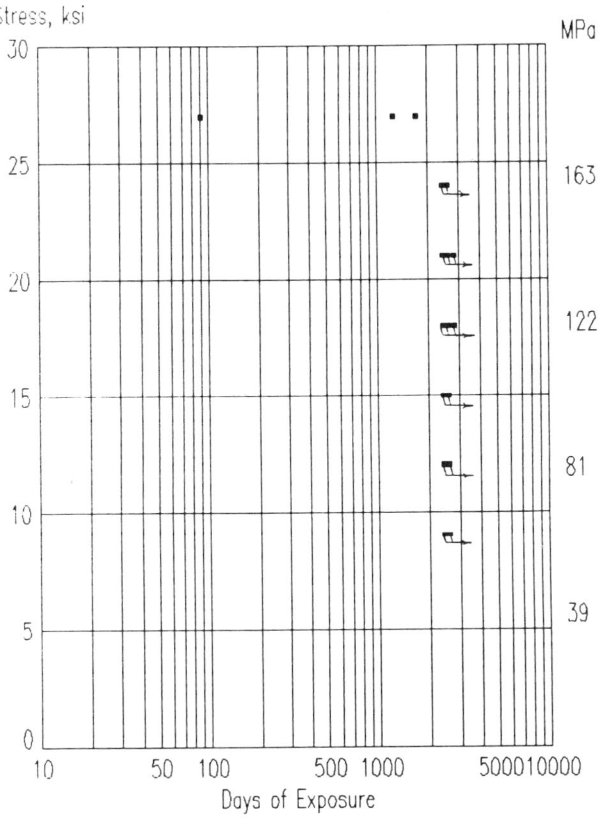

FIGURE 8.26 Stress corrosion of QE22A-T6; rural exposure; tension.

8.10 DAMPING

The damping capacity of magnesium and magnesium alloys is high relative to other metals, as is shown in Table 8.16. Comparison of the stress levels used to determine damping capacity as shown in the table with the data given in Table 8.2 show that, with the exception of alloy K1A, the stress levels used are well below the yield strength. Since the proportional limit is approximately 50% of the yield strength, the stresses are also well below the proportional limit and the strain associated with the stress levels used can be calculated using the elastic modulus.

The capacity to dampen out vibrations is used directly for vibration testing fixtures, and for mounting brackets for electronic equipment sensitive to vibrations. The part shown in Figure 2.15 is typical of such a bracket. In addition to direct usage, damping capacity is valuable for decreasing the magnitude of strain transmitted from one section of a part to another or from one part to another in a structure, both of which lessen the danger of a fatigue failure. Because of high damping capacity, magnesium parts will often live better in vibration environments than will other metals.

TABLE 8.16 Damping Capacity

Alloy	Temper	Specific damping capacity, %, at MPa				
		7.0	13.8	20	25	35
AZ92	F	0.17	0.45	2.09	5.54	
	T4	0.50	1.04	1.29	2.62	3.78
	T6	0.35	0.70	1.64	3.08	4.78
EZ33A	T5		4.88	12.55	18.15	22.42
K1A	F	40.0	48.8	56.0	61.7	66.1
HK31	T6	0.37	0.66	1.12		
HZ32A	T5	1.93	7.81	11.64		
Aluminum alloys						
355	T6		0.51	0.67	1.0	
356	T6	0.3	0.48	0.62	0.82	1.2
Cast iron			5.0	12.2	14.2	16.5

Source: Refs. 20, 21, 22.

TABLE 8.16a Damping Capacity

Alloy	Temper	Specific damping capacity, %, at ksi				
		1.0	2.0	3.0	3.5	5.0
AZ92	F	0.17	0.45	2.09	5.54	
	T4	0.50	1.04	1.29	2.62	3.78
	T6	0.35	0.70	1.64	3.08	4.78
EZ33A	T5		4.88	12.55	18.15	22.42
K1A	F	40.0	48.8	56.0	61.7	66.1
HK31	T6	0.37	0.66	1.12		
HZ32A	T5	1.93	7.81	11.64		
Aluminum alloys						
355	T6		0.51	0.67	1.0	
356	T6	0.3	0.48	0.62	0.82	1.2
Cast iron			5.0	12.2	14.2	16.5

Source: Refs. 20, 21, 22.

REFERENCES

1. Norsk Hydro Bulletin, "Normag; Magnesium, Pure and Alloys."
2. ASTM B 80, "Standard Specification for Magnesium-Alloy Sand Castings," *Annual Book of ASTM Standards*, Section 2, Volume 02.02 (1984).
3. Dow Chemical Bulletin, "Room and Elevated Temperature Properties of Magnesium Casting Alloys," February 1958.
4. U.S. Military Specification MIL-M-46062B, "Magnesium Alloy Castings, High Strength."
5. P. C. Meredith, "QE22A-T6: A Versatile Magnesium Alloy for High Performance Aerospace Applications," Magnesium Elektron Ltd., January 1967.
6. Raymond W. Fenn, Jr. and James A. Gusack, "Effect of Strain-Rate and Temperature on the Strength of Magnesium Alloys," *ASTM Proceedings 58*:685 (1958).
7. *ASM Metals Handbook*, "Properties and Selections: Nonferrous Alloys and Pure Metals," Ninth Edition, Volume 2, 1979.
8. Dow Chemical Company, Private communication.
9. Magnesium Elektron Ltd., "Elektron Magnesium Alloys," 1983.
10. Dow Chemical Company, "Shear, Bearing, and Hardness Data for Cast Magnesium Alloys," Technical Service and Development Letter Enclosure, November 15, 1965.
11. Kirk-Othmer, *Encyclopedia of Chemical Technology*, Volume 14, Third Edition, John Wiley & Sons, Inc., New York, 1981.
12. Dow Chemical Company, "Tables of Fatigue Strength of Sand-cast Magnesium Alloys," TS&D Letter Enclosure, November 19, 1965.
13. Magnesium Elektron Ltd., "Mechanical Properties and Chemical Compositions of Cast Magnesium Alloys," Bulletin 440, March 1981.
14. R. B. Clapper and J. A. Watz, "Determination of Fatigue-Crack Initiation and Propagation in a Magnesium Alloy," *Proceedings of the Second Pacific Area National Meeting*, ASTM, September 1956.
15. Dow Chemical Company, Private communication.
16. S. Suresh and R. O. Ritchie, "Propagation of Short Fatigue Cracks," *International Metals Reviews, 59*(6):445 (1984).
17. Dow Chemical Company, "The Effect of Anodic Coatings on the Fatigue Strength of Magnesium Alloys," TS&D Letter Enclosure, April 6, 1967.
18. P. Klain, "Stress-Relaxation and Stress-Relief of some Magnesium Alloys," *Welding Research Supplement*, September 1955.
19. R. J. M. Payne, "Stress-Relief and Allied Problems in Magnesium Alloy Castings," *Journal of the Institute of Metals, 78*:147 (1950).

20. Stephen C. Erickson, "Magnesium's High Damping Capacity for Automotive Noise and Vibration Attenuation," *Proceedings of the International Magnesium Association*, May 1975.
21. Dow Chemical Company, "Damping Characteristics of some Magnesium Alloys," TS&D Letter Enclosure, August 1, 1957.
22. D. F. Walsh, J. W. Jensen, and J. A. Rowland, Report R. I. 6116, U.S. Bureau of Mines, Rolla, Mo.
23. Magnesium Elektron Ltd., "QH21A," Bulletin 460, September 1979.
24. Magnesium Elektron Ltd., "WE54, a new magnesium casting alloy for use up to 300°C (572 °F)," Bulletin 466, May 1985.
25. C. Sheldon Roberts, *Magnesium and its Alloys*, John Wiley & Sons, Inc., New York, 1960.
26. Colin J. Smithells, *Metals Reference Book*, Butterworths, London and Boston, 1976.
27. Magnesium Elektron Ltd., "Elektron EQ21A: Another casting alloy developed by Magnesium Elektron Ltd.," Bulletin 464, July 1984.

9

Mechanical Properties of Permanent Mold Castings

9.1 GENERAL CHARACTERISTICS

A permanent mold is one made of a material that is re-usable after the casting is removed. While there are a number of materials that can be used, steel is usually chosen. There are three general subheadings of permanent-mold castings, based upon the method of introducing the molten metal to the mold.

In a gravity casting, the molten metal is poured into the mold in the same way as into a sand mold, with gravity used as the force to introduce and distribute the metal. This type of casting is known as a permanent-mold casting in the United States and a gravity die casting in Europe. If sand cores are used, the process is known as a semi-permanent mold casting method.

If a relatively low external pressure is used to force the metal into the mold, the result is known as a low-pressure casting. The pressure is usually applied to the surface of the molten metal, forcing it through a tube into the die cavity.

High-pressure die casting makes use of a high injection pressure to force the metal into the die cavity, a process known simply as die casting in the United States.

9.2 GRAVITY AND LOW-PRESSURE CASTINGS

These methods of preparing a shape are used when production quantities are large enough to justify the cost of the die. Operating costs for the low-pressure process are lower than those for gravity die casting because of higher die efficiency and, usually, higher casting rates. The capital investment is higher than for gravity castings, and therefore costs are sensitive to the quantities of castings required. Castings produced by either process are of high quality with good surface finish.

Since the freezing conditions are not greatly different from those characteristic of sand casting, the property data for sand castings given in Chapter 8 can be used for both gravity and low-pressure castings. As an example, the data in Table 9.1, taken from specifications for permanent mold castings, are no different from those of Table 8.2.

TABLE 9.1 Mechanical Properties of Permanent Mold-Castings

			Minima separate bars					Average from castings				
				MPa		ksi			MPa		ksi	
Alloy	Temper	Temperature, °C	%E	TYS	TS	TYS	TS	%E	TYS	TS	TYS	TS
AM100A	F	25		70	140	10	20					
	T4		6	70	235	10	34	1.5	70	175	10	25
	T6		2	105	235	15	34		95	175	14	25
	T61			115	235	17	34		95	175	14	25
AZ63A	F		4	75	180	11	26					
	T4		7	75	235	11	34	2	70	175	10	25
	T5		2	85	180	12	26					
	T6		3	110	235	16	34	0.8	95	175	14	25
AZ81A	T4		7	75	235	11	34	1.8	70	175	10	25
AZ91C	F			75	160	11	23					
	T4		7	75	235	11	34	1.8	70	175	10	25
	T5		2	85	160	12	23					
	T6		3	110	235	16	34	0.8	100	175	14	25

TABLE 9.1 (*Continued*)

Alloy	Temper	Temperature, °C	Minima separate bars					Average from castings				
				MPa		ksi			MPa		ksi	
			%E	TYS	TS	TYS	TS	%E	TYS	TS	TYS	TS
AZ92A	F			75	160	11	23					
	T4		6	75	235	11	34	2.5	70	175	10	25
	T5			85	160	12	34					
	T6			125	235	18	34		110	175	16	25
EZ33A	T5	25	2	95	140	14	20	0.5	85	105	12	15
		260		55	90	8	13		40	70	6	10
HK31A	T6	25	4	90	185	13	27	1.0	80	160	12	23
		260		90	145	13	21		70	95	10	14
HZ32A	T5	25	4	90	185	13	27	1.0	80	160	12	23
		260		55	90	8	13		40	70	6	10
QE22A	T6	25	2	170	240	25	35	0.5	160	220	23	32
		315		70	90	10	13		55	70	8	10

Source: Refs. 1, 2.

9.3 HIGH-PRESSURE DIE CASTINGS

If a sufficient number of parts are to be produced to amortize the cost of the die, high-pressure die casting is the lowest cost-method for producing metal castings. Parts with excellent definition, with thin walls, and with great complexity can be made, as is shown for many of the applications illustrated in Chapter 2. While some machining may be required, often the cast part is used after only superficial cleaning such as deburring. Surface finishing, such as chemical treatments, painting, or electroplating may be used for specific applications.

There are two major categories of die-casting equipment. With the cold-chamber machine, molten metal is poured into a tube, which is usually horizontal, and a piston in the tube then advanced to push the metal into the die at high speed and with high pressure imposed after the cavity is filled. With the hot-chamber machine, the piston and piston cylinder are immersed in the molten metal, and liquid metal is pushed into the die cavity directly from the reservoir of molten metal. Because higher pressures can be used with the cold- than with the hot-chamber process, larger castings can be produced than by the hot-chamber process. However, because metal efficiency and production rates are higher, castings produced by the hot-chamber process are of lower cost than those produced by the cold-chamber machine. Hot-chamber castings generally have higher quality than cold-chamber castings. Magnesium can be cast on both types so that the process best suited for a specific casting can be selected.

Magnesium has properties that make it particularly suited to producing parts by the high-pressure die-casting process:

1. In almost all cases, the section size of the magnesium can be the same as for either aluminum or zinc. Less weight of metal is required for the part if made of magnesium rather than of aluminum or zinc.

2. Parts are produced by rapid injection of molten metal into a steel die. The limiting factor for production rate is the time required for the cast part to cool to a temperature that will permit extraction from the die. The low volume specific heat of magnesium compared to aluminum results in higher rates of production for magnesium.

3. Magnesium does not react with steel. The molten metal can be handled in easily fabricated steel equipment, the life of the steel die is longer, and less draft is required for magnesium than for aluminum. Either the cold- or hot-chamber process can be used. Aluminum is restricted to the cold-chamber type of machine.

The net result is frequently a lower cost for a magnesium part than for the same part in another material.

9.3.1 Design Considerations

As a general rule of thumb, wall thickness can be 1/100 the distance to the gate. Wall thicknesses as thin as 0.9 mm (0.035 in) have been cast. If minimum wall thickness is important to the design, the die caster should be consulted early to assure that the optimum is obtained.

The Series E Engineering Standards by the American Die Casting Institute [10] should be consulted for details of thickness and flatness tolerances, recommended radii at fillets, and draft tolerances. In many cases the tolerances specified in these standards can be exceeded in practice and the die caster should be consulted before the final design.

As a preliminary guide for what to expect, the thickness, length, and flatness tolerances will be between 0.2 and 0.4% of the linear dimension, provided the linear dimension is at least one inch (25.4 cm). The draft requirement for an interior wall will follow the relation

log draft = 0.48 log wall depth - 1.55

where draft and depth are in inches. The draft for an external wall is one-half that of an interior wall (inch × 25.4 = mm). The draft requirement for cored holes will follow the relation

log draft = 0.52 log hole depth - 1.36

where draft and depth are in inches (inch × 25.4 mm). Many casting configurations can be cast with a zero draft. If this is important to the design, the caster should be consulted. The maximum depth of cored hole that can be cast is related to the diameter of the hole by the relation

Diameter = 0.15 maximum depth + 0.113

where diameter and depth are in inches (inch × 25.4 mm).

Pressed-in or shrink-fit inserts can be used. Because of the potential for stress corrosion if the stresses due to the interference used are too high, it is important to control the amount of the interference. The stress developed depends upon the size of the insert, the insert material, and the size of the magnesium wall surrounding the insert. The graphs of Figures 9.1, 9.2, and 9.3 were developed using the equation [9]:

$$e = b\,\frac{S_M}{E_M}\left[\frac{\left(\dfrac{c^2 + b^2}{c^2 - b^2} + \mu_M\right) + \dfrac{E_M}{E_B}\left(\dfrac{b^2 + a^2}{b^2 - a^2} - \mu_B\right)}{\dfrac{c^2 + b^2}{c^2 - b^2}}\right]$$

where

a = Internal diameter of the insert (mm)
b = Outside diameter of the insert (mm)
c = Outside diameter of the magnesium (mm)
e = Interference (mm)
μ_M = Poisson's ratio of magnesium
μ_B = Poisson's ratio of the insert material
S_M = Stress in the magnesium (MPa)
E_M = Young's modulus of magnesium (GPa)
E_B = Young's modulus of the insert material (GPa)

The curves for Figures 9.1 to 9.3 were calculated for an allowable stress of 7,000 pounds/sq. in, or 48 MPa, which is below the stress corrosion limit shown in Figure 9.8 for AZ91. This same limit is safe also for both AM60A and AS41A. AS21A is not susceptible to stress corrosion. Since the ratios in the curves are dimensionless, any consistent system of units can be used. Any other interference or stress that will result from different dimensions or different materials can be calculated by using the above equation.

It is also possible to use cast-in inserts. In this case it is not possible to calculate the stresses developed since the cooling conditions are an important factor that will be known only for each casting, if even then. However, recommendations for dimensions that have been found to be safe to use are shown in Figure 9.4. The design of the insert is important. Sharp edges or corners that will lead to stress concentrations must be avoided. The best design is an insert that has an oval outside contour which will anchor the insert without appreciable stress concentration.

9.3.2 Alloys Used

Four magnesium alloys are used to produce castings by high-pressure die casting.

AZ91, selected for about 90% of all applications, has three variations: AZ91A, AZ91B, and AZ91D. AZ91A and AZ91B are the same, except that there are wider limits for impurities for AZ91B and thus secondary metal can be used. AZ91D is a high-purity version which has much greater saltwater corrosion resistance than either the A or B variation. See Chapter 13 on corrosion for details. The mechanical properties are the same for all three variations.

AM60A is used where greater toughness and ductility is more important than a slightly lower strength. It has been used for automobile wheels, for example.

AS41A has good resistance to creep at elevated temperatures, and is used where this is important.

AS21A has somewhat higher creep strength than does AS41, but is more difficult to cast and is used to only a small extent.

Mechanical properties for the three major alloys are given in Tables 9.2 through 9.4, and in Figures 9.5 through 9.8.

TABLE 9.2 Mechanical Properties of High-Pressure Die-Cast AZ91A, B, or D

	Temperature, °C				
Property	25	95	150	200	260
%E	3				
TYS (MPa)	160	140	110	85	55
TYS (ksi)	23	20	16	12	8
TS (MPa)	230	220	185	125	75
TS (ksi)	33	32	27	18	11
Modulus of elasticity (GPa)	45	41	39	34	29
Modulus of elasticity (ksi)	6500	5900	5600	5000	4200
Shear strength (MPa)	140				
Shear strength (ksi)	20				
Impact, unnotched (J)	3				
Impact, unnotched (ft-lbf)	2.2				
0.1% Creep strain (MPa)					
1 hr	145	120	35	15	
10 hrs	140	90	25	10	
100 hrs	140	65	15	5	
0.2% Creep strain (MPa)					
100 hrs			25		
1000 hrs		55			
0.1% Creep strain (ksi)					
1 hr	21	17	5	2	
10 hrs	20	13	4	1.5	
100 hrs	20	9	2	0.7	
0.2% Creep strain (ksi)					
100 hrs			3.6		
1000 hrs		8			
Specific damping capacity (%) by cantilever beam					
5 MPa (0.7 ksi) stress	0.8				
15 MPa (2 ksi) stress	1.6				
20 MPa (3 ksi) stress	2.5				

TABLE 9.2 (*Continued*)

Property	Temperature, °C 25	95	150	200	260
Specific damping capacity (%) by torsion					
20 MPa (3 ksi) stress	20				
40 MPa (6 ksi) stress	32				
60 MPa (9 ksi) stress	45				
80 MPa (12 ksi) stress	50				
100 MPa (14 ksi) stress	55				
Stress relaxation (%)					
1 hr			10		
10 hrs			40		
100 hrs			60		
500 hrs			75		

Source: Refs. 1, 2, 4, 5, 6, 7.

TABLE 9.3 Mechanical Properties of High-Pressure Die-Cast AM60A

Property		Value MPa	ksi
%E	8		
TYS		130	19
TS		220	32
Specific damping capacity (%) by torsion			
20 MPa (3 ksi) stress	20		
40 MPa (6 ksi) stress	54		
60 MPa (9 ksi) stress	65		
100 MPa (14 ksi) stress	72		

Source: Refs. 1, 4.

TABLE 9.4 Mechanical Properties of High-Pressure Die-Cast AS41A

	Temperature, °C				
Property	25	95	150	200	260
%E	6				
TYS (MPa)	140	115	105	60	
TYS (ksi)	20	17	15	9	
TS (MPa)	210	160	130	90	
TS (ksi)	30	23	19	13	
Rotating beam fatigue limits (MPa)					
10^5 cycles	110	80	70		
10^6 cycles	90	75	60		
10^7 cycles	90	75	50		
Rotating beam fatigue limits (ksi)					
10^5 cycles	16	12	10		
10^6 cycles	13	11	9		
10^7 cycles	13	11	7		
Stress relaxation (%)					
1 hr			10		
10 hrs			35		
100 hrs			50		
500 hrs			80		
0.2% Creep strain					
100 hrs (MPa)			40		
1000 hrs (MPa)		50			
0.2% Creep strain					
100 hrs (ksi)			6		
1000 hrs (ksi)		7			
Specific damping capacity (%) by torsion					
20 MPa (3 ksi) stress	20				
40 MPa (6 ksi) stress	54				
60 MPa (9 ksi) stress	60				
80 MPa (12 ksi) stress	65				
100 MPa (14 ksi) stress	70				

Source: Refs. 1, 4, 6, 7.

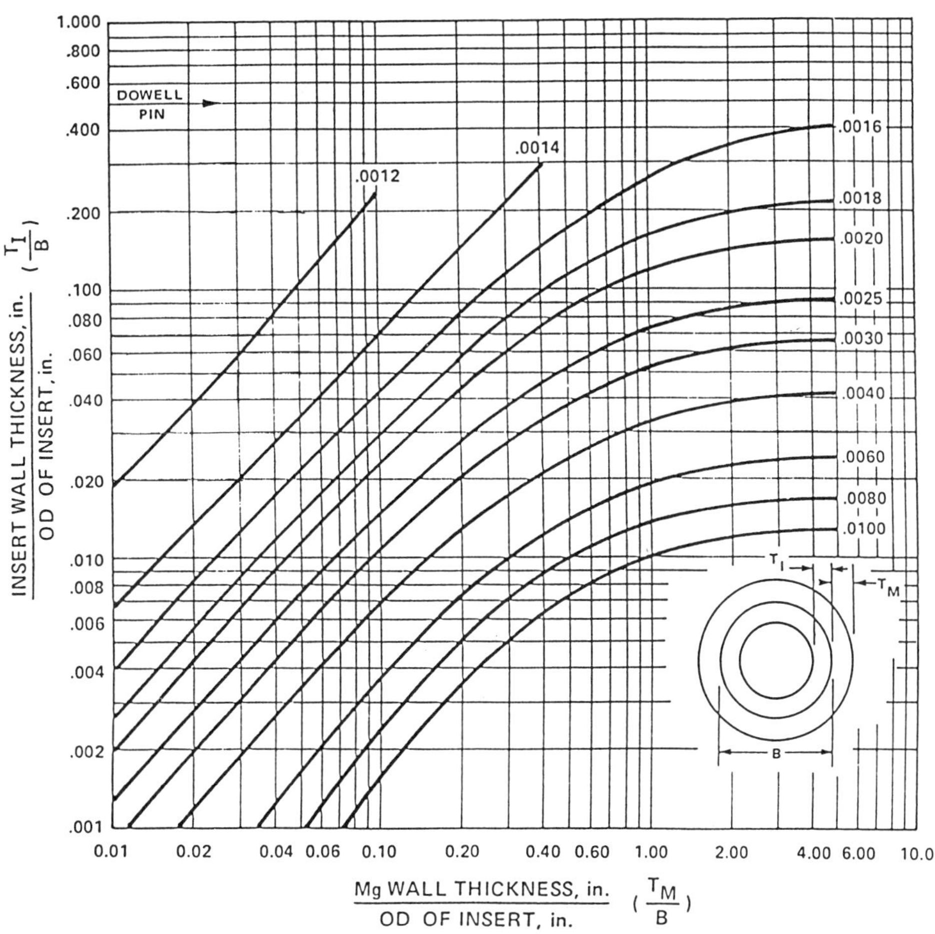

FIGURE 9.1 Interferences, in inches, that will result in a stress in the magnesium of 7 ksi when a steel insert is used.

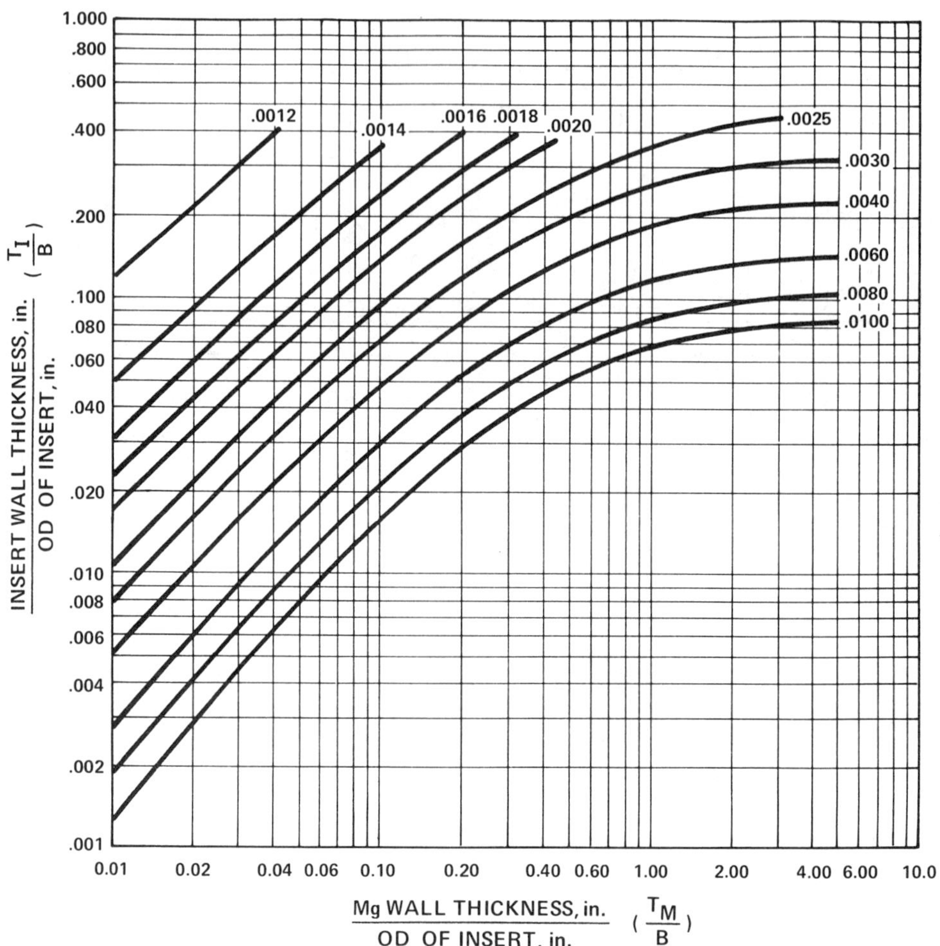

FIGURE 9.2 Interferences, in inches, that will result in a stress in the magnesium of 7 ksi when a brass insert is used.

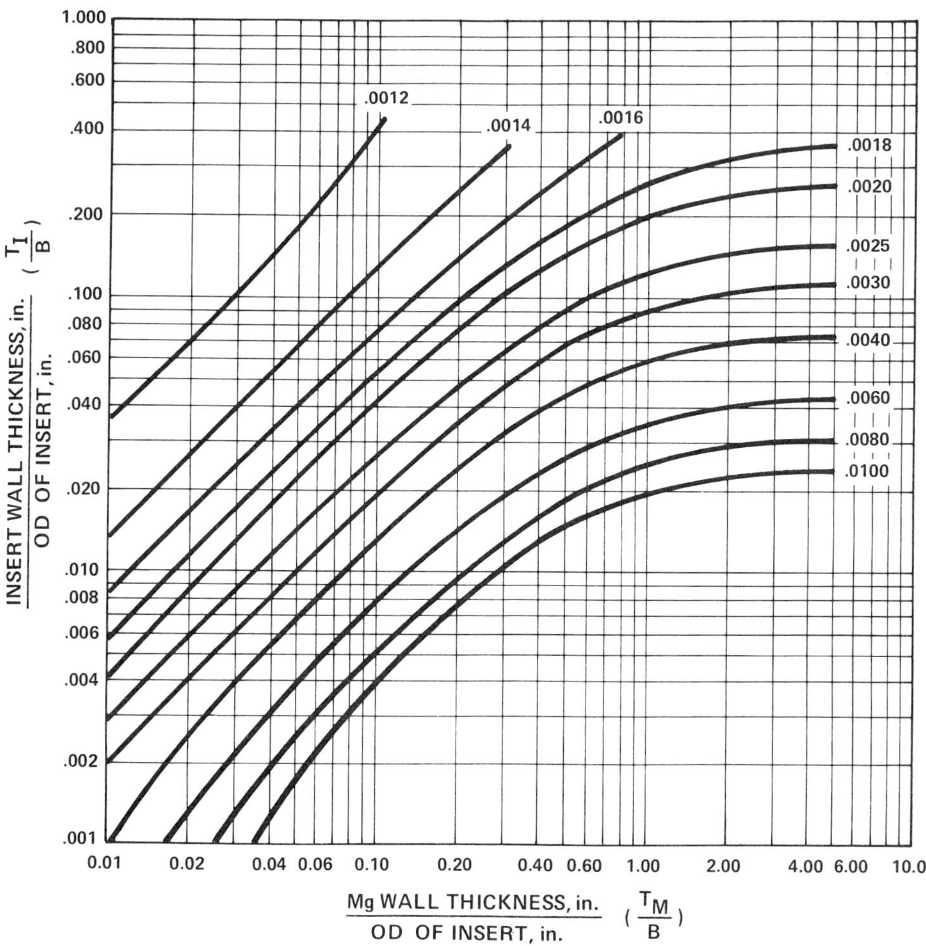

FIGURE 9.3 Interferences, in inches, that will result in a stress in the magnesium of 7 ksi when a Babbitt insert is used.

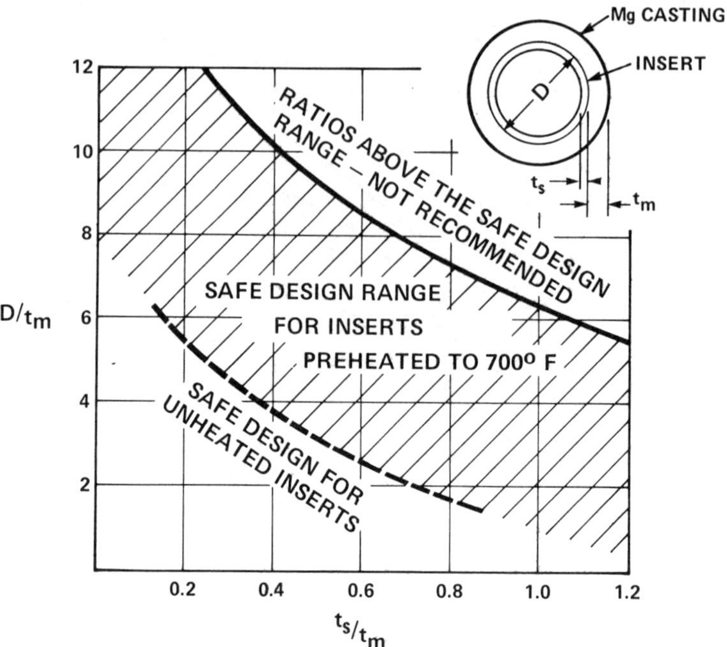

FIGURE 9.4 Recommended designs for interferences for cast-in steel inserts.

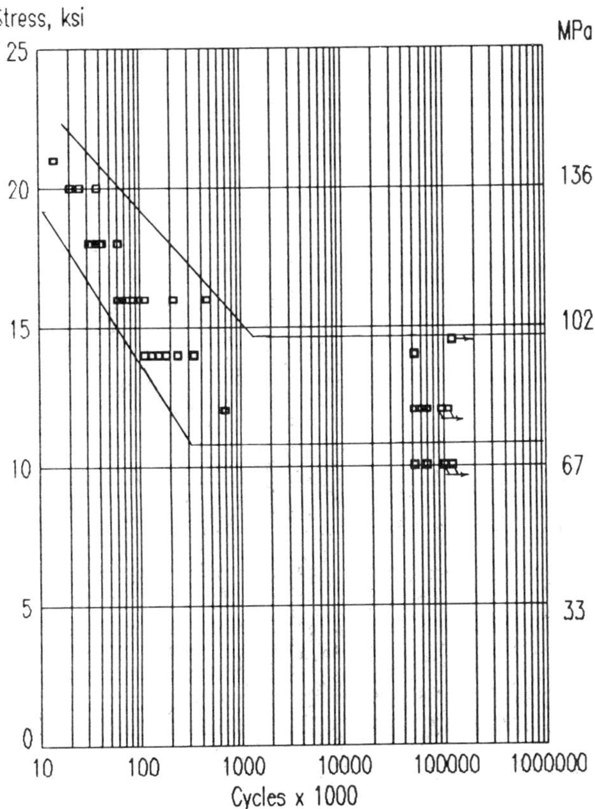

FIGURE 9.5 Rotating-beam fatigue results for high-pressure die-cast AZ91B. Separately-cast test bars with the as-cast surface left on the bars.

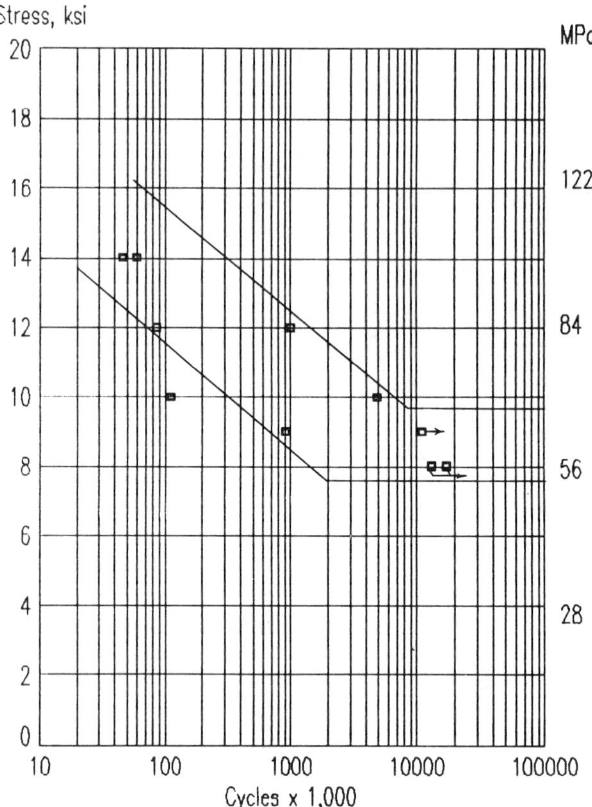

FIGURE 9.6 Axial-load fatigue results for high-pressure die-cast AZ91B. Separately-cast test bars with the as-cast surface left on the bars.

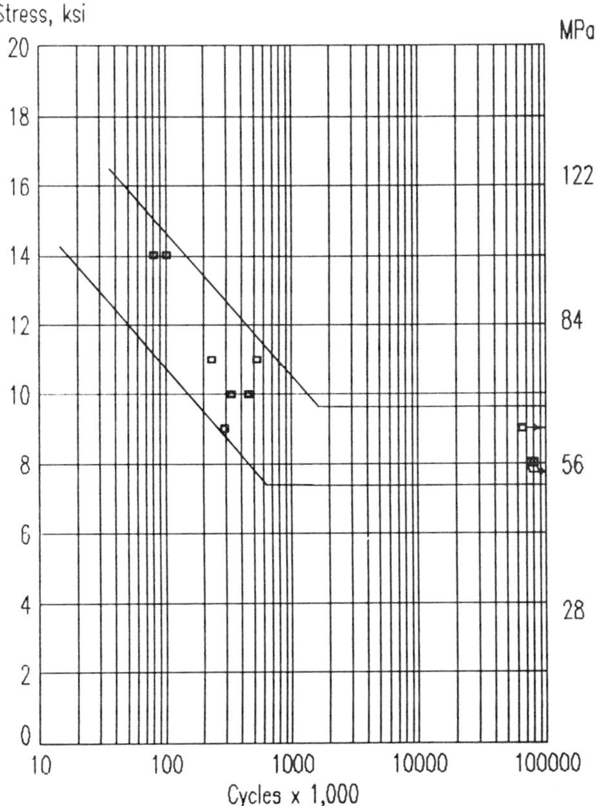

FIGURE 9.7 Plate-bending fatigue results for high-pressure die-cast AZ91B. Determined on cast panels with only the edges machined.

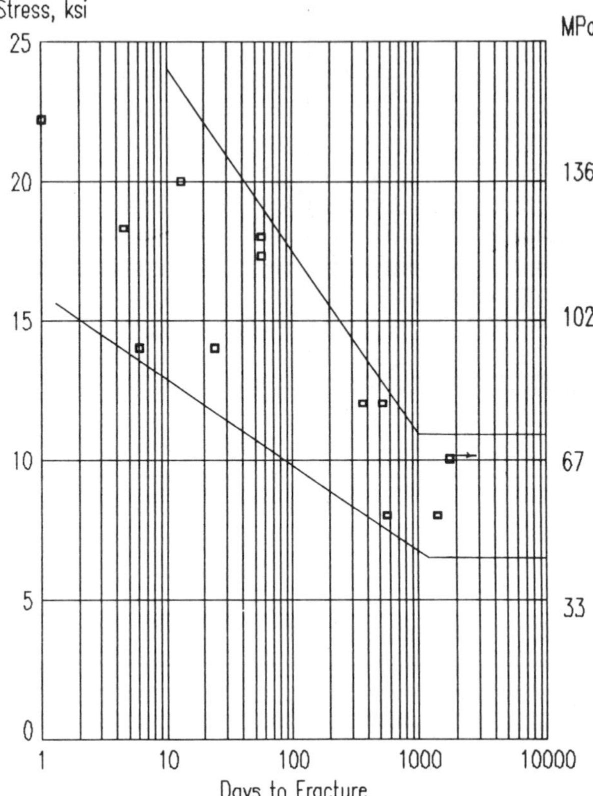

FIGURE 9.8 Stress corrosion of AZ91 high-pressure die-cast (rural exposure).

REFERENCES

1. ASTM Specification B 199, "Standard Specification for Magnesium-Alloy Permanent Mold Castings," *Annual Book of ASTM Standards*, Vol. 02.02, Sec. 2, June 1984.
2. Federal Specification QQ-M-55C, *Magnesium Alloy, Permanent and Semi-Permanent Mold Castings*, December 1, 1970.
3. Dow Chemical Company, *Magnesium Die Casting Manual*, 1980.
4. ASTM Standard B94, "Standard Specification for Magnesium-Alloy Die Castings," *Annual Book of ASTM Standards*, Vol. 02.02, Sec. 2, 1984.
5. Dow Chemical Company, "HM11XA Die Casting Alloy," Technical Service and Development Letter Enclosure, November 27, 1957.
6. Dow Chemical Company, "AZ91A-F Die Castings. R. R. Moore Fatigue Curves," TS&D Letter Enclosure, March 3, 1957.
7. Anders Vickberg and Rolf Ericsson, "Magnesium in the Volvo LCP 2000," *SAE Technical Paper 850418*, February 1985.
8. George S. Foerster, "New Developments in Magnesium Die Casting," *Proceedings of the International Magnesium Association*, 1976.
9. H. N. Hill, "Formulas for Calculating Press Fits in Aluminum Rings," *Product Engineering*, November 1942.
10. American Die Casting Institute, Series E Engineering Standards, Number 402, English units; Number 403, SI units; American Die Casting Institute, Des Plaines, Illinois.

10

Mechanical Properties of Extrusions

10.1 GENERAL

Magnesium alloys are easily extruded into either solid or hollow complex sections. Extrusion speeds, and hence cost, are dependent on the alloy but are comparable to the speeds typical of aluminum alloys of the same general strength level. Extruded magnesium sections are often competitive in cost per volume with the same sections in aluminum.

Alloys used, as with cast alloys, fall in two major groupings: those with and those without zirconium. Most of those without zirconium contain aluminum and zinc, with some manganese added for improved corrosion resistance. One alloy, M1A, contains only manganese. Those alloys containing no more than three percent aluminum can be extruded at high speeds and are commonly used for low- and medium-strength applications. ZM21A, with no aluminum, can be extruded at especially high speeds and is the lowest-cost extrusion alloy available. ZK60A, containing 6% zinc and zirconium, is a high-strength, but more costly, alloy. HM31A, containing thorium, is used for its good strength at elevated temperatures.

10.2 DESIGN CONSIDERATIONS [5]

Although very complex shapes can be extruded, the designer should make every effort to attain maximum simplicity of design, in order to minimize cost and maximize properties.

The most important consideration in shape design for ease of extrusion is symmetry, preferably about both axes. Radii should be generous. Sharp corners will amplify hot-shortness which requires a lowering of extrusion speed and thus an increase in cost. In addition, sharp corners in the die are points of stress concentration which can lower the life of the die. Generous fillet radii on the magnesium part are important for reducing stress concentrations in service.

The optimum width-to-thickness ratio is less than 20. Higher ratios require very generous tolerances.

In Figure 10.1, A is easily extrudable since the part is symmetrical about both axes. The design shown in B has lost its symmetry about one axis, but is still extrudable since the wall thickness about the holes is uniform. However, that shown in C, with its large mass of metal on one side, is extremely difficult to extrude with good tolerances on the hole dimensions.

A wedgelike shape such as the one in Figure 10.2A is extremely difficult to extrude. The metal tends to flow more rapidly through the thicker end resulting in a curved extrusion. It is almost impossible to fill the thin edge. Some relief can be obtained by adding metal to the knife edge as shown in Figure 10.2B.

Long, thin legs projecting from thick sections, as shown in Figure 10.3 should be avoided. The legs should be as short and as thick as possible. The situation shown in Figure 10.3C is especially difficult since symmetry is lost about both axes.

When there are slots such as shown in Figure 10.4, the length should not exceed three times the width. The slot shown as being in the extrusion is a tongue on the die which can be easily broken if too slender. Nonsymmetrical slots, such as in Figure 10.4C should be avoided.

Extrudability of the shape shown in Figure 10.5 can be greatly improved by adding fillets and thickening the web as shown by the dotted line. A web much thinner than the flanges makes it difficult to meet flatness and angularity tolerances.

The integrally-stiffened flooring section shown in Figure 10.6 illustrates the desirability of symmetry. The nonsymmetrical section shown in the upper part of the sketch is extremely difficult to extrude with good angularity tolerances, while the section in the lower part is easily extruded.

The producibility of thin hollow shapes like those shown in Figure 10.7 is greatly improved by the addition of one or more stiffening ribs as indicated by the dotted lines. Such ribs are effective where the width is three or more times the depth and maximum depth is 50 mm (2 in) or less. If the addition of stiffeners is not possible for the application intended, the wall thickness may have to be increased if flatness or contour tolerances are to be met.

Tolerances as generally accepted by industry are detailed in magnesium specifications, such as ASTM B107 [6]. Tighter tolerances can sometimes be met, depending upon the section complexity, the alloy, and the property requirements. If especially strict tolerances are important, the extruder should be consulted early in the design stage.

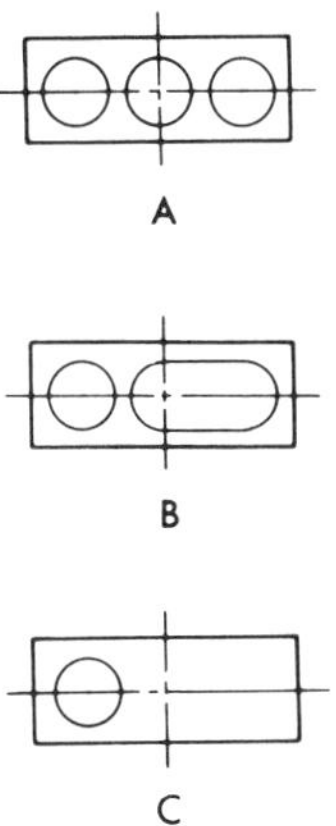

FIGURE 10.1 Examples of the importance of symmetry for ease of extrusion. A is symmetrical about both axes and is easily extruded. B has lost symmetry about one axis, but wall thicknesses are uniform and it is extrudable. C has symmetry about neither axis, and extrusion is very difficult.

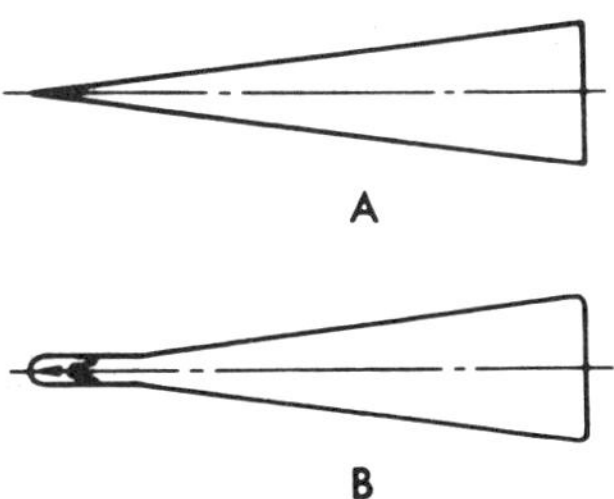

FIGURE 10.2 The sharp point shown in A is almost impossible to fill, and the extrusion will tend to curve as it comes out of the die. Adding a bulb as shown in B improves the extrudability.

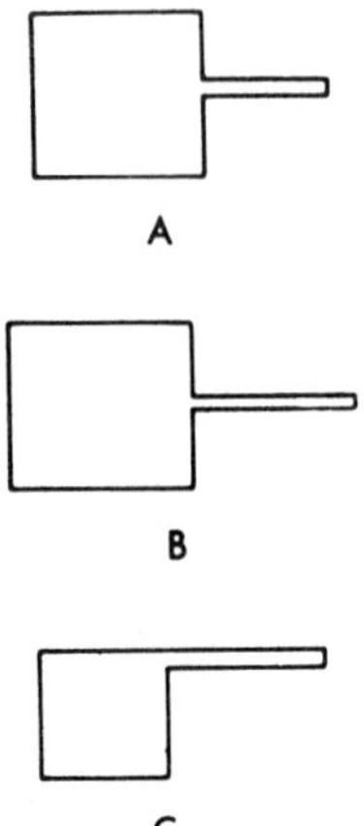

FIGURE 10.3 Legs as shown here should be as short and as thick as possible. Long, thin legs are very difficult to extrude. The section shown in C is extremely difficult to extrude since symmetry has been lost about both axes.

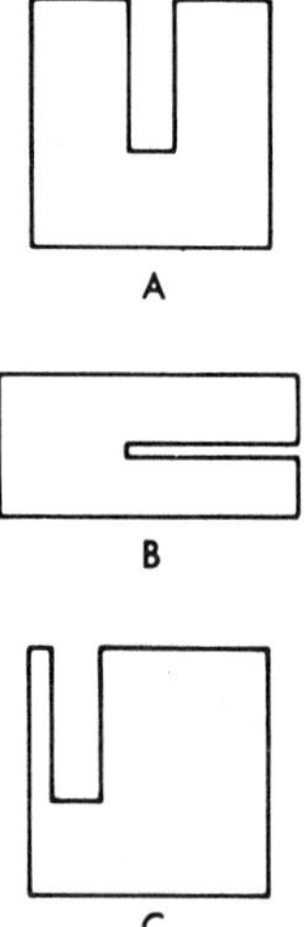

FIGURE 10.4 Slot length should not exceed three times the slot width. The section shown in C, with no symmetry about either axis, is very difficult to extrude.

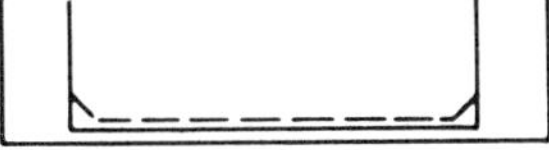

FIGURE 10.5 The section shown by the dotted lines is much to be preferred over the section shown by the solid lines. If the web is much thinner than the flanges, flatness and angularity tolerances are difficult to meet.

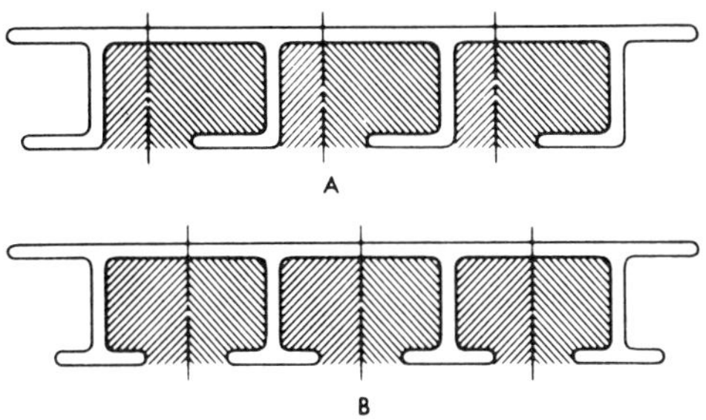

FIGURE 10.6 The section shown in A is not symmetrical and is difficult to extrude. That shown in B is symmetrical and easy to extrude.

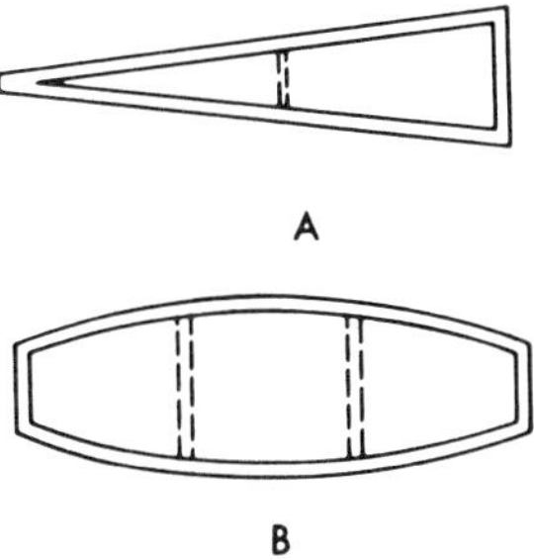

FIGURE 10.7 Stiffening ribs as shown by the dotted lines improve the extrudability of hollow sections.

10.3 TENSILE PROPERTIES

10.3.1 Room Temperature

Typical and minimum properties at room temperature for extruded sections are shown in Tables 10.1 through 10.3. Note that the compressive yield strength is lower than the tensile yield strength. Twinning is more active in compression than in tension because of the preferred orientation developed during extrusion. Since less force is required to initiate yielding by twinning than by slip, the compressive yield strength is low relative to the tensile yield strength. Twinning is inhibited by a fine grain size, resulting in a higher ratio of compressive yield to tensile yield strength for those alloys that do have a fine grain size.

Properties of extrusions depend not only on the alloy composition, but also on any extrusion conditions that affect the temperature of the section as it emerges from the die. In general, the higher the temperature of the extrusion, the lower will be the strength. This is especially true of the compressive yield strength. The two major conditions that will raise the temperature of the extrusion are shape complexity and speed of extrusion. As the shape becomes more complex, it is necessary to raise the starting temperature for extrusion in order to push the billet through the die. As the speed of extrusion increases, the generated heat increases and the temperature rises. Thus, as an example, tubes have lower strength than bars. The typical properties in the tables are those that will be obtained for most extrusions under normal operating conditions. The properties actually obtained in a specific case may be higher or lower, depending upon shape and speed of extrusion. The minimum properties are those that can be met with any section that can be extruded.

Stress-Strain Curves

Relations between stress and strain for a number of extrusion alloys are shown in Figures 10.8 through 10.16. Data at both room and elevated temperatures are included for the high-temperature alloy HM31A-T5. Although formulae dealing with the design of structures assume in most cases that stress is proportional to strain, it is necessary for many designs to consider those portions of the stress-strain curve beyond the proportional limit. A useful concept is the reduced modulus, making use of either the tangent or secant modulus at a given stress level [10-14]. Curves showing these values, as derived from the stress-strain curve, are given in Figures 10.17 through 10.23.

10.3.2 The Effect of Temperature

The effect of the temperature of testing on the tensile properties of extrusions are given in Table 10.4. The room temperature properties in Table 10.4 do not agree exactly with those in Table 10.1. The properties in the first table are typical; those in the second are specific to the extrusion made to study the effect of temperature.

There is an apparent reversal of the ratio of compressive yield strength to the tensile yield strength as the temperature becomes high. This is a testing artifact, related to specimen end problems during the measurement of the compressive yield strength. The compressive yield strength can be considered equal to the tensile yield strength at the elevated temperatures.

As discussed in Chapter 8, the value measured for the elastic modulus depends upon any relaxation that may occur during the test. Since relaxation increases as testing speed decreases and as testing temperature increases, the value found is sensitive to the testing conditions. Essentially no relaxation occurs at room temperature in the time used for testing, so that the modulus measured for room temperature is the true one. At elevated temperatures, relaxation does occur during normal testing speeds. The values given for the modulus at elevated temperature in Table 10.4 were determined using the standard testing speeds.

10.3.3 The Effect of Exposure to Elevated Temperatures

Table 10.5 gives information on exposure at elevated temperatures on the properties of alloys at room temperature and at the temperature of exposure.

10.3.4 The Effect of Speed of Testing

Strength values for 5 orders of magnitude of increasingly rapid strain rates are given in Table 10.6 [9].

Isochronous stress-strain curves for HM31A-T5 are given in Figures 10.24 through 10.27.

Creep data for five extrusion alloys are in Table 10.7.

TABLE 10.1 Tensile Properties of Bars Rods and Shapes at Room Temperature (SI Units)

Alloy	Temper	Least dimension, mm	Area, cm^2	Typical %E	Typical TYS (MPa)	Typical CYS (MPa)	Typical TS (MPa)	Minimum %E	Minimum TYS (MPa)	Minimum CYS (MPa)	Minimum TS (MPa)
AZ10A	F	<6.4		10	145	70	240				
		6.4-38		10	150	75	204				
AZ31B or C	F	<6.4		14	195	105	260	7	145		240
		6.4-38		15	200	95	260	7	150	85	240
		38-64		14	195	95	260	7	150	85	235
		64-127		15	195	95	260	7	150	85	220
AZ61A	F	<6.4		17	230		315	8	145		260
		6.4-64		16	230	130	310	9	165	95	275
		64-127		15	215	145	310	7	150	95	275
AZ80A	F	<6.4		12	250		340	9	195		295
		6.4-38		11	250		340	8	195	115	295
		38-64		11	240		340	6	195	115	295
		64-127		9	250		330	4	185	115	290
	T5	<6.4		8	260	234	380	4	205		325
		6.4-38		7	275	240	380	4	230	195	330
		38.1-64		6	270	220	365	4	230	185	330
		64-127		6	260	215	345	2	205	180	310
HK31A	F	<10						8	200		280
		10-100						8	225		305

HM31A	T5		<6.4	10	270	185	305	4	180	130	225
			6.4-26	10	270	160	305	4	180	105	255
M1A	F			12	180	125	225				
ZC71A	F	0-130						7	160		240
	T5							5	200		250
	T6			5	330	141	356	3	300		325
ZH11A	F			18	147		263				
ZK10A	F			13	208		293				
ZK30A	F	2.54 cm diameter		18	239	213	309				
ZK60A	F		<13	14	260	230	340	5	215	185	295
			13-19	14	255	195	340	5	215	180	295
			19-32	14	250	185	340	5	215	170	295
			32-258	9	255	160	330	6	215	140	295
	T5		<13	11	305	250	365	4	250	205	310
			13-19	12	295	215	360	4	250	195	310
			19-65					6	235	160	310
			65-160					6	235	150	310
			160-260					6	215	140	295
ZM21A	F			11	162		255	17	185		255

Source: Refs. 1, 2, 3, 18, 23, 24.

TABLE 10.1a Tensile Properties of Bars Rods and Shapes at Room Temperature (English Units)

Alloy	Temper	Least dimension, in	Area, in^2	Typical				Minimum			
					ksi				ksi		
				%E	TYS	CYS	TS	%E	TYS	CYS	TS
AZ10A	F	<0.25		10	21	10	35				
		0.25-1.5		10	22	11	35				
AZ31B	F	<0.25		14	28	15	38	7	21		35
		0.25-1.5		15	29	14	38	7	22	12	35
		1.5-2.5		14	28	14	38	7	22	12	35
		2.5-5.0		15	28	14	38	7	20	10	32
AZ61A	F	<0.25		17	33		46	8	21		38
		0.25-2.5		16	33	19	45	9	24	14	40
		2.5-5.0		15	31	21	45	7	22	14	40
AZ80A	F	<0.25		12	36		49	9	28		43
		0.25-1.5		11	36		49	8	28	17	43
		1.5-2.5		11	35		49	6	28	17	43
		2.5-5.0		9	36		48	4	27	17	42
	T5	<0.25		8	38	34	55	4	30		47
		0.25-1.5		7	40	35	55	4	33	28	48
		1.5-2.5		6	39	32	53	4	33	27	48
		2.5-5.0		6	38	31	50	2	30	26	45
HK31A	F	<0.40						8	29		41
		0.40-4.0						8	33		44

HM31A	T5		<1.0	10	39	27	44	4	26	19	37
			1.0-4.0	10	39	23	44	4	26	15	37
M1A	F	<5.0		12	26	18	37				
ZC71A	F	<5.0						7	23		35
	T5							5	29		36
	T6			5	48	20	52	3	44		47
ZH11A	F			18	21		38				
ZK10A	F			13	30		42				
ZK30A	F	1 in diameter		18	35	31	45				
ZK60A	F		<2.0	14	38	33	49	5	31	27	43
			2.0-3.0	14	37	28	49	5	31	26	43
			3.0-5.0	14	36	27	49	5	31	25	43
			5.0-40.0	9	37	23	48	6	31	20	43
	T5		<2.0	11	44	36	53	4	44	30	45
			2.0-3.0	12	43	31	52	4	36	26	45
			3.0-5.0	14	42	30	51	4	36	25	45
			5.0-10.0					6	34	23	45
			10.0-25.0					6	34	22	45
			25.0-40.0					6	31	20	43
ZM21A	F			11	23		37	17	27		37

Source: Refs. 1, 2, 3, 18, 23, 24.

TABLE 10.2 Tensile Properties of Extruded Tubing at Room Temperature (SI Units)

Alloy	Temper	Outside diameter, mm	Wall thickness, mm	Typical				Minimum			
					MPa				MPa		
				%E	TYS	CYS	TS	%E	TYS	CYS	TS
AZ10A	F	150 mx.	0.7-6.4	8	145	70	230				
AZ31B	F	150 max.	0.7-6.4	16	165	85	250	8	110	70	220
		150 max.	6.4-19	12	165	85	250	4	100	70	220
AZ61A	F	150 max.	0.7-19	14	165	110	285	7	110	75	250
M1A	F			9	145		240				
ZK10A	F			7	193		278				
ZK60A	F	76 max.	0.7-6.4	13	240	175	325	5	195	140	275
	T5	76 max.	0.7-6.4	11	275	205	345	4	260	180	315
		76-215	2.4-30	12	270	180	340	4	230	145	305

Source: Refs. 1, 2, 23.

TABLE 10.2a Tensile Properties of Extruded Tubing at Room Temperature (English Units)

				Typical				Minimum			
					ksi				ksi		
Alloy	Temper	Outside diameter, in	Wall thickness, in	%E	TYS	CYS	TS	%E	TYS	CYS	TS
AZ10A	F	6 max.	0.28-0.250	8	21	10	33				
AZ31B	F	6 max.	0.28-0.25	16	24	12	36	8	16	10	32
		6 max.	0.25-0.75	12	24	12	36	4	16	10	32
AZ61A	F	6 max.	0.28-0.75	14	24	16	41	7	16	11	36
M1A	F			9	21		35				
ZK10A	F			7	28		40				
ZK60A	F	3 max.	0.28-0.25	13	35	25	47	5	28	20	40
	T5	3 max.	0.28-0.25	11	40	30	50	4	38	26	46
		3-8.5	0.094-0.19	12	39	26	49	4	33	21	44

Source: Refs. 1, 2, 23.

TABLE 10.3 Tensile Properties of Extruded Semi-hollow and Hollow Shapes at Room Temperature (SI Units)

		Typical				Minimum			
			MPa				MPa		
Alloy	Temper	%E	TYS	CYS	TS	%E	TYS	CYS	TS
AZ10A	F	8	145	70	230				
AZ31B	F	16	165	85	250	8	140	70	220
AZ61A	F	14	165	110	285	7	110	75	250
ZK60A	F	12	235	175	315	5	195	140	275
	T5	11	275	200	345	4	260	180	315

TABLE 10.3a Tensile Properties of Extruded Semi-hollow and Hollow Shapes at Room Temperature (English Units)

		Typical				Minimum			
			ksi				ksi		
Alloy	Temper	%E	TYS	CYS	TS	%E	TYS	CYS	TS
AZ10A	F	8	21	10	33				
AZ31B	F	16	24	12	36	8	20	10	32
AZ61A	F	14	24	16	41	7	16	11	36
ZK60A	F	12	34	25	46	5	28	20	40
	T5	11	40	29	50	4	38	26	46

Source: Refs. 1, 2.

TABLE 10.4 Effect of Temperature on Tensile Properties of Extrusions (SI Units)

Alloy	Temper	Temperature, °C	%E	MPa TYS	CYS	TS	GPa, modulus
AZ31B	F	-175	6	325		395	
		-215	10	295		310	
		-75	10	295		300	
		0	11	220		285	
		25	12	205	95	275	45
		100	21	145	95	235	
		150	39	105	95	170	
		200	42	65	80	110	
		250	57	40	60	75	
		300	64	15	30	45	
AZ61A	F	-175	5	320		355	
		-125	9	300		350	
		-75	12	275		340	
		0	15	240		335	
		25	16	230	130	325	45
		100	23	180	130	285	
		150	40	140	130	225	
		200	42	95	115	155	
		250	45	70	80	100	
		300	64	40	50	65	
AZ80A	F	-75	8	270		385	
		25	11	250		340	45
		95	18	220		305	41
		150	26	175		240	38
		200	35	120		200	32
		260	57	75		110	
HM31A	F	-175	4	305		350	
		-125	5	295		365	
		-75	6	280		340	
		0	10	250		300	
		25	12	240	185	285	45
		100	23	200	180	225	
		150	30	180	170	195	
		200	32	160	160	170	
		250	27	140	140	150	
		300	22	120	120	125	

TABLE 10.4 (*Continued*)

Alloy	Temper	Temperature, °C	%E	MPa TYS	MPa CYS	MPa TS	GPa, modulus
HM31A	T5	-175	7	325		400	
		-125	7	330		365	
		-75	8	305		360	
		0	9	280		320	
		25	10	265	185	305	45
		100	16	205	160	230	43
		150	25	215	205	235	41
		200	32	175	175	205	40
		250	32	130	135	150	39
		300	29	110	115	130	38
M1A	F	25	12	180		255	45
		95	16	145		200	
		150	21	110		145	
		200	27	85		115	
		315	53	35		60	
ZH11A	F (hard)	20	4	380[a]		385	
		250	13	170[a]		195	
	F (soft)	20	31	125[a]		220	
		250	54	80[a]		105	
ZK30A	F	20	18	255		309	45
		100	33	162		182	40
		200	56	46		127	22
		250	71	11		100	12
		300	90			63	
ZK60A	T5	-175	3	405		470	
		-125	4	320		455	
		-75	5	370		440	
		0	9	330		395	
		25	11	305	250	360	45
		100	43	205	200	250	41
		150	53	140	155	175	36
		200	78	85	95	110	
		250		35	45	55	

[a]0.1% deviation from the modulus line.
Source: Refs. 2, 4, 5, 23, 24.

TABLE 10.4a Effect of Temperature on Tensile Properties of Extrusions (English Units)

Alloy	Temper	Temperature, °C	%E	ksi TYS	ksi CYS	ksi TS	ksi, modulus
AZ31B	F	-175	6	47		57	
		-125	10	43		45	
		-75	10	37		44	
		0	11	32		41	
		25	12	30	14	40	6500
		100	21	21	14	34	
		150	39	15	14	25	
		200	42	9	12	16	
		250	57	6	9	11	
		300	64	2	4	7	
AZ61A	F	-175	5	46		51	
		-125	9	44		51	
		-75	12	40		49	
		0	15	35		49	
		25	16	33	19	47	6500
		100	23	26	19	41	
		150	40	20	19	33	
		200	42	14	17	22	
		250	45	10	12	15	
		300	64	6	7	9	
AZ80A	F	-75	8	39		56	
		25	11	36		49	6500
		95	18	32		44	5950
		150	26	25		35	5500
		200	35	17		29	4650
		260	57	11		16	
HM31A	F	-175	4	44		51	
		-125	5	43		53	
		-75	6	41		49	
		0	10	36		44	
		25	12	35	27	41	6500
		100	23	29	26	33	
		150	30	26	25	28	
		200	32	23	23	25	
		250	27	20	20	22	
		300	22	17	17	18	

TABLE 10.4a (*Continued*)

Alloy	Temper	Temperature, °C	%E	ksi TYS	ksi CYS	ksi TS	ksi, modulus
HM31A	T5	-175	7	47		58	
		-125	7	48		53	
		-75	8	44		52	
		0	9	41		46	
		25	10	38	27	44	6500
		100	16	30	23	33	6250
		150	25	31	30	34	5950
		200	32	25	25	30	5800
		250	32	19	20	22	5650
		300	29	16	17	19	5500
M1A	F	25	12	26		37	6500
		95	16	21		29	
		150	21	16		21	
		200	27	12		17	
		315	53	5		9	
ZH11A	F (hard)	20	4	55[a]		56	
		250	13	25[a]		28	
	F (soft)	20	31	18[a]		32	
		250	54	12[a]		15	
ZK30A	F	20	18	37		45	6500
		100	33	23		26	5800
		200	56	7		18	3200
		250	71	2		15	1740
		300	90			9	
ZK60A	T5	-175	3	59		68	
		-125	4	46		66	
		-75	5	54		64	
		0	9	48		57	
		25	11	44	36	52	6500
		100	43	30	29	36	5950
		150	53	20	22	25	5200
		200	78	12	14	16	
		250		5	7	8	

[a]0.1% deviation from the modulus line.
Source: Refs. 2, 4, 23, 24.

TABLE 10.5 The Effect of Temperature Exposure on Tensile Properties

Alloy	Exposure °C	Exposure Hours	Test temperature, °C	%E	MPa TYS	MPa CYS	MPa TS	ksi TYS	ksi CYS	ksi TS
HM31A-T5	315	0	20	10	270	185	305	39	27	44
		1		10	270	185	305	39	27	44
		10		10	270	170	305	39	25	44
		100		10	250	170	290	36	25	42
		0	315	27	105	110	125	15	16	18
		1		30	105	110	125	15	16	18
		10		33	95	105	115	14	15	17
		100		35	95	105	110	14	15	16
	370	0	20	10	270	185	305	39	27	44
		1		11	255	170	290	37	25	42
		10		11	240	160	255	35	23	37
		100		11	215	145	270	31	21	39
		0	370	35	75	85	90	11	12	13
		1		39	70	75	90	10	11	13
		10		44	60	70	85	9	10	13
		100		48	60	70	75	9	10	11
	425	0	20	10	270	185	305	39	27	44
		1		13	205	150	270	30	22	39
		10		13	180	130	255	26	19	37
		100		13	150	115	235	22	17	34
		0	425	52	50	50	60	7	7	9
		1		58	35	35	50	5	5	7

TABLE 10.5 (*Continued*)

Alloy	Exposure °C	Exposure Hours	Test temperature, °C	%E	MPa TYS	MPa CYS	MPa TS	ksi TYS	ksi CYS	ksi TS
		10		64	20	20	35	3	3	5
		100		70	20	20	30	3	3	4
	480	0	20	10	270	185	305	39	27	44
		1		15	165	130	240	24	19	35
		10		15	130	105	220	19	15	32
		100		15	125	90	205	18	13	30
		0	480	70	20	20	30	3	3	4
		1		78	15	15	20	2	2	3
		10		86	5	5	15	1	1	2
		100		95	5	5	15	1	1	2
ZK60A-T5	100	0	20	10	275	240	350	40	35	51
		16		10	275	240	350	40	35	51
		48		10	275	240	350	40	35	51
		192		10	275	240	350	40	35	51
		500		10	275	240	350	40	35	51
		1000		10	275	240	350	40	35	51
	100	0	100	41	190	215	245	28	31	36
		16		37	200	210	245	29	30	36
		48		38	190	225	245	28	33	36
		192		38	200	220	250	29	32	36
		500		36	200	225	260	29	33	38
		1000		34	200	220	255	29	32	37

120	0	20	10	275	240	350	40	35	51
	16		10	270	240	345	39	35	50
	48		10	290	240	350	42	35	51
	192		10	275	245	345	40	36	36
	500		8	280	245	360	41	36	52
	1000		10	290	245	355	42	36	51
120	0	120	48	165	190	210	24	28	30
	16		44	165	200	210	24	29	30
	48		45	160	210	210	23	30	30
	192		43	165	205	220	24	30	32
	500		45	165	205	210	24	30	30
	1000		45	160	205	205	23	30	30
150	0	20	10	275	240	350	40	35	51
	16		10	280	240	345	41	35	50
	48		10	290	235	345	42	34	50
	192		12	240	230	340	35	33	49
	500		10	260	230	335	38	33	49
	1000		12	255	220	325	37	32	47
150	0	150	52	130	160	175	19	23	25
	16		45	135	170	175	20	25	25
	48		52	135	180	170	20	26	25
	192		55	130	165	165	19	24	24
	500		56	130	165	165	19	24	24
	1000		56	120	160	165	17	23	24

Source: Refs. 7, 8.

TABLE 10.6 Effect of Strain Rate and Temperature on the Strength of Extrusions (SI Units)

Alloy	Test temperature, °C	TYS (MPa) at strain/min				TS (MPa) at strain/min			
		0.005	0.050	0.50	5.0	0.005	0.050	0.50	5.0
AZ31B-F	25	160	170	175	180	265	265	265	265
	95	135	140	150	160		225	230	240
	150	90	100	115	130		165	180	200
	200	60	75	90	105		105	125	150
	260	40	55	65	80		65	85	115
	315	20	35	45	60		40	55	75
	370	10	25	30	45		30	40	60
	425		15	25	30		15	25	40
	480		10	15	25		10	15	25
AZ61A-F	25	170	175	175	180	310	310	310	310
	95	155	160	165	170	275	275	275	275
	150	115	130	140	150		195	230	245
	200	75	100	115	130		115	155	190
	260	50	70	80	110		75	100	140
	315	25	45	50	70		45	65	90
	370	10	25	35	55		25	40	65
	425		10	25	40		15	30	40
	480		5	15	25		10	15	25
AZ80A-T5	25	210	220	240	255	335	335	335	335
	95	165	185	205	225		285	295	305
	150	110	130	155	175		180	215	250
	200	65	90	120	145		110	145	185
	260	35	60	75	110	65	95	135	

	315	20	40	55	85		45	65	95
	370	10	25	35	55		30	40	65
	425		15	30	40		15	30	45
	480		5	15	20		10	15	20
HM31A-F	25	245	245	250	250	260	265	275	290
	95	210	210	215	230	205	215	230	245
	150	175	180	190	210	180	185	195	215
	200	155	155	160	180	160	160	170	185
	260	140	140	145	160	140	140	145	160
	315	90	105	115	125	90	110	120	130
	370	50	60	70	90	65	65	75	95
	425	20	25	30	55	35	35	40	60
	480	5	10	15	35	10	10	15	35
M1A-F	25	170	180	190	200	230	250	275	295
	95	135	150	165	175		180	205	230
	150	85	105	135	165		120	140	175
	200	50	70	90	120		80	95	125
	260	35	45	60	85		50	65	90
	315	25	30	40	55		35	50	65
	370	20	25	30	35		25	35	45
	425		20	20	25		20	25	30
	480		10	15	20		15	20	20

TABLE 10.6 (*Continued*)

Alloy	Test temperature, °C	TYS (MPa) at strain/min				TS (MPa) at strain/min			
		0.005	0.050	0.50	5.0	0.005	0.050	0.50	5.0
ZK60A-T5	25	275	290	310	325		355	365	370
	95	185	215	250	280		260	285	315
	150	115	145	180	210		170	200	240
	200	45	90	120	155		100	130	170
	260	15	35	55	105		40	70	125
	315	10	15	30	60		15	35	75
	370	10	10	20	40		15	25	45
	425		10	15	25		10	15	30
	480		10	15	25		10	15	25

Source: Ref. 9.

TABLE 10.6a Effect of Strain Rate and Temperature on the Strength of Extrusions (English Units)

Alloy	Test temperature °C	TYS (ksi) at strain/min				TS (ksi) at strain/min			
		0.005	0.050	0.50	5.0	0.005	0.050	0.50	5.0
AZ31B-F	25	23	25	25	26	38	38	38	38
	95	20	20	22	23		33	33	35
	150	13	14	17	19		24	26	29
	200	9	11	13	15		15	18	22
	260	6	8	9	12		9	12	17
	315	3	5	7	9		6	8	11
	370	1	4	4	7		4	6	9
	425		2	4	4		2	4	6
	480		1	2	4		1	2	4
AZ61A-F	25	25	25	25	26	45	45	45	45
	95	22	23	24	25	40	40	40	40
	150	17	19	20	22		28	33	36
	200	11	14	17	19		17	22	28
	260	7	10	12	16		11	14	20
	315	4	7	7	10		7	9	13
	370	1	4	5	8		4	6	9
	425		1	4	6		2	4	6
	480		1	2	4		1	2	4
AZ80A-T5	25	30	32	35	37	49	49	49	49
	95	24	27	30	33		41	43	44
	150	16	19	22	25		26	31	36
	200	9	13	17	21		16	21	27
	260	5	9	11	16		9	14	29

TABLE 10.6a (*Continued*)

Alloy	Test temperature, °C	TYS (ksi) at strain/min				TS (ksi) at strain/min			
		0.005	0.050	0.50	5.0	0.005	0.050	0.50	5.0
	315	3	6	8	12		7	9	14
	370	1	4	5	8		4	6	9
	425		2	4	6		2	4	7
	480		1	2	3		1	2	3
HM31A-F	25	36	36	36	36	38	38	40	42
	95	30	30	31	33	30	31	33	36
	150	25	26	28	30	26	27	28	31
	200	22	22	23	26	23	23	25	27
	260	20	20	21	23	20	20	21	23
	315	13	15	17	18	13	16	17	19
	370	7	9	10	13	9	9	11	14
	425	3	4	4	8	5	5	5	9
	480	1	1	2	5	1	1	2	5

M1A-F	25	25	26	28	29	33	36	40	43
	95	20	22	24	25		26	30	33
	150	12	15	20	24		17	20	25
	200	7	10	13	17		12	14	18
	260	5	7	9	12		7	9	13
	315	4	4	6	8		5	7	9
	370	3	4	4	5		4	5	7
	425		3	3	4		3	4	4
	480		1	2	3		2	3	3
ZK60A-T5	25	40	42	45	47		51	53	54
	95	27	31	36	41		38	41	46
	150	17	21	26	30		25	29	35
	200	7	13	17	22		14	19	25
	260	2	5	8	15		6	10	18
	315	1	2	4	9		2	5	11
	370	1	1	3	6		2	4	7
	425		1	2	4		1	2	4
	480		1	2	4		1	2	4

Source: Ref. 9.

TABLE 10.7 Creep of Extrusion Alloys (SI Units)

Alloy	Temperature, °C	Stress (MPa) at stated total extension for indicated hours															
		0.1%				0.2%				0.5%				1.0%			
		1	10	100	500	1	10	100	500	1	10	100	500	1	10	100	500
AZ31B-F	95	40	35	30	25	70	60	50	40	110	95	85	70	130	115	105	90
	120	35	30	20	15	60	50	35	30	90	75	60	50	110	95	75	60
	150	30	20	10	5	50	35	20	15	75	55	35	30	85	70	50	35
	175	20	15			35	20			55	40			70	55		
AZ61A-F	95	40	35	30	20	75	70	50	40	140	125	95	85	170	150	130	110
	120	35	30	15	5	60	50	30	20	110	90	55	40	140	110	75	55
	150	25	15			50	30	5		85	55	20	5	110	75	35	15
	175	15				30	10			60	30			75	40		
HM31A-T5	200	40	40	40				75				115					
	260	40	40	40				70				95					
	315	35	35	35				50				65					
	370			10				15				15					
ZK30A-F[a]	150							26	19		43	33	27			39	32
	200	31	15	14	8		24	20	17		28	26	24				
ZK06A-T5	95					55	35	20	15	105	70	40	30	140	60	60	40
	120					35	20	5		70	40	15		95	60	30	
	150					20	5	5		35	15	10					

[a] Creep estension only.

Source: Refs. 5, 23.

TABLE 10.7a Creep of Extrusion Alloys (English Units)

Alloy	Temperature, °C	Stress (ksi) at stated total extension for indicated hours															
		0.1%				0.2%				0.5%				1.0%			
		1	10	100	500	1	10	100	500	1	10	100	500	1	10	100	500
AZ31B-F	95	6	5	4	4	10	9	7	6	16	14	12	10	19	17	15	13
	120	5	4	3	2	9	7	5	4	13	11	9	7	16	14	11	9
	150	4	3	1	1	7	5	3	2	11	8	5	4	12	10	7	5
	175	3	2			5	3			8	6			10	8		
AZ61A-F	95	6	5	4	3	11	10	7	6	20	18	22	12	25	22	19	16
	120	5	4	2	1	9	7	4	3	16	13	8	6	20	16	11	8
	150	4	2			7	4	1		12	8	3	1	16	11	5	2
	175	2				4	1			9	4			11	6		
HM31A-T5	200	6	6	6				11			17						
	260	6	6	6				10			14						
	315	5	5	5				7			9						
	370			1				2			2						
ZK30A-F[a]	150							4	3		6	5	4			6	5
	200	4	2	2	1		4	3	2		4	4	3				
ZK60A-T5	95					8	5	3	2	15	10	6	4	20	9	9	6
	120					5	3	1		10	6	2		14	9	4	
	150					3	1	1		5	2	1		8	4	1	

[a]Creep extension only.

Source: Refs. 5, 23.

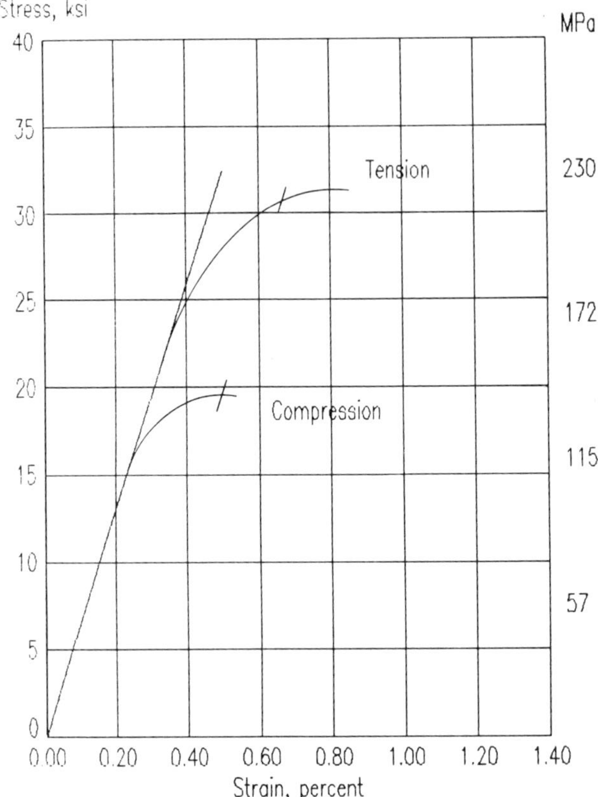

FIGURE 10.8 Stress-strain curves for AZ31B-F extrusions at room temperature.

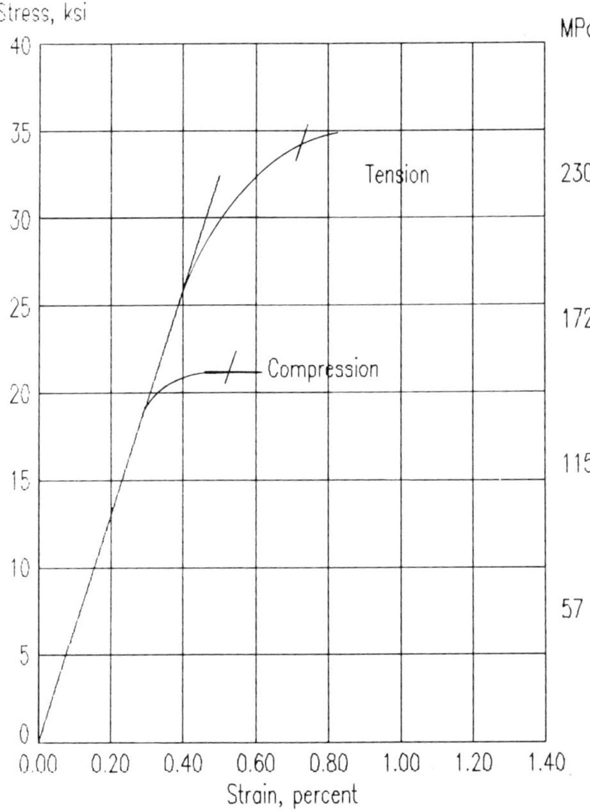

FIGURE 10.9 Stress-strain curves for AZ61A-F extrusions at room temperature.

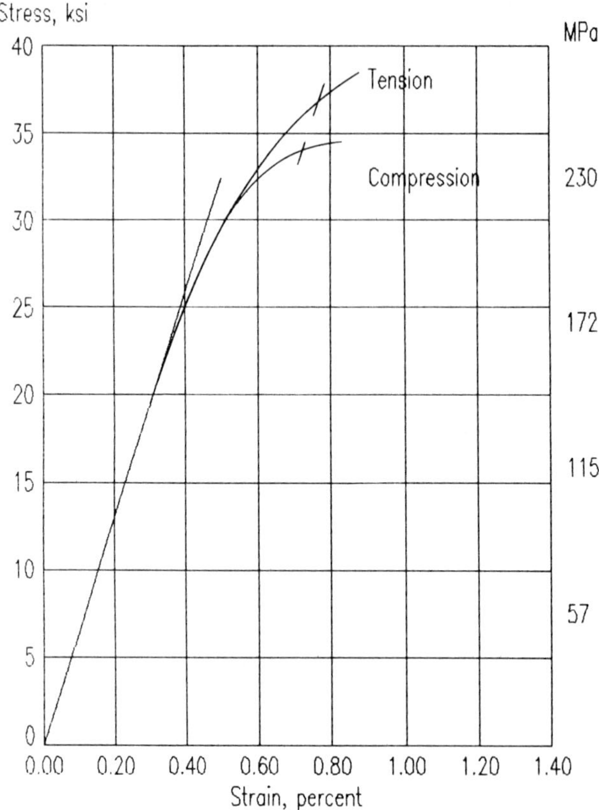

FIGURE 10.10 Stress-strain curves for AZ80A-T5 extrusions at room temperature.

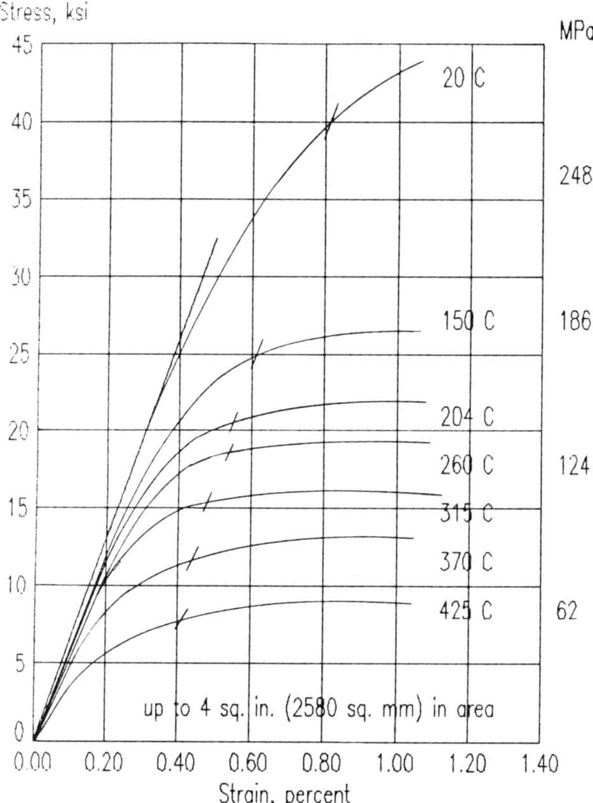

FIGURE 10.11 Tensile stress-strain curves for HM31A extrusions at room and elevated temperature.

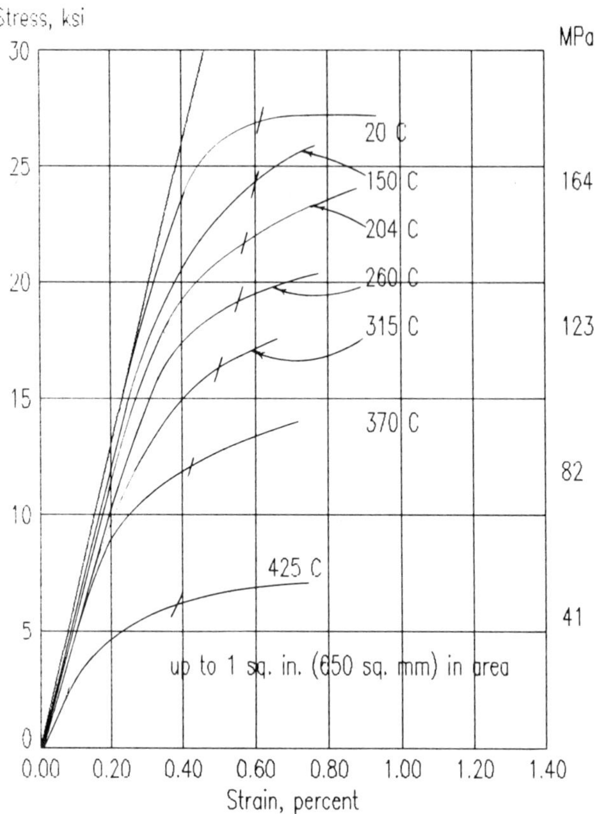

FIGURE 10.12 Compressive stress-strain curves for small HM31A-T5 extrusions at room and elevated temperatures.

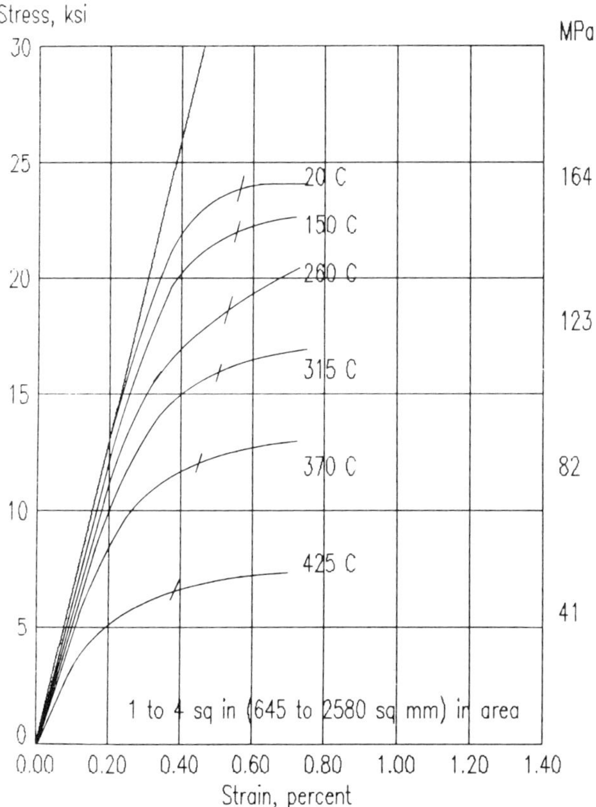

FIGURE 10.13 Compressive stress-strain curves for large HM31A-T5 extrusions at room and elevated temperatures.

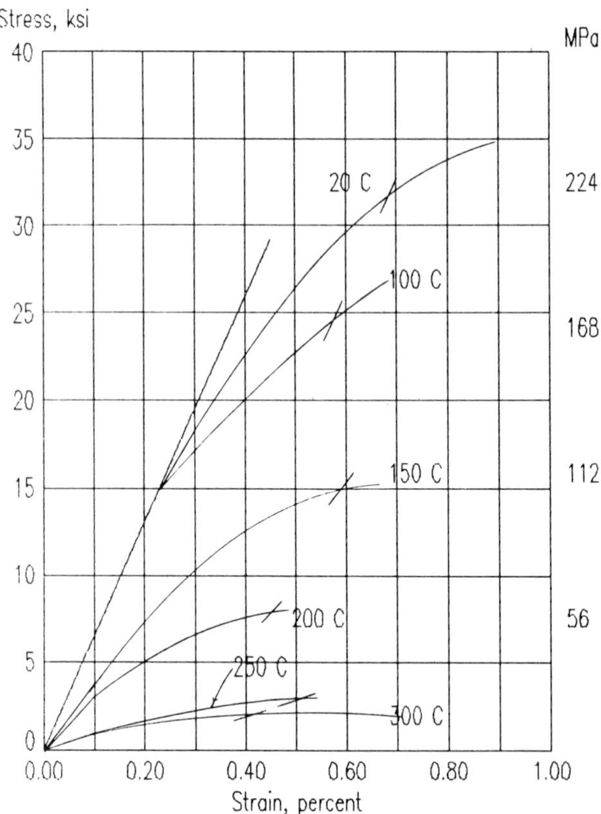

FIGURE 10.14 ZK30A-T5 Extrusion tensile stress-strain curves.

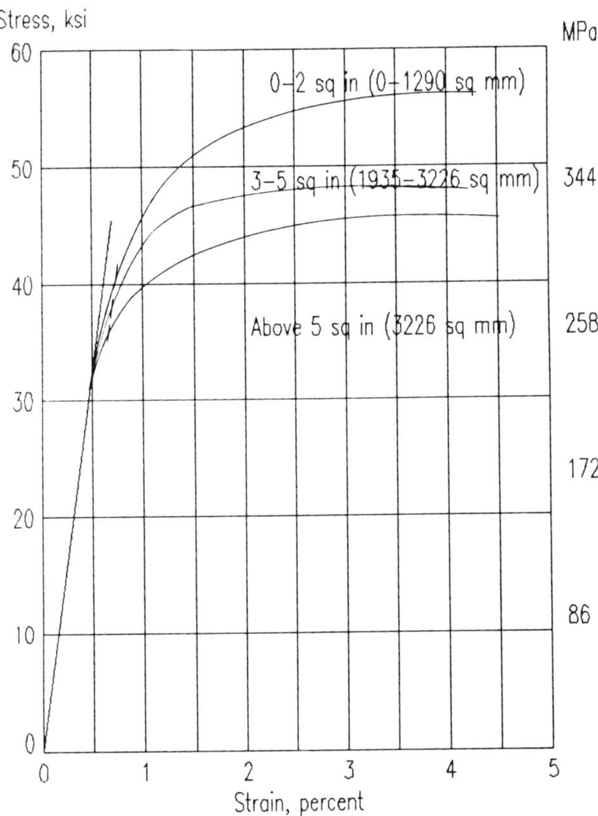

FIGURE 10.15 Tensile stress-strain curves for ZK60A-T5 extrusions of varying size at room temperature.

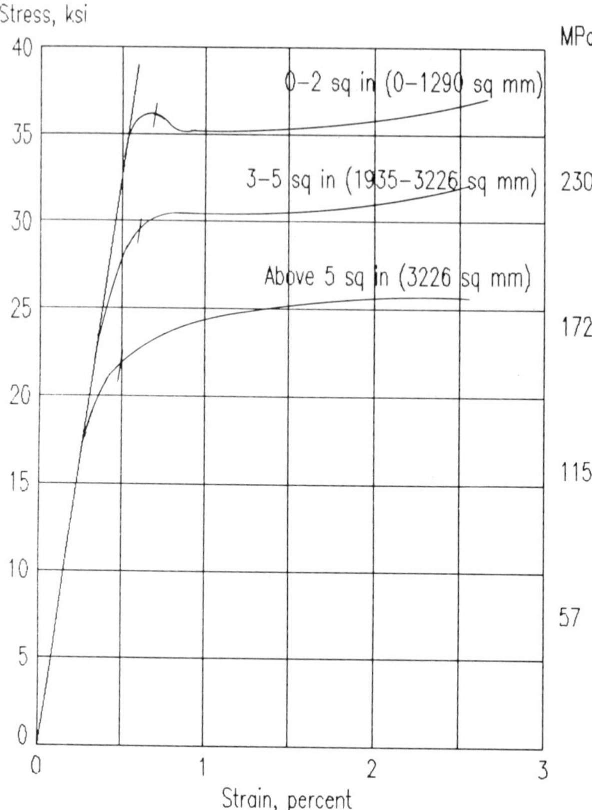

FIGURE 10.16 Compressive stress-strain curves for ZK60A-T5 extrusions of varying size at room temperature.

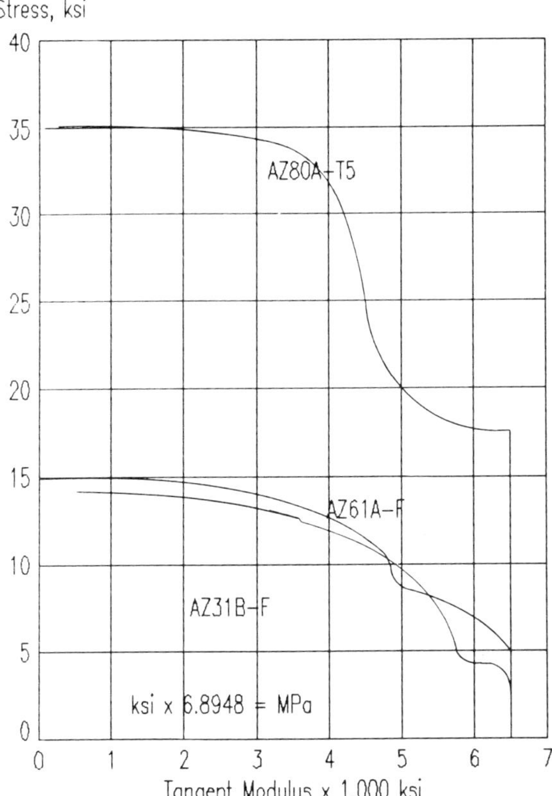

FIGURE 10.17 Compressive tangent modulus for AZ31B-F, AZ61A-F, and AZ80A-T5 extrusions based upon minimum properties at room temperature.

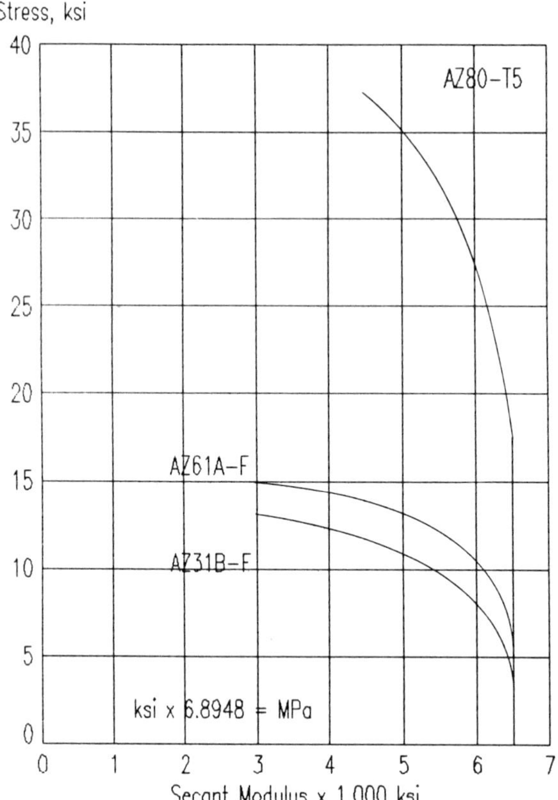

FIGURE 10.18 Compressive secant modulus for AZ31B-F, AZ61A-F, and AZ80-T5 extrusions based upon minimum properties at room temperature.

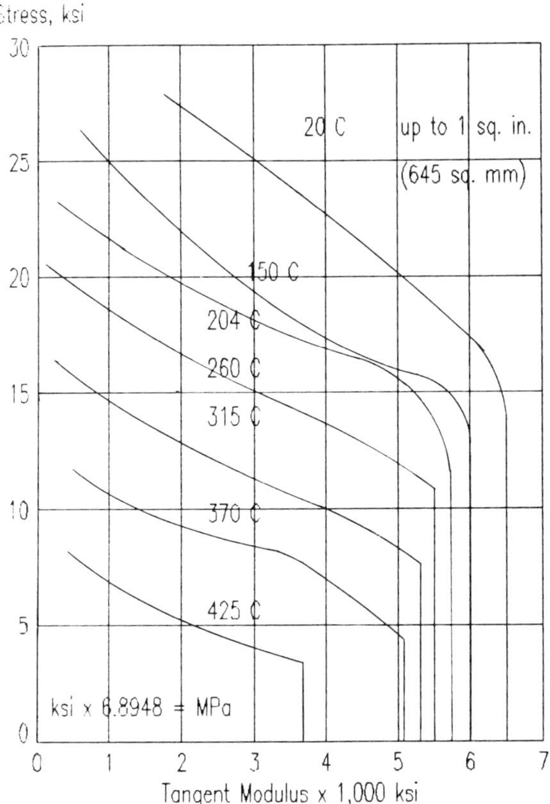

FIGURE 10.19 Compressive tangent modulus for HM31A-T5 extrusions at room and elevated temperatures.

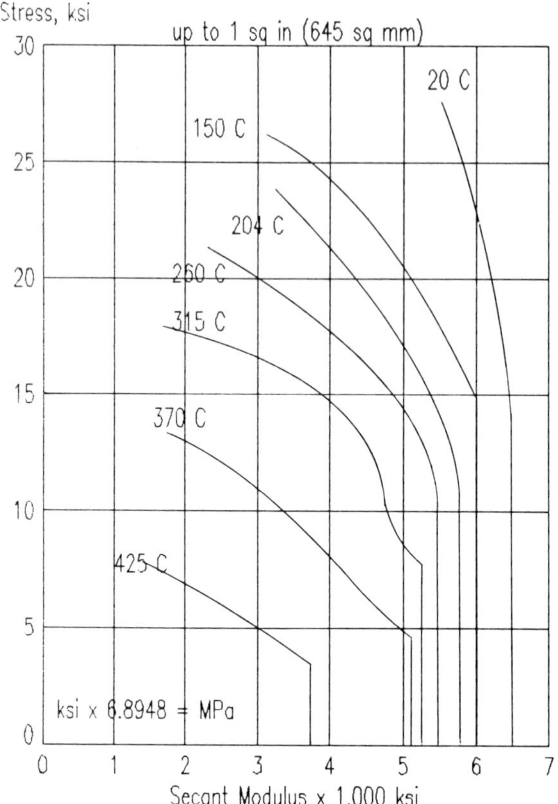

FIGURE 10.20 Compressive secant modulus for HM31A-T5 extrusions at room and elevated temperatures.

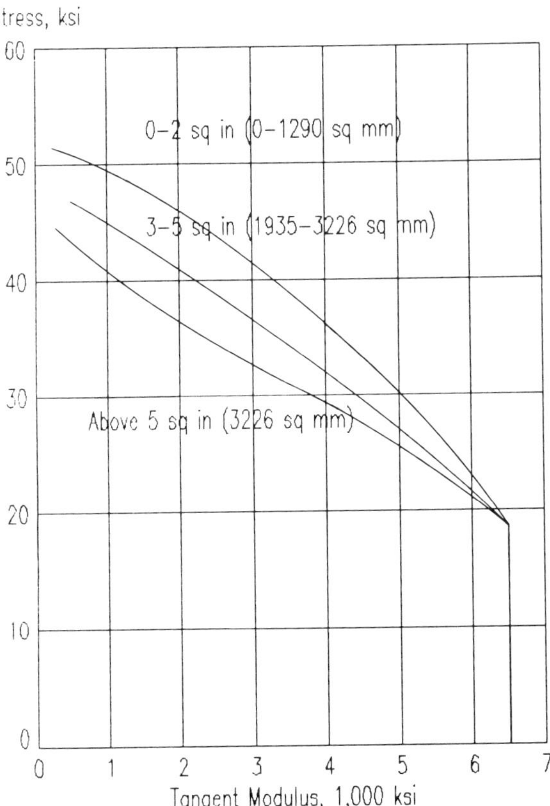

FIGURE 10.21 Tensile tangent modulus for ZK60A-T5 extrusions at room temperature.

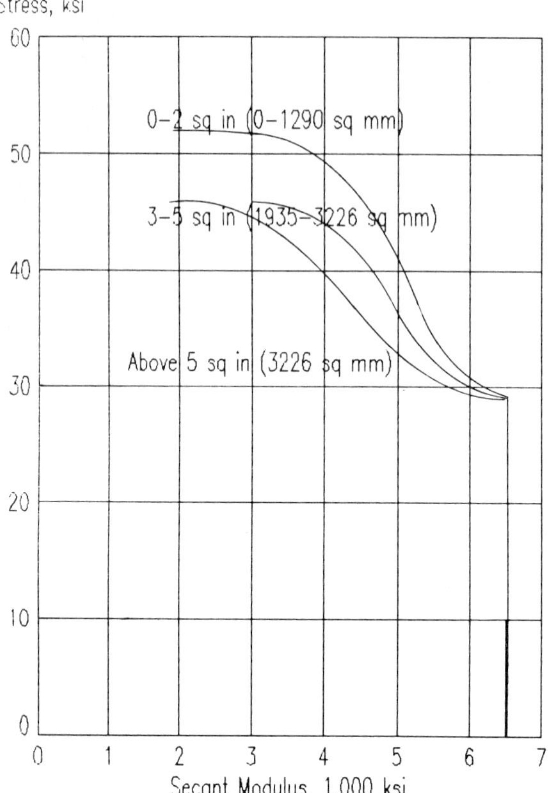

FIGURE 10.22 Tensile secant modulus for ZK60A-T5 extrusions at room temperature.

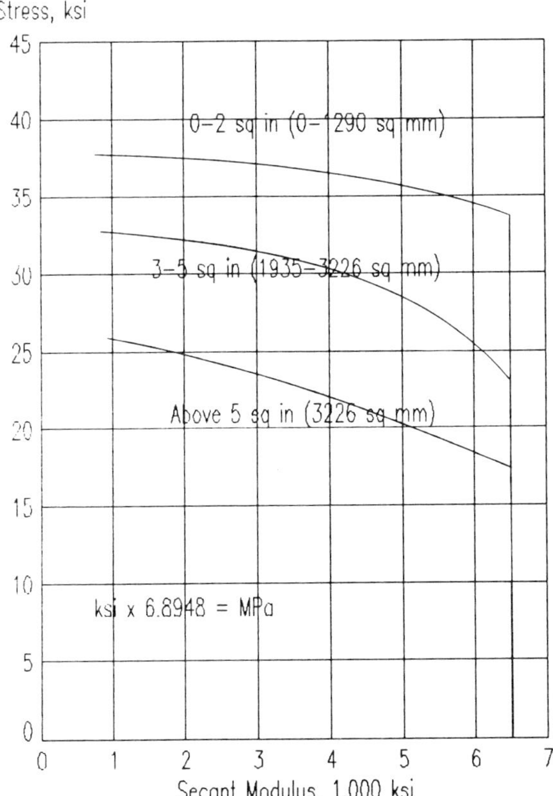

FIGURE 10.23 Compressive secant modulus for ZK60A-T5 extrusions at room temperature.

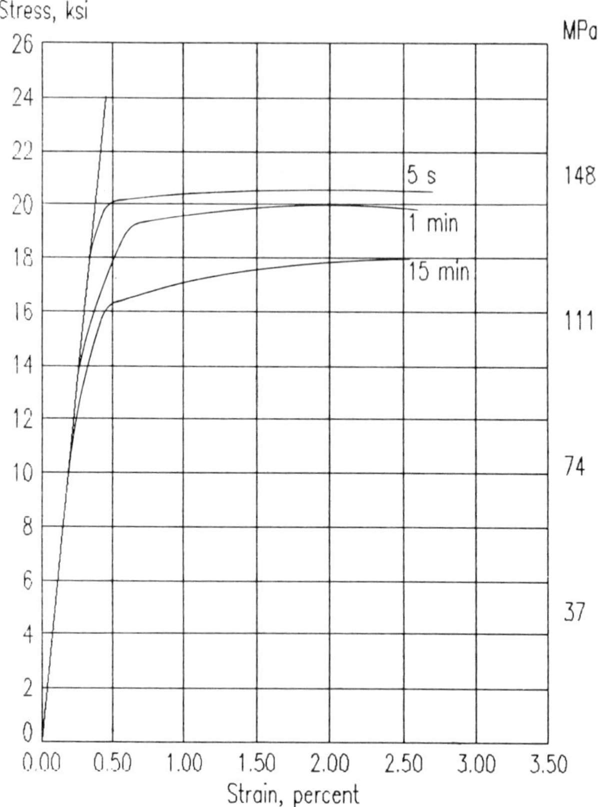

FIGURE 10.24 Isochronous stress-strain curves for HM31A-T5 extrusions at 260°C.

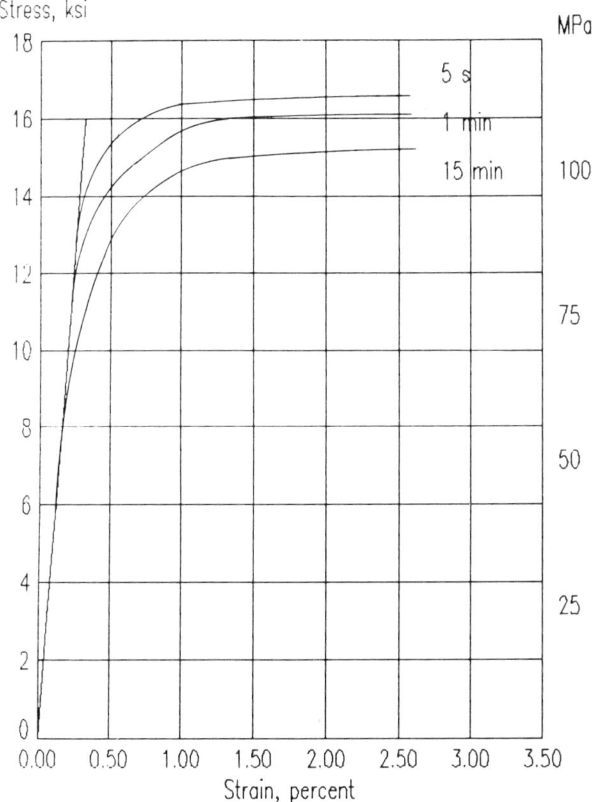

FIGURE 10.25 Isochronous stress-strain curves for HM31A-T5 extrusions at 315°C.

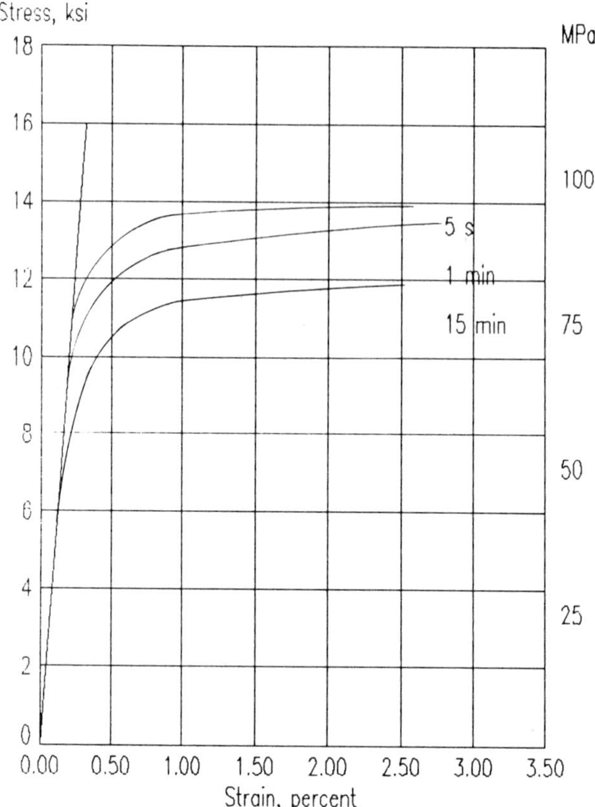

FIGURE 10.26 Isochronous stress-strain curves for HM31A-T5 extrusions at 370°C.

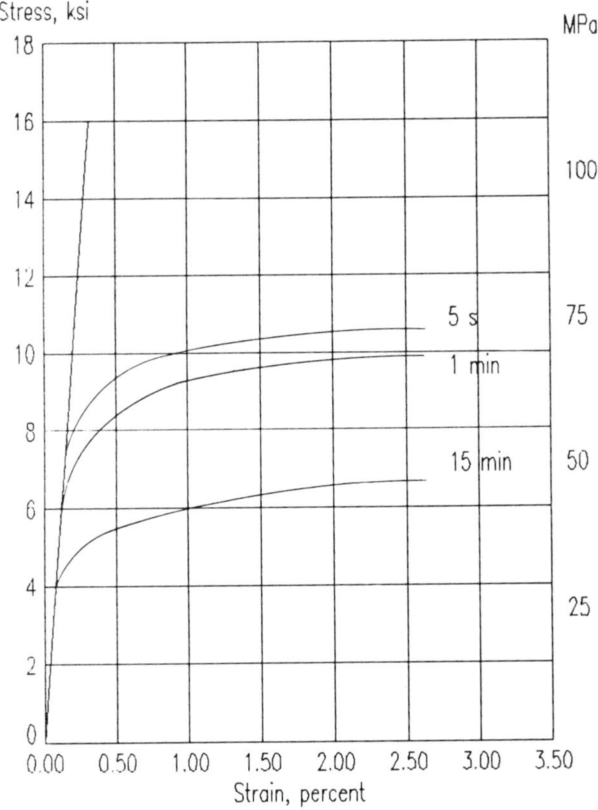

FIGURE 10.27 Isochronous stress-strain curves for HM31A-T5 extrusions at 425°C.

10.4 TOUGHNESS

Three measures of toughness are given in Table 10.8. The ratio of notched to unnotched tensile strength is an indication of the reaction of the metal to stress concentration when a tensile stress is applied. The Charpy impact values measure the ability to absorb energy in the presence of stress concentrations. K_{Ic}, a material constant, is the stress intensity at the head of an infinitely sharp notch, under conditions of plane strain, for which failure is catastrophic. It can be used to calculate the largest flaw size that can be tolerated for a given geometry and value of applied stress. The values for AZ80A are probably good estimates of the true valus of K_{Ic}, but the values for ZK30A and ZK60A are suspiciously high. The specimens used for determining the values were probably not large enough to result in true plane strain during the test. Enough plastic deformation occurred at the tip of the crack to lower significantly the stress intensity before failure.

TABLE 10.8 Notch Sensitivity and Toughness of Extruded Alloys (SI Units)

Alloy	Temper	Temperature, °C	Tensile strength (MPa)			Notched charpy, joules	K_{Ic} MPa $cm^{1/2}$
			Unnotched	Notched[a]	Ratio		
AZ31B	F	25				4.6	280
AZ61A	F	25				6.0	300
AZ80A	F	25	315	235	0.75	1.8	290
		-195	420	170	0.40		
	T5	25	345	155	0.45	1.4	160
		-195	445	100	0.22		
	T6	25				1.4	
HM31A	T5	25	300	260	0.87		
		-78	355	265	0.75		
		-195	410	270	0.66		
ZK30A	F	25				5.4	460[b]
ZK60A	T5	25	355	340	0.96	4.6	345
		0				3.0	
		-78				3.0	
		-195	510	310	0.61		

[a]The notched specimen has a reduced cross section of 1.524 × 25.4 mm; a 60° V notch; a notched width of 17.78 mm; and a notch root radius of 0.0076 mm.

[b]Value is for J_{Ic} since specimen was too small for an accurate K_{Ic} value; the true K_{Ic} is lower.

Source: Refs. 15, 16, 17, 18, 22.

TABLE 10.8a Notch Sensitivity and Toughness of Extruded Alloys (English Units)

Alloy	Temper	Temperature, °C	Tensile strength (ksi)			Notched charpy, joules	K_{Ic} MPa $cm^{1/2}$
			Unnotched	Notched[a]	Ratio		
AZ31B	F	25				3.4	25.5
AZ61A	F	25				4.4	27.3
AZ80A	F	25	46	34	0.75	1.3	26.4
		-195	61	25	0.40		
	T5	25	50	22	0.45	1.4	14.75
		-195	65	14	0.22		
	T6	25				1.4	
HM31A	T5	25	44	38	0.87		
		-78	51	38	0.75		
		-195	59	39	0.66		
ZK30	F	25				4.0	41.8[b]
ZK60A	T5	25	51	49	0.96	3.4	31.4
		0				2.2	
		-78				2.2	
		-195	74	45	0.61		

[a]The notched specimen has a reduced section of 0.06 in × 1 in; a 60° V notch; a 0.700 in notched width; and a notch root radius of 0.0003 in.

[b]Value is for J_{Ic} since specimen was too small for an accurate K_{Ic} value; true value for K_{Ic} is lower.

Source: Refs. 15, 16, 17, 18, 22.

10.5 BEARING AND SHEAR STRENGTH

Bearing strength is the resistance to a stress applied by a pin in a hole. Bearing yield strength is defined here as the stress at which the hole elongation deviates 2% from the previously linear relation between hole diameter and stress. It is sensitive to the distance of the hole to the edge of the plate section, and therefore that distance (E) must be expressed as a ratio to the hole diameter (D) for any set of data. Bearing ultimate strength is that stress needed to break the pin out of the section.

Both bearing strengths and shear strengths are clearly related to the strengths of the material determined in a normal tensile test. One would expect the bearing yield strength to be related to the compressive yield strength, and the bearing ultimate and shear strengths to the tensile strength. The Dow Chemical Company measured the bearing and shear properties of extrusions made from AZ31B, AZ61A, AZ80A, HM31A, ZK21A, and ZK60A and compared the results to the compressive yield and the tensile strengths [15]. The correlations found are good enough to enable an estimation of bearing and shear strengths from tensile data, both at room and elevated temperatures.

The relations are linear; the following equations are such that all bearing or shear data for a given tensile property are higher than the value calculated using the equations. Thus the numbers obtained are conservative. Those calculated by using minima properties can be considered as allowables for design purposes.

$$\text{Bearing yield strength} = 1.14\ \text{CYS} + 98.4 \text{ for } E/D = 2.0\ (\text{MPa})$$
$$= 0.97\ \text{CYS} + 77.0 \text{ for } E/D = 1.5\ (\text{MPa})$$

$$\text{Bearing ultimate strength} = 1.54\ \text{TS} - 53.0 \text{ for } E/D = 2.0\ (\text{MPa})$$
$$= 1.20\ \text{TS} - 35.7 \text{ for } E/D = 1.5\ (\text{MPa})$$

where
E = distance to the edge
D = hole diameter

$$\text{Shear strength} = 0.65\ \text{TS at room temperature}$$
$$= 0.79\ \text{TS at } 95°\text{C}$$
$$= 1.0\ \text{TS at } 150°\text{C}$$

As an illustration, Table 10.9 was prepared using these relationships and the values in Tables 10.1 to Table 10.3. The assumption is made that the equations apply to all magnesium alloys.

TABLE 10.9 Bearing and Shear Strengths of Extrusions (SI Units (MPa); E/D = 2.0)

Alloy	Form	Bearing typical		Bearing minimum		
		Yield	Ultimate	Yield	Ultimate	Shear
AZ10A-F	Solid	185	315			155
	Tube	180	300			150
	Hollow	180	300			150
AZ31B-F	Solid	205	345	195	315	170
	Tube	195	330	180	285	165
	Hollow	195	330	180	285	165
AZ61A-F	Solid	245	425	205	370	200
	Tube	225	385	185	330	185
	Hollow	225	385	185	330	185
AZ80A-T5	Solid	370	530	320	325	220
HK31A-T5	Solid				415	200[a]
HM31A-T5	Solid	280	415	218	340	200
M1A-F	Solid	240	340			165
	Tube		315			155
ZC71A-F	Solid				315	155[a]
-T5					330	165[a]
-T6		260	495		450	210[a]
ZK10A-F	Solid		400			
	Tube		375			
ZK30A-F	Solid	340	420			
ZK60A-T5	Solid	385	510	330	425	235
	Tube	305	470	270	430	220
	Hollow	325	480	305	430	225
ZM21A-F	Solid				340	165[a]

[a]Minimum rather than typical.
Source: See text.

TABLE 10.9a Bearing and Shear Strengths of Extrusions (English Units (ksi); E/D = 2.0)

		Bearing typical		Bearing minimum		
Alloy	Form	Yield	Ultimate	Yield	Ultimate	Shear
AZ10A-F	Solid	27	46			22
	Tube	26	44			22
	Hollow	26	44			22
AZ31B-F	Solid	30	50	28	46	25
	Tube	28	48	26	41	24
	Hollow	28	48	26	41	24
AZ61A-F	Solid	36	62	30	54	29
	Tube	33	56	27	48	27
	Hollow	33	56	27	48	27
AZ80A-T5	Solid	54	77	46	47	32
HK31A-T5	Solid				60	29[a]
HM31A-T5	Solid	41	60	32	49	29
M1A-F	Solid	35	49			24
	Tube		46			22
ZC71A-F	Solid				46	22[a]
-T5					48	24[a]
-T6		38	72		65	30[a]
ZK10A-F	Solid		58			
	Tube		54			
ZK03A-F	Solid	49	61			
ZK60A-T5	Solid	56	74	48	62	34
	Tube	44	68	39	62	32
	Hollow	47	70	44	62	33
ZM21A-F	Solid				49	24[a]

[a]Minimum rather than typical.
Source: See text.

10.6 COLUMNS

A column is a compression member whose length is long in comparison with its transverse dimensions, which tends to fail by buckling rather than by direct compression. If the stress at which buckling commences is below the proportional limit, the column is considered long, and the buckling stress follows Euler's formula [19]:

$$P/A = \pi^2 E/(l/k)^2$$

where
- P = load
- A = area of cross section
- E = modulus of elasticity
- l = column effective pin-ended length
- k = least radius of gyration

All materials follow this formula for long columns. A short column is one for which the critical compressive stress at which buckling starts is higher than the proportional limit. In this case, the Euler formula will not apply and it is necessary to develop empirical relations for all alloys. Schuette [20] has shown that hyperbolic formulae fit data for magnesium alloys in the short column range, and proposed the following:

$$P/A = (180\ F_{cy}^{1/2})/(l/k) \qquad (\text{Max } P/A = 0.90F_{cy}) \text{ for M1A alloy}$$

$$P/A = (2900\ F_{cy}^{1/4})/(l/k)^{1.5} \qquad (\text{Max } P/A = F_{cy}) \text{ for AZ31B, AZ61A and AZ80A-F}$$

$$P/A = (3300\ F_{cy}^{1/4})/(l/k)^{1.5} \qquad (\text{Max } P/A = 0.96\ F_{cy}) \text{ for AZ80A-T5 and ZK60A}$$

where
- P = load
- A = area of column cross section
- F_{cy} = compressive yield strength
- l = column effective pin-ended length
- k = minimum radius of gyration of column cross section

Figure 10.28 and 10.29 are curves obtained by using the above equations [5].

When columns with cross sections of elements relatively thin in comparison to their widths are loaded, these elements may buckle individually before a load is reached that would cause the column to buckle as a whole. The failure that occurs at a stress higher than that required to initiate local failure is known as crippling. Local buckling, or crippling, may be determined by calculating the crippling loads for each individual plate element and then summing these loads. This sum divided by the combined area of the individual elements gives the average crippling stress for the entire section. Figures 10.30 through 10.36 give data for alloys on local buckling, and Figure 10.37 an example of how the curves may be used [5].

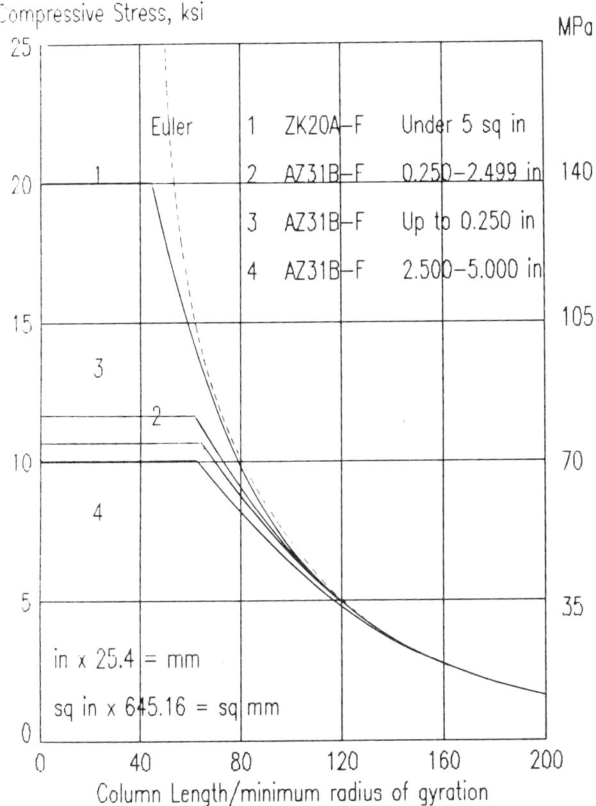

FIGURE 10.28 Column curves for AZ31B-F and ZK21A-F extrusions at minimum property levels. L' = column effective pin-ended length; r = minimum radius of gyration of column cross section; ksi × 6.8948 = MPa.

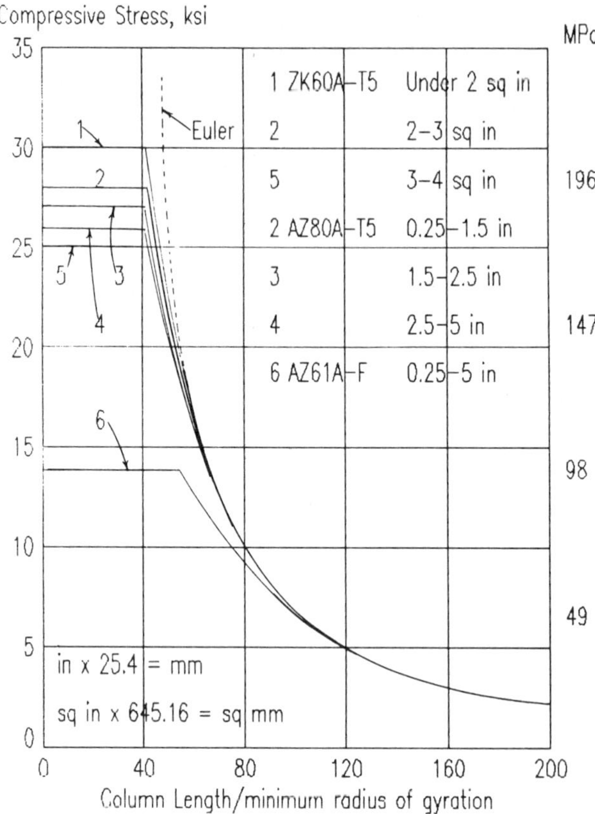

FIGURE 10.29 Column curves for AZ61A-F, AZ80A-T5, and ZK60A-T5 extrusions at minimum property levels. ksi × 6.8948 = MPa.

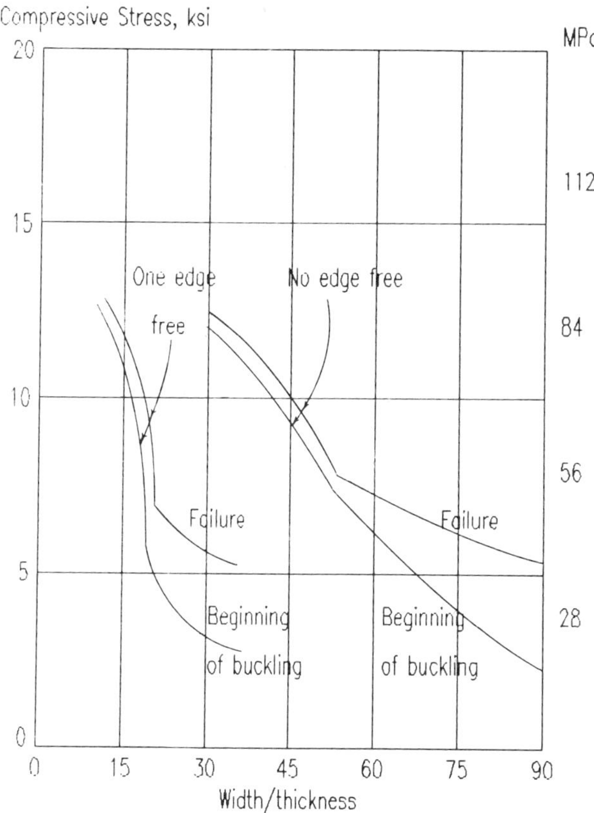

FIGURE 10.30 Crippling curves for AZ31B-F extrusions at minimum property levels; ksi × 6.8948 = MPa.

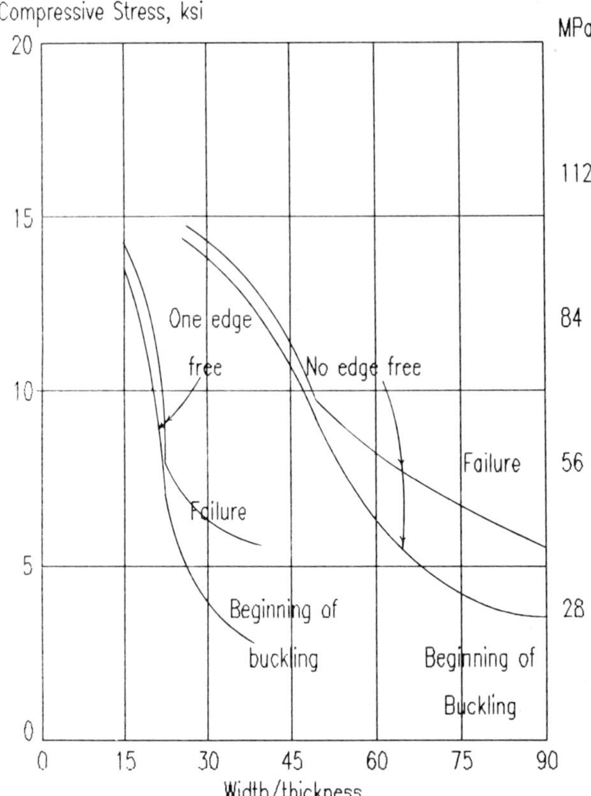

FIGURE 10.31 Crippling curves for AZ61A-F extrusions at minimum property levels; ksi × 6.8948 = MPa.

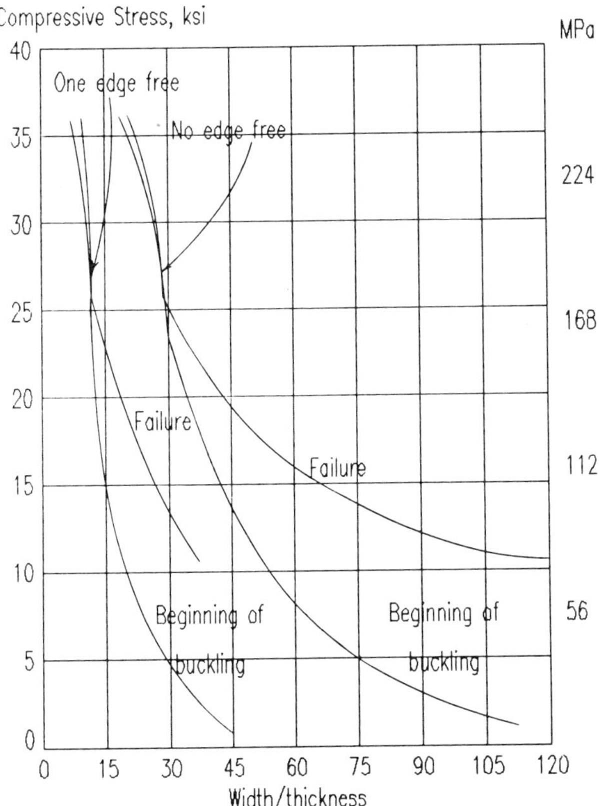

FIGURE 10.32 Crippling curves for AZ80A-T5 extrusions; ksi × 6.8948 = MPa.

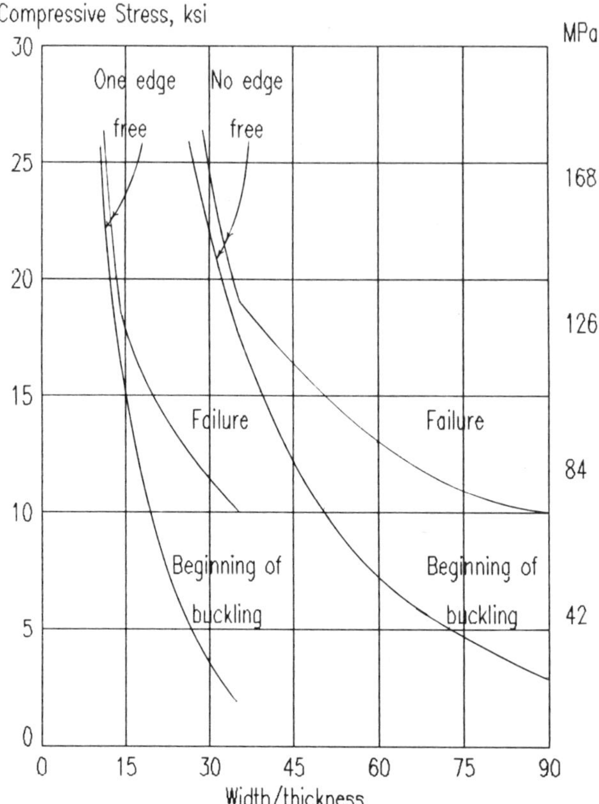

FIGURE 10.33 Crippling curves for HM31A-T5 extrusions at 20°C, ksi × 6.8948 = MPa.

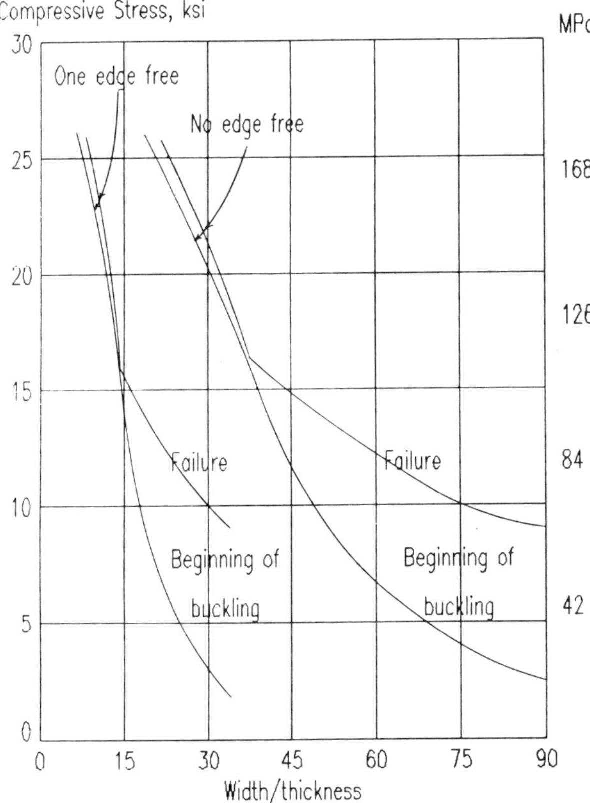

FIGURE 10.34 Crippling curves for HM31A-T5 extrusions at 150°C; ksi × 6.8948 = MPa.

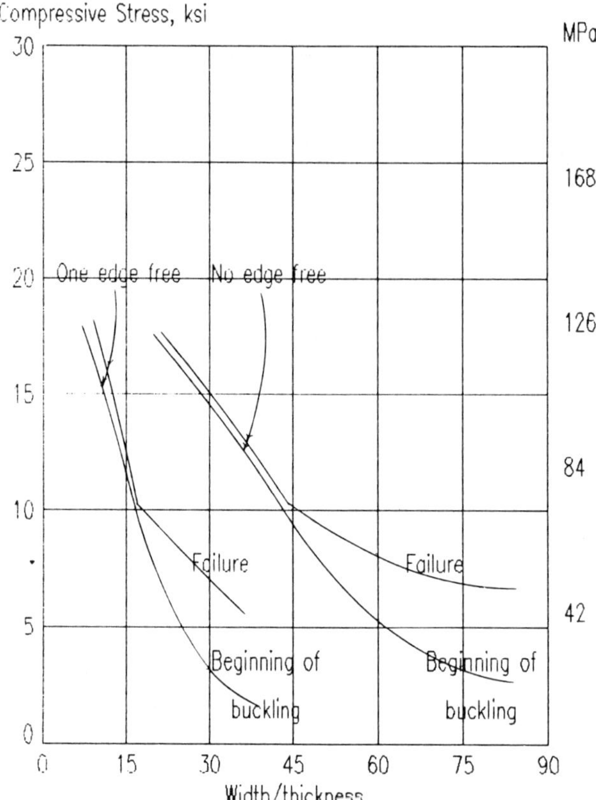

FIGURE 10.35 Crippling curves for HM31A-T5 extrusions at 315°C; ksi × 6.8948 = MPa.

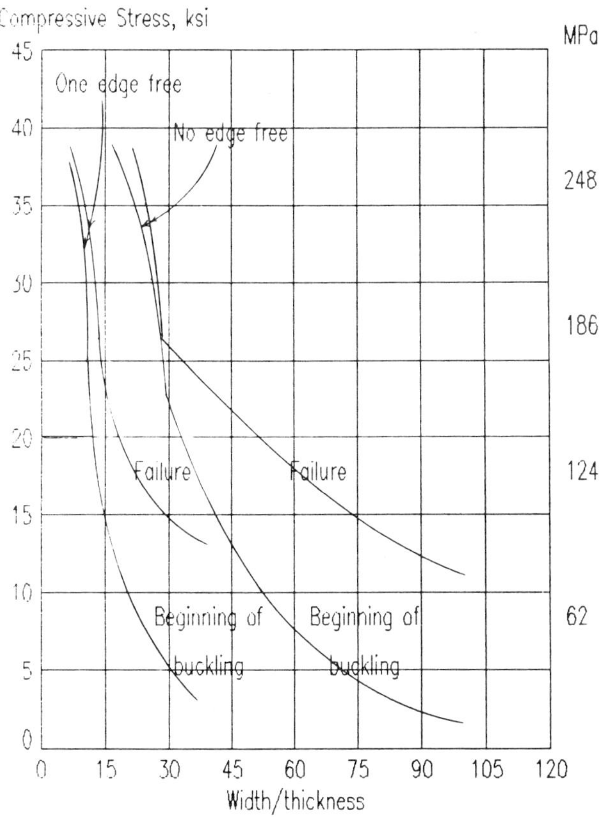

FIGURE 10.36 Crippling curves for ZK60A-T5 extrusions at minimum property levels; ksi × 6.8948 = MPa.

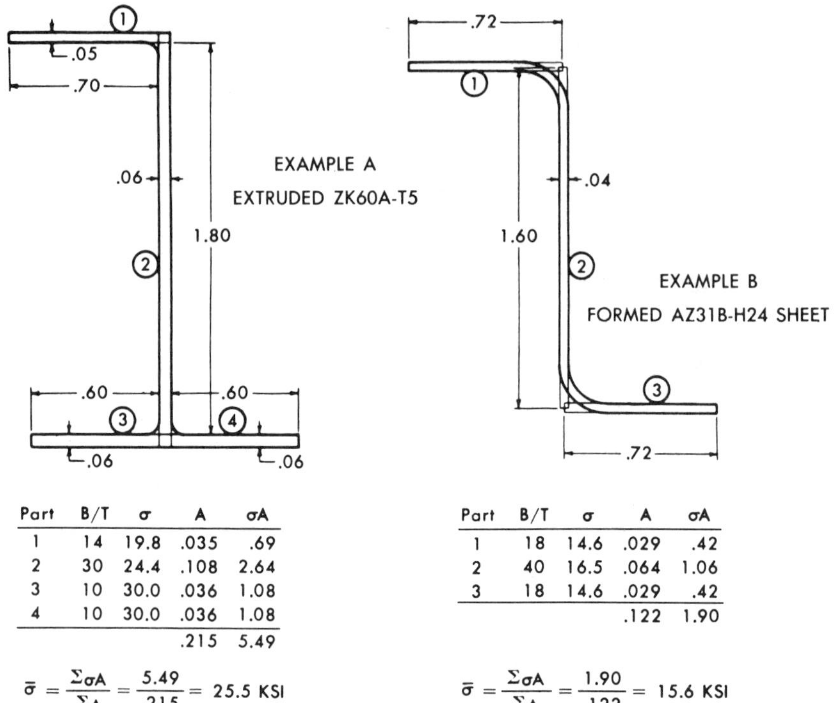

Part	B/T	σ	A	σA
1	14	19.8	.035	.69
2	30	24.4	.108	2.64
3	10	30.0	.036	1.08
4	10	30.0	.036	1.08
			.215	5.49

$$\bar{\sigma} = \frac{\Sigma \sigma A}{\Sigma A} = \frac{5.49}{.215} = 25.5 \text{ KSI}$$

Part	B/T	σ	A	σA
1	18	14.6	.029	.42
2	40	16.5	.064	1.06
3	18	14.6	.029	.42
			.122	1.90

$$\bar{\sigma} = \frac{\Sigma \sigma A}{\Sigma A} = \frac{1.90}{.122} = 15.6 \text{ KSI}$$

FIGURE 10.37 Examples of crippling strength calculation for extruded sections; ksi × 6.8948 = MPa.

10.7 FATIGUE

Information on the fatigue properties of many extrusion alloys with a variety of surface conditions are given in Table 10.10. As a rough rule of thumb, the standard deviation of fatigue data is 14 MPa (2 ksi). All of the information in Table 10.10 were obtained on specimens tested while exposed to normal atmospheric conditions.

Corrosion fatigue is the simultaneous action of corrosion and alternating stress on an alloy, leading to a lower fatigue limit than is the case with fatigue alone. Table 10.11 shows some data obtained by the Dow Chemical Company [15]. The lowering of the fatigue limit in the presence of water condensed on the speciman is substantial. Note that the samples in the desiccator have a slightly higher limit than the polished samples in Table 10.10, which were tested in a normal laboratory atmosphere. It is clear that the principal active agent lowering the limit is the presence of acidic water. Condensed water will absorb CO_2 from the atmosphere to result in carbonic acid; absorbed SO_2 results in sulfurous acid. The acid in a large body of water such as that used for the full immersion will be less concentrated than in small condensed droplets. While acidified water is the most effective in lowering the fatigue limit, even non-acidified water (argon + water) has an effect. Coating the magnesium so that no condensed water comes in contact with the metal completely eliminates the effect [15]. The effect of condensed water on the fatigue properties of bare magnesium should be considered in any design, with the preferable solution being to assure the metal is free from condensed water during service.

TABLE 10.10 Fatigue Strengths of Extrusions (SI Units)

Alloy	Type of test	R value	Surface or stress conc.	Fatigue limit (MPa) at cycles				
				10^5	10^6	10^7	5×10^7	10^8
AZ31B-F	Rotating beam	-1	Polished	180	160	145		130
	Plate bending	-1	Etched	105	95	90		85
	Axial load	0.25	Polished	175	160	150		
AZ61A-F	Rotating beam	-1	Polished	185	170	155		140
			2.0				130	
	Plate bending	-1	Extruded	125	95	85		
	Axial load	0.25	Polished	170	140	130		
AZ80A-F	Rotating beam	-1	Polished	190	175	160		150
	Plate bending	0.25	Extruded	140	100	90		
HM31A-T5	Rotating beam	-1	Polished	125	105	90		95
			2.0	90	55	45		40
M1A-F	Rotating beam	-1	Polished	107	88	85	83	
			1.8	76	54	50	48	

ZH11A-F	Rotating beam	-1	Polished	100	83	76	74	
			1.8	73	51	48	46	
ZK30A-F	Rotating beam	-1	Polished	152	131	124	124	
			2.0	124	90	90	90	
			3.0	90	62	48	48	
ZK60A-T5	Rotating beam	-1	Polished	210	170	150		140
			Anodized	210	170	150		140
	Plate bending	0	Machined	170	135	135		135
	Axial load	0	Machined	285	250	235		
			Extruded	215	180	160		
			1.8	155	125	110		
	Axial load	0.25	Polished	310	295	290		
			Machined	315	290	270		
			Extruded	240	200	185		
			1.8	180	140	130		

Source: Refs. 5, 17, 15, 23, 24.

TABLE 10.10a Fatigue Strengths of Extrusions (English Units)

Alloy	Type of test	R value	Surface or stress conc.	Fatigue limit (ksi) at cycles				
				10^5	10^6	10^7	5×10^7	10^8
AZ31B-F	Rotating beam	-1	Polished	26	23	21		19
	Plate bending	-1	Etched	15	14	13		12
	Axial load	0.25	Polished	25	23	22		
AZ61A-F	Rotating beam	-1	Polished	27	25	22		20
			2.0				19	
	Plate bending	-1	Extruded	18	14	12		
	Axial load	0.25	Polished	25	20	19		
AZ80A-F	Rotating beam	-1	Polished	28	25	23	22	
	Plate bending	0.25	Extruded	20	14	13		
HM31A-T5	Rotating beam	-1	Polished	18	15	13		
			2.0	13	8	7		6
M1A-F	Rotating beam	-1	Polished	16	13	12	12	
			1.8	11	8	7	7	

ZH11A-F	Rotating beam	-1	Polished	15	12	11	11	
			1.8	11	7	7	7	
ZK30A-F	Rotating beam	-1	Polished	22	19	18	18	
			2.0	18	13	13	13	
			3.0	13	9	7	7	
ZK60A-T5	Rotating beam	-1	Polished	30	25	22		20
			Anodized	30	25	22		20
	Plate bending	0	Machined	25	20	20		20
	Axial load	0	Machined	41	36	34		
			Extruded	31	26	23		
			1.8	22	18	16		
	Axial load	0.25	Polished	45	43	42		
			Machined	46	42	39		
			Extruded	35	29	27		
			1.8	26	20	19		

Source: Refs. 5, 17, 15, 23, 24.

TABLE 10.11 Effect of Corrosion on Fatigue Properties of Extrusions (Axial Load; R = 0.25)

Alloy	Atomsphere of test	Fatigue limit at cycles, MPa, 10^5	MPa, 10^6	MPa, 10^7	ksi, 10^5	ksi, 10^6	ksi, 10^7
Fatigue limits							
AZ31B-F	Desiccator	195	170	165	28	25	24
	Water immersion		145	145		21	21
	Condensed water in air	130	90	85	19	13	12
ZK60A-T5	Desiccator	260	255	250	38	37	36
	Condensed water in air	160	85	80	23	12	12

Life of AZ31B-F at a constant stress of 140 MPa (20 ksi)

Atmosphere of test	Cycles to failure
Desiccator	$>10^8$
Condensed water in:	
air	6×10^4
oxygen	3.5×10^5
nitrogen	6×10^5
argon	2×10^6
argon + CO_2	10^5
argon + SO_2	10^5
argon + ammonia	5×10^5
air + ammonia	5×10^5
air + ammonia + SO_2	10^5

Source: Ref. 15.

10.8 STRESS CORROSION

Stress-corrosion data for some extrusion alloys are given in Figures 10.38 through 10.42 [15]. These data were obtained by exposing test bars to a rural atmosphere while under constant tension, a test that simulates service expectations better than short-time tests.

The alloys containing aluminum should not be designed so that stresses close to the yield point must be sustained for long periods. Residual stresses from operations such as welding or machining must be relieved by a stress-relief heat treatment. Recommended schedules are given in Chapter 4.

The two alloys not containing aluminum, HM31A and ZK60A, both show superior performance to the aluminum-containing series. Each sustained a constant tension stress between 100 and 140 MPa (15 and 20 ksi) for about seven years without failure. These two are good representatives of all nonaluminum containing alloys, showing there is in practice no reason to have concern about stress corrosion during service with these alloys.

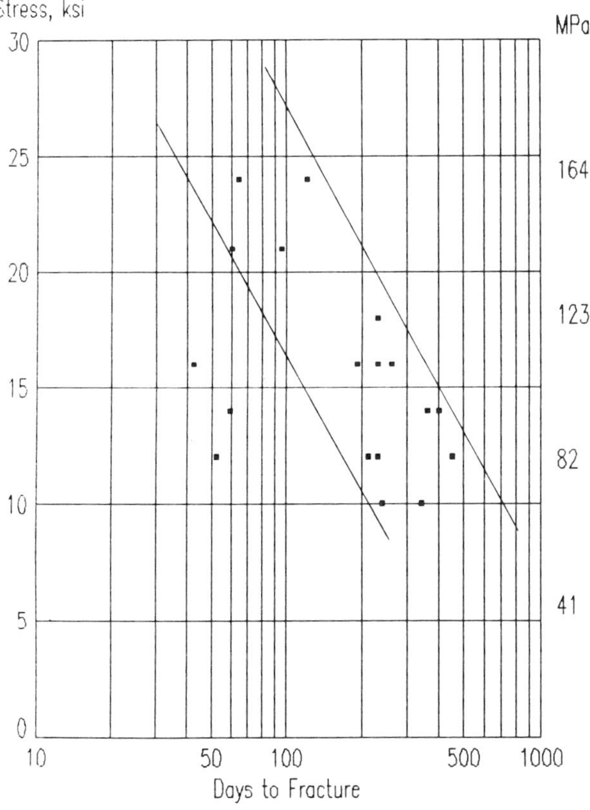

FIGURE 10.38 Stress corrosion of AZ31B-F extrusion at rural exposure under constant tension.

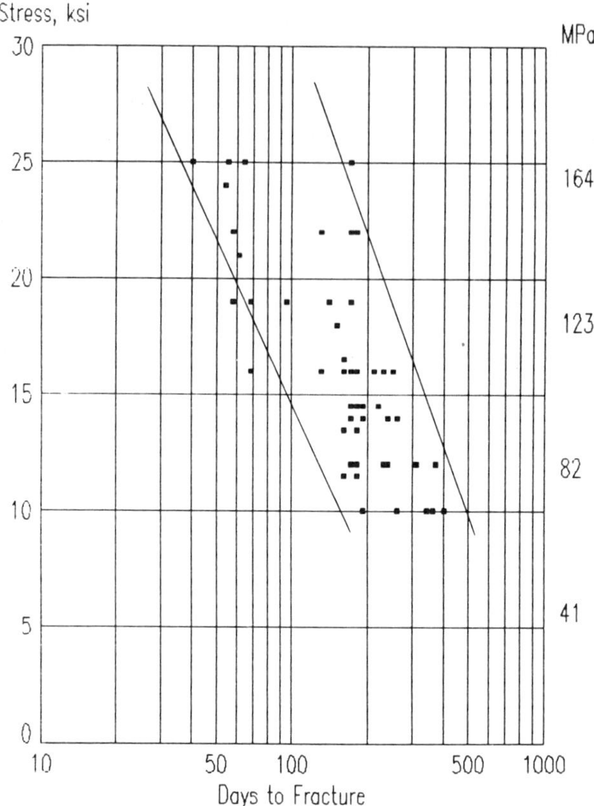

FIGURE 10.39 Stress corrosion of AZ61A-F extrusion at rural exposure under constant tension.

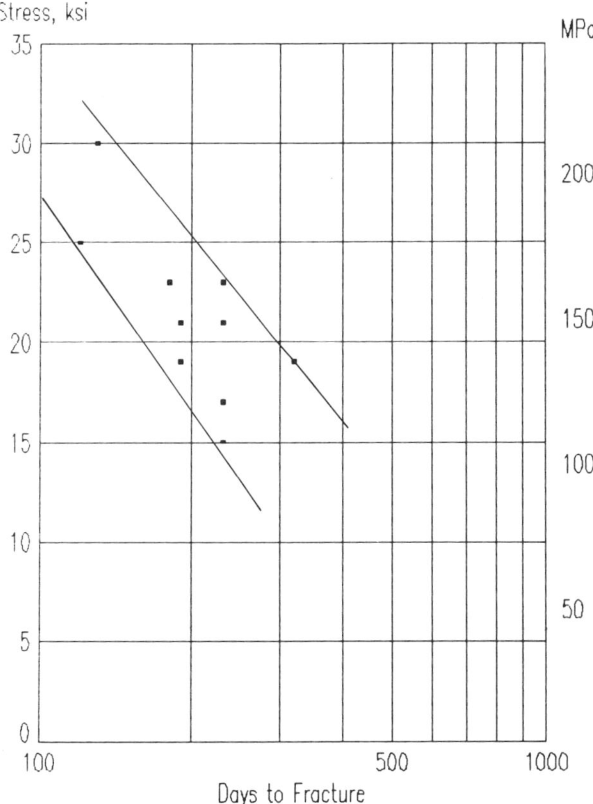

FIGURE 10.40 Stress corrosion of AZ80A-T5 extrusion at rural exposure under constant tension.

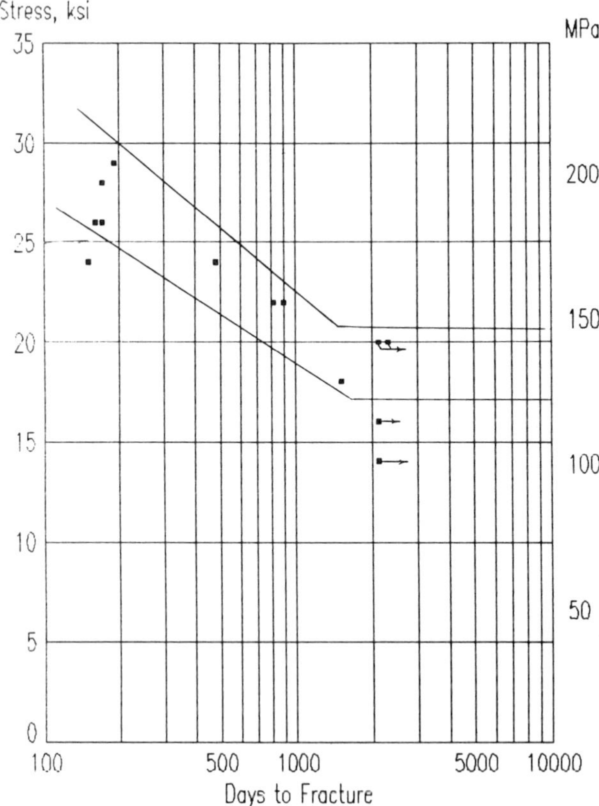

FIGURE 10.41 Stress corrosion of HM31A-T5 extrusion at rural exposure under constant tension.

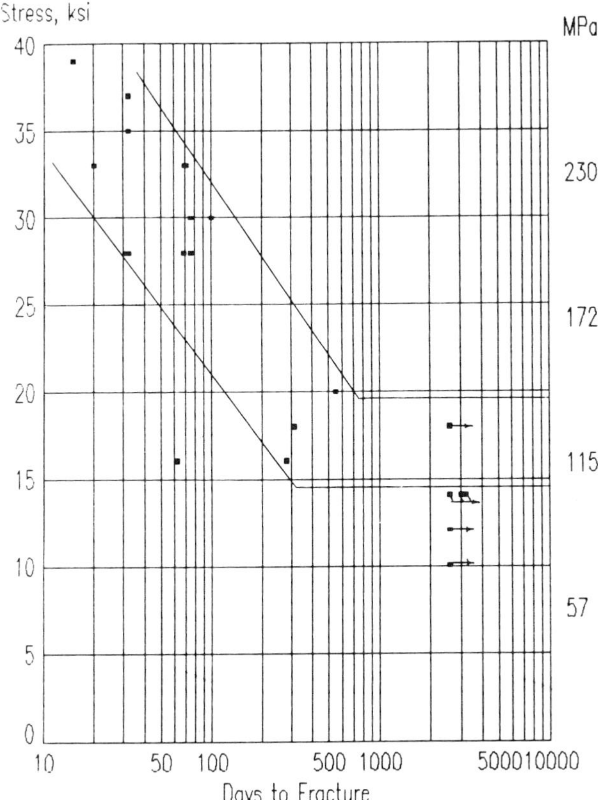

FIGURE 10.42 Stress corrosion of ZK60A-T5 extrusion at rural exposure under constant tension.

10.9 DAMPING

Information on the damping capacity of two extrusion alloys is given in Table 10.12.

TABLE 10.12 Specific Damping Capacity (%) of Two Extrusion Alloys (Cantilever Beam Tests)

	Bending stress, MPa				
Alloy	7.0	13.8	20	25	35
M1A-F	0.35	1.28	2.22	3.14	3.92
AZ31B-F	1.04	1.57	2.04	2.38	2.72

Source: Ref. 21.

REFERENCES

1. Dow Chemical Company, *Magnesium Mill Products*, 1965.
2. American Society for Metals, "Properties of Magnesium Alloys," *Metals Handbook*, Vol. 2, 9th Edition, American Society for Metals, Metals Park, Ohio, 1979.
3. Metals Progress, March 1972.
4. Dow Chemical Company, "Low Temperature Properties of Wrought Magnesium Alloys," Technical Service and Development Letter Enclosure, March 23, 1962.
5. Dow Chemical Company, "Magnesium in Design," Bulletin 141-213, 1967.
6. ASTM Specification B107, "Standard Specification for Magnesium-Alloy Extruded Bars, Rods, Shapes, Tubes, and Wire," *Annual Book of ASTM Standards*, Vol. 02.02, Sec. 2, 1984.
7. Dow Chemical Company, "HM31A Magnesium Alloy Extrusions," TS&D Letter Enclosure, October 28, 1965.
8. Dow Chemical Company, "ZK60A Magnesium Alloy Extrusions," TS&D Letter Enclosure, November 29, 1963.
9. Raymond W. Fenn, Jr. and James A. Gusack, "Effect of Strain-Rate and Temperature on the Strength of Magnesium Alloys," *ASTM Proceedings 58*:685 (1958).
10. D. S. Wolford, "Significance of the Secant and Tangent Modulus of Elasticity in Structural Design," *J. Aeronautical Sciences, 10*(6):160 (1943).
11. E. Z. Stowell, "Unified Theory of Plastic Buckling of Columns and Plates," NACA Technical Note #1556.

12. E. H. Schuette, "Column Curves for Magnesium-Alloy Sheet," *J. Aeronautical Sciences, 16*(5):301 (1949).
13. E. H. Schuette, "Buckling of Curved Sheet in Compression and its Relation to the Secant Modulus," *J. Aeronautical Sciences, 15*(1):18 (1949).
14. E. H. Schuette and J. C. McDonald, "Prediction and Reduction to Minimum Properties of Plate Compressive Curves," *J. Aeronautical Sciences, 15*(1):23 (1948).
15. Dow Chemical Company, Private communication.
16. Dow Chemical Company, "ZK60A Magnesium Alloy Extrusions," TS&D Letter Enclosure, November 29, 1963.
17. Magnesium Elektron Ltd., "Mechanical Properties and Chemical Compositions of Wrought Magnesium Alloys," Bulletin 441, March 1983.
18. J. P. Doan and J. C. McDonald, "The Notch Sensitivity in Static and Imapct Loading of Some Magnesium-Base and Aluminum-Base Alloys," *ASTM, 46* (1946).
19. Mott Sauders, *The Engineer's Companion*, p. 129, John Wiley & Sons, Inc., New York, 1966.
20. E. H. Schuette, "Hyperbolic Column Formulas for Magnesium-Alloy Extrusions," *J. Aeronautical Sciences, 15*:523 (1948).
21. Dow Chemical Company, "Damping Characteristics of Some Magnesium Alloys," TS&D Letter Enclosure, August 1, 1957.
22. Magnesium Elektron Ltd., "Elektron Magnesium Alloy," Bulletin 400, September 1983.
23. Colin J. Smithells, *Metal Reference Book*, Butterworths, London and Boston, 1976.
24. Magnesium Elektron Ltd., Private communication.

11

Mechanical Properties of Sheet and Plate

Magnesium alloys can be rolled in the form of sheet or plates on standard equipment. The most generally-used alloy is AZ31B which has a good combination of strength, ductility, and corrosion resistance. Since its combination of tensile properties and stress corrosion tendency is such that stress relief is required after welding, ZE10 is used where avoidance of stress relief after welding is required. ZM21 has seen some limited use as a sheet alloy. Three thorium-containing alloys, ZH11A, HK31A, and HM21A, have been developed for use at high-temperature. ZK10A and ZK30A are the two Zn-Zr type of alloy that are used as rolled products.

11.1 TENSILE PROPERTIES

The tensile properties of rolled magnesium alloys depend upon the chemical composition, the grain size, and the amount of strain hardening present. The smaller the grain size and the greater the amount of strain hardening, the higher are the tensile strengths. Although alloy HM21A is strongest in the T8 temper, in general there is little benefit derivable from heat treating of flat products. Since both grain size and strain hardening change upon exposure to elevated temperature, the effect of such exposure is important and must be known. Temperature exposure during fabrication must be controlled to maintain the desired property level.

Defects that affect properties include surface contamination, inclusions and surface scratches, all of which affect ductility and fatigue more than other properties. The seriousness of a defect depends upon its size, location, and orientation.

Table 11.1 gives typical and minimum property values for magnesium rolled products at a number of thicknesses for each alloy. Since the degree of work hardening is important to the development of properties, the strengths in general are higher at the thinner gauges. Tooling plate is made of an alloy similar to AZ31B and is produced to very close flatness tolerances in order to make it suitable for the construction of tooling and jigs. Tooling plate has outstanding dimensional stability over time.

11.1.2 The Effect of Temperature

Tensile values over a range of temperatures for the alloys are given in Table 11.2. In general, strength decreases and ductility increases as the temperature is raised. There is no brittle point, such as is common for many steel alloys, even at very low temperatures. The data for Table 11.2 were obtained at normal testing speeds.

Stress-strain curves are shown for both room and elevated temperatures in Figures 11.1 through 11.18 [5,10,19]. These can be used to determine a yield strength at any offset from the modulus line, to estimate the proportional limit, to estimate the elastic modulus at high temperatures, and to determine tangent and secant moduli. The latter are shown for the more common alloys in Figures 11.30 through 11.35.

The effect of time and temperature of exposure to elevated temperatures on properties is shown in Table 11.3. The fact that the ductility increases with all alloys with increasing exposure is an indication that the principal effect is removal of strain hardening. An increasing grain size would be accompanied by a decrease in ductility. These data can be used to design fabricating procedures more critically than by using the recommendations in Chapter 5 as well as for designing for applications which involve long exposure to elevated temperatures during service.

11.1.3 The Effect of Strain Rate

TABLE 11.4 gives information on the effect of rapid strain rates on the tensile properties of rolled products. The higher the rate of strain, the higher are the strengths, with the effect becoming more pronounced at higher temperatures.

As the strain rate becomes slower than normal, the strengths decrease, with the effect again being more pronounced as the temperature becomes higher. Figures 11.19 through 11.29 are isochronous stress-strain curves showing the effect of testing times up to 100 hours at various temperatures [5,6].

Creep strengths for a number of temperatures and times are shown in Table 11.5. The superiority of the thorium-containing alloys is apparent.

TABLE 11.1 Tensile Properties of Sheet and Plate at Room Temperature (SI Units)

			Typical				Minimum			
				MPa				MPa		
Alloy	Temper	Thickness, mm	%E	TYS	CYS	TS	%E	TYS	CYS	TS
AZ31B	0	0.41-1.52	21	150	110	255	12	125	85	220
		1.53-1.72	21	150	110	255	12	105	85	220
		1.73-12.7	21	150	90	250	12	105	70	220
		12.8-50.8	17	150	85	250	10	105	70	220
		50.8-76.2	17	145	75	250	9	105	55	220
	H24	0.41-6.32	15	220	180	290	6	200	165	270
		6.33-9.50	17	200	160	275	8	180	140	260
		9.51-12.7	19	185	130	270	8	165	110	255
		12.8-25.4	17	165	110	260	8	150	90	250
		25.5-50.8	14	160	95	255	8	140	70	235
		50.9-76.2	16	145	85	255	8	125	60	235
	H26	6.35-9.50	16	205	165	275	6	185	150	270
		9.51-11.1	13	195	150	275	6	180	145	260
		11.2-12.7	13	195	150	275	6	180	125	260
		12.8-19.1	10	195	130	275	6	170	117	255
		19.2-25.4	10	180	125	270	6	160	110	255
		25.5-38.1	10	170	110	260	6	150	105	240
		38.2-50.8	10	170	105	260	6	145	95	240
Tooling plate		6.35-150	10	130	70	240				

TABLE 11.1 (*Continued*)

			Typical				Minimum			
				MPa				MPa		
Alloy	Temper	Thickness, mm	%E	TYS	CYS	TS	%E	TYS	CYS	TS
HK31A	0	0.41-6.35	23	140	95	230	12	125	85	205
		6.36-12.7	20	130	90	230	13	110	70	205
		12.8-25.4	17	125	90	230	13	105	70	205
		25.5-76.2	17	115	90	220	13	95	70	200
	H24	0.41-3.18	9	205	160	260	4	180	140	235
		3.19-6.35	10	205	160	255	4	165	150	235
		6.36-25.4	14	215	170	270	4	170	140	235
		25.5-76.2	10	200	150	255	4	170	115	230
HM21	T8	0.41-6.35	11	160	130	250	6	125	105	230
		6.36-12.7	11	185	160	255	6	145	140	220
		12.8-25.4	12	180	145	240	6	145	115	205
		25.5-50.8	11	170	140	230	6	145	105	205
		50.9-76.2	10	170	125	230	6	145	95	205
	T81	3.18-2.15					4	170	150	235

M1A			6	100		232			
ZE10A	0	0.41-1.52	23	160	110	230	15	125		205
		1.52.6.35	23	140	110	230	15	125		205
		6.36-12.7	18	110	105	215	12	85		200
	H24	0.41-0.86	12	200	180	260	6	170	150	250
		0.87-4.78	12	180	170	255	6	150	140	235
		4.79-12.7	8	165	160	235	6	140	140	215
ZH11A	0		6	125		235				
	H24		12	170		260				
ZK10A			10	178	154	263				
ZK30A			8	185	154	270				
ZM21A	0		13	131		232				
	H24						8	180		275

Source: 1, 2, 3, 4, 18, 20.

TABLE 11.1a Tensile Properties of Sheet and Plate at Room Temperature (English Units)

			Typical				Minimum			
				ksi				ksi		
Alloy	Temper	Thickness, in	%E	TYS	CYS	TS	%E	TYS	CYS	TS
AZ31B	0	0.016-0.060	21	22	16	37	12	18	12	12
		0.061-0.249	21	22	16	37	12	15	12	32
		0.250-0.500	21	22	13	36	12	15	10	32
		0.501-2.000	17	22	12	36	10	15	10	32
		2.001-3.000	17	21	11	36	9	15	8	32
	H24	0.016-0.249	15	32	26	42	6	29	24	39
		0.250-0.374	17	29	23	40	8	26	20	38
		0.375-0.500	19	27	19	39	8	24	16	37
		0.501-1.000	17	24	16	38	8	22	13	36
		1.001-2.000	14	23	14	37	8	20	10	34
		2.001-3.000	16	21	12	37	8	18	9	34
	H26	0.250-0.374	16	30	24	40	6	27	22	39
		0.375-0.438	13	28	22	40	6	26	21	38
		0.439-0.500	13	28	22	40	6	26	18	38
		0.501-0.750	10	28	19	40	6	25	17	40
		0.751-1.000	10	26	18	39	6	23	16	37
		1.001-1.500	10	25	16	38	6	22	15	35
		1.501-2.000	10	24	15	38	6	21	14	35
Tooling plate		0.250-5.90	10	19	10	35				
HK31A	0	0.016-0.250	23	20	14	33	12	18	12	30
		0.251-0.500	20	19	13	33	12	16	10	30

		0.501-1.000	17	18	13	33	12	15	10	30
		1.001-3.000	17	17	13	32	12	14	10	29
	H24	0.016-0.125	9	30	23	38	4	26	20	34
		0.126-0.250	10	28	23	37	4	24	22	34
		0.251-1.000	14	34	25	39	4	25	20	34
		1.001-3.000	10	29	22	37	4	25	17	33
HM21A	T8	0.016-0.250	11	23	19	35	6	18	15	33
		0.251-0.500	12	27	23	37	6	21	20	32
		0.501-1.000	12	26	21	30	6	21	17	30
		1.001-2.000	10	25	20	33	6	21	15	25
		2.001-3.000	10	25	18	33	6	21	14	30
	T81	0.125-0.312					4	25	22	34
ZE10A	0	0.016-0.060	23	23	16	33	15	18		30
		0.061-0.250	23	20	16	33	15	18		30
		0.251-0.500	18	16	15	31	12	12		29
	H14	0.016-0.034	12	29	26	38	6	25	22	36
		0.035-0.188	12	26	25	37	6	22	20	34
		0.189-0.500	8	24	23	34	6	20	20	31
M1A			6	15		34				
ZH11A	0		6	18		34				
	H24		12	25		38				
ZK10A			10	26	22	38				
ZK30A			8	27	22	39				
ZM21A	0		13	19	34					
	H24						8	26		40

Source: Refs. 1, 2, 3, 4, 18, 20.

TABLE 11.2 Effect of Temperature on the Tensile Properties of Sheet and Plate (SI Units)

			Longitudinal				Transverse			
				MPa				MPa		
Alloy	Temper	Temperature, °C	%E	TYS	CYS	TS	%E	TYS	CYS	TS
AZ31B sheet	0	-253	7	205		435				
		-197	9	185		370				
		-77	22	175		290				
		23	26	145		255				
	H24	-196	4	235		400				
		-78	6	210		320				
		21	14	220	165	285				
		100	30	145	135	205				
		150	58	90	110	150				
		204	82	55	75	90				
		260	92	35	50	55				
		315	136	15		40				
		370	140	15		30				
HK31A sheet	0	-251	9	195		380	10	195		380
		-196	10	185		360	11	185		360
		-78	14	165		295	15	165		295
		-18	18	150		260	19	150		260
		20	23	140	95	230	24	140	105	230
		95	40	125	95	170	40	125	95	170
		120	46	110	95	150	46	110	95	150
		150	50	105	90	140	50	105	90	140

		200	50	90	90	110	50	90	90	110
		260	34	75	75	105	34	75	75	105
		315	60	50	50	85	60	50	50	85
		370	>100	15	15	40	>100	15	15	40
	H24	-251	6	240	170	400	9	250	180	405
		-196	6	240	170	370	9	250	180	380
		-78	7	230	165	315	11	235	170	325
		-18	8	220	165	285	13	230	170	290
		20	11	205	160	260	15	215	165	270
		95	16	180	160	205	20	185	165	215
		120	18	165	150	185	22	170	150	195
		150	20	160	150	180	23	165	160	185
		200	21	145	145	165	23	150	150	170
		260	19	115	115	140	19	125	125	145
		315	70	50	50	85	70	55	55	90
		345	>100	20	20	55	>100	20	20	60
		370	>100	15	15	40	>100	15	15	50
HK31A plate	H24	24	12	200	150	255				
		150	30	150		165				
		200	28	150		150				
		260	22	115		130				
		315	36	75		110				
HM21A sheet	T8	-250	6	215		365	12	185		370
		-196	6	215		345	12	185		350
		-78	7	205		295	13	180		305
		-18	9	200		270	15	170		275
		20	11	195	145	250	17	165	140	240
		95	14	170	145	195	22	150	140	285

TABLE 11.2 (*Continued*)

Alloy	Temper	Temperature, °C	Longitudinal				Transverse			
				MPa				MPa		
			%E	TYS	CYS	TS	%E	TYS	CYS	TS
		120	17	160	145	170	25	140	140	165
		150	20	145	145	160	30	130	140	150
		200	30	125	140	130	32	115	130	125
		260	25	110	115	115	22	105	110	110
		315	15	90	95	205	15	90	95	95
		370	50	55	60	75	50	55	60	75
		425	100	20	30	35	100	20	30	35
		480	>100	5	5	15	>100	5	5	15
plate		24	11	205	165	255	14	195	170	270
		95	16	185	165	200	19	170	170	215
		120	22	170	165	180	25	160	160	195
		150	30	160	160	160	32	145	150	170
		200	36	130	140	140	36	125	130	150
		260	33	115	125	125	33	110	115	130
		315	25	95	105	110	25	90	95	115
		370	40	55	60	85	40	55	60	85
		425	50	20	30	40	50	20	30	40
		480	>100	5	5	15	>100	5	5	15

ZE10A sheet	0	20	24	175	130	240	26	180	125	240
		120	65	125		150	55	110		140
		200	60	35		85	60	40		85
	H24	20	17	190	175	260	24	215	185	275
		120	59	100		150	55	110		160
		200	64	35		70	60	40		70
							Modulus (GPa)			
ZH11A sheet		20	10	181		266	45			
		100	10	179		224	41			
		150	11	176		201	41			
		200	15	165		171	40			
		250	20	124		134	40			
		300	27	73		96	34			
		350	38	17		56	29			
ZK30A sheet		20	10	195		270	45			
		100	33	120		165	40			
		150	42	74		116	33			
		200	51			76				
		250	59			49				

Source: Refs. 1, 5, 6, 7, 8, 9, 18, 20.

TABLE 11.2a Effect of Temperature on the Tensile Properties of Sheet and Plate (English Units)

			Longitudinal				Transverse			
				ksi				ksi		
Alloy	Temper	Temperature, °C	%E	TYS	CYS	TS	%E	TYS	CYS	TS
AZ31B sheet	0	-253	7	30		63				
		-197	9	27		54				
		-77	22	25		42				
		23	26	21		37				
	H24	-196	4	34		58				
		-78	6	30		46				
		21	14	32	23	41				
		100	30	21	20	30				
		150	58	13	16	22				
		204	82	8	11	13				
		260	92	5	7	8				
		315	136	2		6				
		370	140	2		4				
HK31A sheet	0	-251	9	28		55	10	28		55
		-196	10	27		52	11	27		52
		-78	14	24		43	15	24		43
		-18	18	22		38	19	22		38
		20	23	20	14	33	24	20	15	33
		95	40	18	14	25	40	18	15	25
		120	46	16	14	22	46	16	14	22
		150	50	15	13	20	50	15	13	20

		200	50	13	13	16	50	13	13	16
		260	34	11	11	15	34	11	11	15
		315	60	7	7	12	60	7	7	12
		370	>100	2	2	6	>100	2	2	6
	H24	-251	6	35	25	58	9	36	26	59
		-196	6	35	25	54	9	36	26	55
		-78	7	33	24	46	11	34	25	47
		-18	8	32	24	41	13	33	25	42
		20	11	30	23	38	15	30	24	39
		95	16	26	23	20	20	27	24	31
		120	18	24	22	27	22	25	22	28
		150	20	23	22	26	23	24	23	27
		200	21	21	21	24	23	22	22	25
		260	29	17	17	20	19	18	18	21
		315	70	7	7	12	70	8	8	13
		345	>100	3	3	8	>100	3	3	9
		370	>100	2	2	6	>100	2	2	7
HK31A plate	H24	24	12	29	22	37				
		150	30	22		24				
		200	28	22		22				
		260	22	17		19				
		315	36	11		16				
HM21A sheet	T8	-250	6	31		53	12	27		54
		-196	6	31		50	12	27		51
		-78	7	30		43	13	26		44
		-18	9	29		39	15	25		40
		20	11	28	21	36	17	24	20	35
		95	14	25	21	28	22	22	20	27

TABLE 11.2a (*Continued*)

Alloy	Temper	Temperature, °C	Longitudinal				Transverse			
				ksi				ksi		
			%E	TYS	CYS	TS	%E	TYS	CYS	TS
		120	17	23	21	25	25	20	20	24
		150	20	21	21	23	30	19	20	22
		200	30	18	20	19	32	17	19	18
		260	25	16	17	17	22	15	16	16
		315	15	13	14	15	15	13	14	14
		370	50	8	9	11	50	8	9	11
		425	100	3	4	5	100	3	4	5
		480	>100	1	1	2	>100	1	1	2
plate		24	11	30	24	37	14	28	25	39
		95	16	27	24	29	19	25	25	31
		120	22	25	24	26	25	23	23	28
		150	30	23	23	23	32	21	22	25
		200	36	19	20	20	36	18	19	22
		260	33	17	18	18	33	16	17	19
		315	25	14	15	16	25	13	14	17
		370	40	8	9	12	40	8	9	12
		425	50	3	4	6	50	3	4	6
		480	>100	1	1	2	>100	1	1	2

ZE10A sheet	0	20	24	25	19	35	26	26	18	35
		120	65	18		22	55	16		20
		200	60	5		12	60	6		12
	H24	20	17	28	25	38	24	31	27	40
		120	59	15		22	55	16		23
		200	64	5		10	60	6		10
							Modulus (ksi)			
ZH11A sheet		20	10	26		39	6500			
		100	10	26		32	5950			
		150	11	26		29	5950			
		200	15	24		25	5800			
		250	20	18		19	5800			
		300	27	11		14	4950			
		350	38	2		8	4200			
ZK30A sheet		20	10	28		39	6500			
		100	33	17		24	5800			
		150	42	11		17	4800			
		200	51			11				
		250	59			7				

Source: Refs. 1, 5, 6, 7, 8, 9, 18, 20.

TABLE 11.3 The Effect of Time and Temperature of Exposure on Tensile Properties of Sheet

Alloy	Exposure °C	Exposure Hours	Test temperature, °C	%E	MPa TYS	MPa CYS	MPa TS	ksi TYS	ksi CYS	ksi TS
AZ31B-H24	95	0	20	13	220	165	280	32	24	41
		16		13	220	165	280	32	24	41
		48		13	220	165	280	32	24	41
		192		13	220	165	280	32	24	41
		500		13	220	165	280	32	24	41
		1000		13	220	165	280	32	24	41
		16	95	37	170	155	230	25	22	33
		48		37	170	155	230	25	22	32
		192		37	170	155	230	25	22	33
		500		37	170	155	230	25	22	33
		1000		37	170	155	230	25	22	33
	120	0	20	13	220	165	280	32	24	41
		16		13	220	165	280	32	24	41
		48		13	220	165	280	32	24	41
		192		13	220	165	280	32	24	41
		500		13	220	165	280	32	24	41
		1000		17	220	165	280	32	24	41
		16	120	50	140	150	190	20	22	28
		48		50	140	150	190	20	22	28
		192		50	140	150	190	20	22	28
		500		50	140	150	190	20	22	28
		1000		50	140	150	190	20	22	28

150	0	20	13	220	165	280	32	24	41
	16		15	215	165	275	31	24	40
	48		16	200	165	275	29	24	40
	192		18	195	150	275	28	22	40
	500		18	200	155	270	29	22	39
	1000		20	205	145	270	30	21	39
	16	150	52	110	150	160	16	22	23
	48		55	110	130	160	16	19	23
	192		58	110	130	150	16	19	22
	500		63	110	130	150	16	19	22
	1000		64	105	125	145	15	18	21
200	0	20	13	220	165	280	32	24	41
	16		21	175	130	255	25	19	37
	48		21	175	130	260	25	19	38
	192		21	175	130	260	25	19	38
	500		21	175	130	260	25	19	38
	1000		21	175	130	260	25	19	38
	16	200	73	70	85	90	10	12	13
	48		73	70	85	90	10	12	13
	192		73	70	85	90	10	12	13
	500		73	70	85	90	10	12	13
	1000		73	70	85	90	10	12	13
260	0	20	13	220	165	280	32	24	41
	16		21	160	120	255	23	17	37
	48		22	160	120	255	23	17	37
	192		22	160	120	255	23	17	37
	500		21	160	120	255	23	17	37
	1000		22	160	120	255	23	17	37

TABLE 11.3 *(Continued)*

Alloy	Exposure °C	Exposure Hours	Test temperature, °C	%E	MPa TYS	MPa CYS	MPa TS	ksi TYS	ksi CYS	ksi TS
		16	260	70	50	60	60	7	9	9
		48		90	50	60	60	7	9	9
		192		90	50	60	60	7	9	9
		500		90	50	60	60	7	9	9
		1000		90	50	60	60	7	9	9
	315	0	20	13	220	165	280	32	24	41
		16		18	155	110	250	22	16	36
		192		22	155	110	250	22	16	36
		500		22	155	110	250	22	16	36
		16	315	113	30	40	40	4	6	6
		192		113	30	40	40	4	6	6
		500		113	30	40	40	4	6	6
HM31A-H23	150	0	20	8	200	170	255	29	25	37
		5000		8	200	170	255	29	25	37
		0	150	20	160	160	180	23	23	26
		500		20	160	160	180	23	23	26
	200	0	20	8	200	170	255	29	25	37
		5000		8	200	170	255	29	25	37
		0	200	21	145	150	165	21	22	24
		500		29	130	150	150	19	22	22
		5000		29	130	150	150	19	22	22

	260	0	20	8	200	170	255	29	25	37
		1000		15	200	150	255	29	22	37
		5000		16	195	145	250	28	21	36
HM31A-H24	315	0	20	8	200	170	255	29	25	37
		1		11	200	165	250	29	24	36
		100		20	185	130	250	27	19	36
		1000		20	170	125	240	25	18	35
		5000		20	165	110	235	24	16	34
HM21A-T8	315	0	20	7	170		230	25		33
		5000		7	170		230	25		33
		0	315	15	85	90	95	12	13	14
		5000		15	85	90	95	12	13	14
	370	0	20	7	170		230	25		33
		500		9	165		220	24		32
		0	370	50	55	60	75	8	9	11
		100		50	55	60	75	8	9	11
		1000		112	35	40	50	5	6	7
	425	0	20	7	170		230	25		33
		100		12	125		205	18		30

Source: Refs. 5, 10.

TABLE 11.4 Effect of Strain Rate and Temperature on the Strength of Sheet (SI Units)

Alloy	Test temperature, °C	TYS (MPa) at strain/min				TS (MPa) at strain/min			
		0.005	0.050	0.50	5.0	0.005	0.050	0.50	5.0
AZ31B-O	24	150	150	150	155	250	250	255	255
	95	125	125	130	145	185	200	215	225
	150	90	95	100	120	130	150	165	185
	200	65	75	85	95	65	85	115	145
	260	40	55	60	80	40	55	70	105
	315	20	35	45	65	25	35	50	70
	370	5	25	30	50	15	25	35	70
	425		15	20	30		15	25	30
	480			15	20			15	20
AZ31B-H24	24	205	210	215	220	275	285	280	280
	95	160	175	185	200	195	215	235	255
	150	100	125	140	155	100	140	170	200
	200	45	65	95	130	50	70	100	160
	260	25	45	65	85	35	45	65	100
	315	15	30	45	65	20	30	45	70
	370	5	20	30	50	15	20	30	50
	425		10	20	30		10	20	35
	480			15	20			15	20
HK31A-O	24	145	150	155	150	195	210	225	240
	95	125	130	135	145	145	160	175	190
	150	110	115	115	130	135	135	140	150
	200	90	100	105	125	110	110	110	125
	260	85	85	90	105	100	100	100	105

	315	60	65	70	85	60	80	90	95
	370	15	35	45	65	15	35	45	70
	425		15	30	45		15	30	45
	480			20	30			20	30
HK31A-H24	24	200	205	210	210	240	245	250	250
	95	185	190	190	195	195	205	215	225
	150	165	170	175	175	165	175	185	195
	200	135	145	155	165	140	155	165	175
	260	115	125	135	145	115	130	140	150
	315	45	85	105	125	50	90	110	125
	370	10	30	50	95	15	35	65	95
	425		10	25	50		15	30	85
	480			15	35			20	35
HM21A-T8	24	170	170	170	170	210	220	230	240
	95	150	155	155	160	170	175	190	200
	150	130	135	145	150	135	145	155	170
	260	95	100	105	125	100	100	110	125
	315	80	85	90	100	85	90	95	105
	370	40	60	70	85	45	60	75	90
	425	15	25	40	60	15	30	50	65
	480			10	20			15	35

Source: Ref. 11.

TABLE 11.4a Effect of Strain Rate and Temperature on the Strength of Sheet (English Units)

Alloy	Test temperature, °C	TYS (ksi) at strain/min				TS (ksi) at strain/min			
		0.005	0.050	0.50	5.0	0.005	0.050	0.50	5.0
AZ31B-O	24	22	22	22	22	36	36	37	37
	95	18	18	19	21	27	29	31	33
	150	13	14	15	17	19	22	24	27
	200	9	11	12	14	9	12	17	21
	260	6	8	9	12	6	8	10	15
	315	3	5	7	9	4	5	7	10
	370	1	4	4	7	3	4	5	10
	425		2	3	4		2	4	4
	480			2	3			2	3
AZ31B-H24	24	30	30	31	32	40	41	41	41
	95	23	25	27	29	28	31	34	37
	150	15	18	20	22	15	20	25	29
	200	7	9	14	19	7	10	15	23
	260	4	7	9	12	5	7	9	15
	315	2	4	7	9	3	4	7	10
	370	1	3	4	7	2	3	4	7
	425		1	3	4		1	3	5
	480			2	3			2	3
HK31A-O	24	21	22	22	23	28	30	33	35
	95	18	19	20	21	21	23	25	28
	150	16	17	17	19	20	20	20	22
	200	13	15	15	18	16	16	16	18
	260	12	12	13	15	15	15	15	15

	315	9	9	10	12	9	12	13	14
	370	2	5	7	9	2	5	7	10
	425		2	4	7		2	4	7
	480			3	4			3	4
HK31A-H24	24	29	30	30	30	35	36	36	36
	95	27	28	28	28	28	30	31	33
	150	24	25	25	25	24	25	27	28
	200	20	21	22	24	20	22	24	25
	260	17	18	20	21	17	19	20	22
	315	7	12	15	18	7	13	16	18
	370	1	4	7	14	2	5	9	14
	425		1	4	7		2	4	12
	480			2	5			3	5
HM21A-T8	24	25	25	25	25	30	32	33	35
	95	22	22	22	23	25	25	28	29
	150	19	20	21	22	20	21	22	25
	260	14	15	15	18	15	15	16	18
	315	12	12	13	15	12	13	14	15
	370	6	9	10	12	7	9	11	13
	425	2	4	6	9	2	4	7	9
	480			1	3			2	5

Source: Ref. 11.

TABLE 11.5 Creep Strength of Sheet Alloys (SI Units)

					MPa (in %)								
					Stress for total extension					Stress for creep extension			
Alloy	Temper	Exposure, hrs[a]	Temperature, °C	Test duration, hrs	0.1	0.2	0.5	1.0	2.0	0.01	0.03	0.05	0.1
AZ31B	0	0	95	1	40	70	105	110					
				10	35	60	95	100					
				100	30	55	85	90					
				500	30	50	70	75					
			120	1	35	55	85	95					
				10	30	50	75	85					
				100	15	35	60	70					
				500	15	20	50	60					
			150	1	30	50	70	85					
				10	15	30	50	60					
				100	5	15	30	35					
				500	3	5	20	25					
			175	1	15	30	50	55					
				10	5	15	30	35					
AZ31B	H24	0	95	1	40	60	105	130					
				10	30	50	75	95					
				100	15	30	50	70					
				500	15	20	35	50					
			120	1	30	40	60	90					
				10	15	30	40	55					
				100	5	15	30	35					
				500	5	10	15	20					

			150	1	15	30	40	55
				10	5	15	20	35
				100	3	5	10	15
				500		3	5	5
			175	1	5	10	30	40
				10	3	5	10	15
HK31A	0	0	200	100		70	75	
			260	100		10	15	
HK31A	H24	0	150	100	50	85	140	180
		500				75	130	
		1000				75	130	
		5000				75	130	
				500	45	70	130	175
				1000		70	125	
		0	200	1	40	80	110	120
				10	40	40	90	105
				100	25	40	65	75
		500				30	50	
		1000				30	45	
		5000				30	45	
		0		500	25	35	50	60
				1000		35	50	
HM21A	T8	0	200	1	40	80	115	120
				10	40	80	115	120
				100	40	80	95	95
			260	1	40	70	100	105
				10	40	60	85	90
				100	40	50	60	65

TABLE 11.5 (*Continued*)

					MPa (in %)								
				Test	Stress for total extension					Stress for creep extension			
Alloy	Temper	Exposure, hrs[a]	Temperature, °C	duration, hrs	0.1	0.2	0.5	1.0	2.0	0.01	0.03	0.05	0.1
			315	1	35	50	60	65					
				10	30	40	50	55					
				100	30	35	40	45					
			370	1	20	25	35	40					
				10	20	20	30	30					
				100	15	20	25	25					
ZH11A			120	10								130	145
				100									138
				1000									128
			150	10								132	138
				100								124	132
				1000								103	124
			250	30s					110				
				60s					95				
				100						46			
				1000							46		
			350	30s			20	32	48				
				60s			18	28	40				
				600s				15	20				

[a]Hours of exposure at the testing temperature before testing.
Source: Refs. 5, 6, 7, 18, 19.

TABLE 11.5a Creep Strength of Sheet Alloys (English Units)

					ksi (in %)								
				Test	Stress for total extension					Stress for creep extension			
Alloy	Temper	Exposure, hrs[a]	Temperature, °C	duration, hrs	0.1	0.2	0.5	1.0	2.0	0.01	0.03	0.05	0.1
AZ31B	0	0	95	1	6	10	15	16					
				10	5	9	14	15					
				100	4	8	12	13					
				500	4	7	10	11					
			120	1	5	11	12	14					
				10	4	7	11	12					
				100	2	5	9	10					
				500	2	3	7	9					
			150	1	4	7	10	12					
				10	2	4	7	9					
				100	1	2	4	5					
				500	0.5	1	2	4					
			175	1	2	4	7	8					
				10	1	2	4	5					
AZ31B	H24	0	95	1	6	9	15	19					
				10	4	7	11	14					
				100	2	4	7	10					
				500	2	3	5	7					
			120	1	4	6	9	13					
				10	2	4	4	11					
				100	1	2	4	5					
				500	1	1	2	3					

TABLE 11.5a (*Continued*)

					ksi (in %)								
					Stress for total extension					Stress for creep extension			
Alloy	Temper	Exposure, hrs[a]	Temperature, °C	Test duration, hrs	0.1	0.2	0.5	1.0	2.0	0.01	0.03	0.05	0.1
			150	1	2	4	6	8					
				10	1	2	3	5					
				100	0.5	1	1	2					
				500		0.5	1	1					
			175	1	1	1	4	6					
				10	0.5	1	1	2					
HK31A	0	0	200	100		10	11						
			260	100		1	2						
HK31A	H24	0	150	100	7	12	20	26					
		500				11	19						
		1000				11	19						
		5000				11	19						
				500	7	10	19	25					
				1000		10	18						
		0	200	1	6	12	16	17					
				10	6	6	13	15					
				100	4	6	9	11					
		500				4	7						
		1000				4	7						
		5000				4	7						
		0		500	4	5	7	9					
				1000		5	7						

HM21A	T8	0	200	1	6	12	16	17					
				10	6	12	15	17					
				100	6	12	14	14					
			260	1	6	10	15	15					
				10	6	9	12	13					
				100	6	7	9	9					
			315	1	5	7	9	9					
				10	4	6	7	8					
				100	4	5	6	7					
			370	1	3	4	5	6					
				10	3	3	4	4					
				100	2	3	4	4					
ZH11A			120	10								19	21
				100									20
				1000									18
			150	10								19	20
				100								18	19
				1000								15	18
			250	30s					16				
				60s					14				
				100						6.7			
				1000							6.7		
			350	30s			2.9	4.6	7				
				60s			2.6	4	5.8				

[a]Hours of exposure at the testing temperature before testing.

Source: Refs. 5, 6, 7, 18, 19.

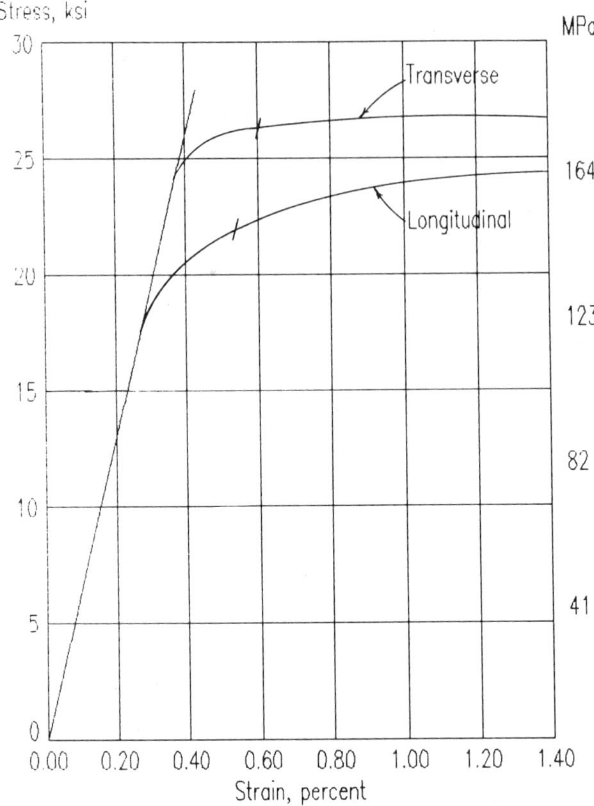

FIGURE 11.1 Tensile stress-strain curves for AZ31B-O sheet at room temperature.

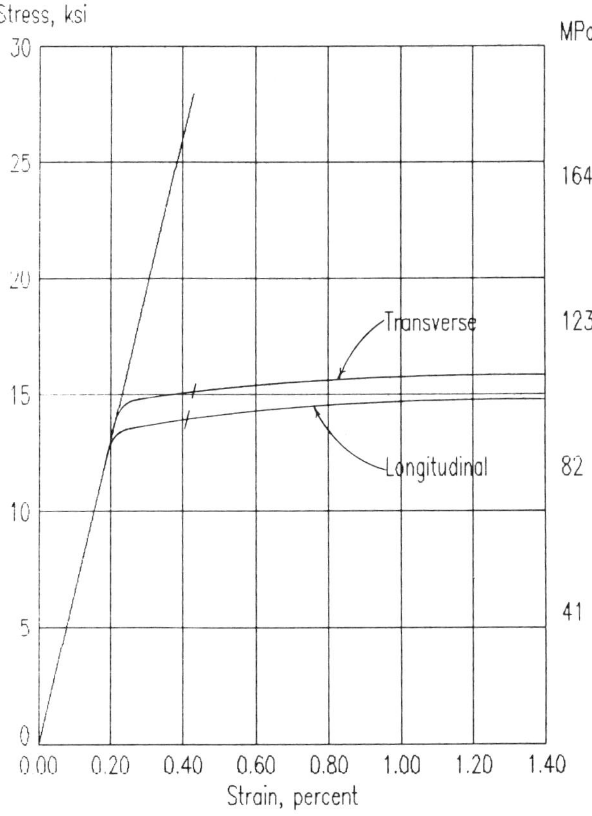

FIGURE 11.2 Compressive stress-strain curves for AZ31B-O sneet at 20°C.

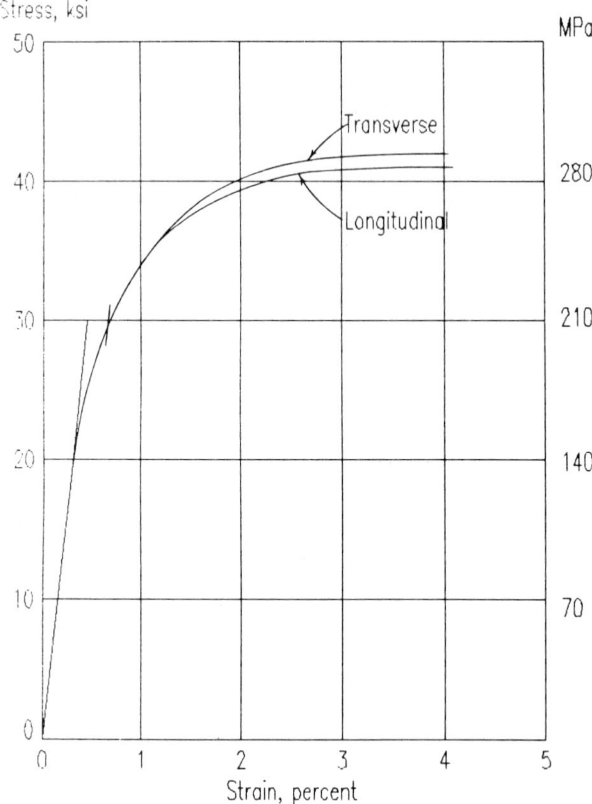

FIGURE 11.3 Tensile stress-strain curves for AZ31B-H24 sheet at 20°C.

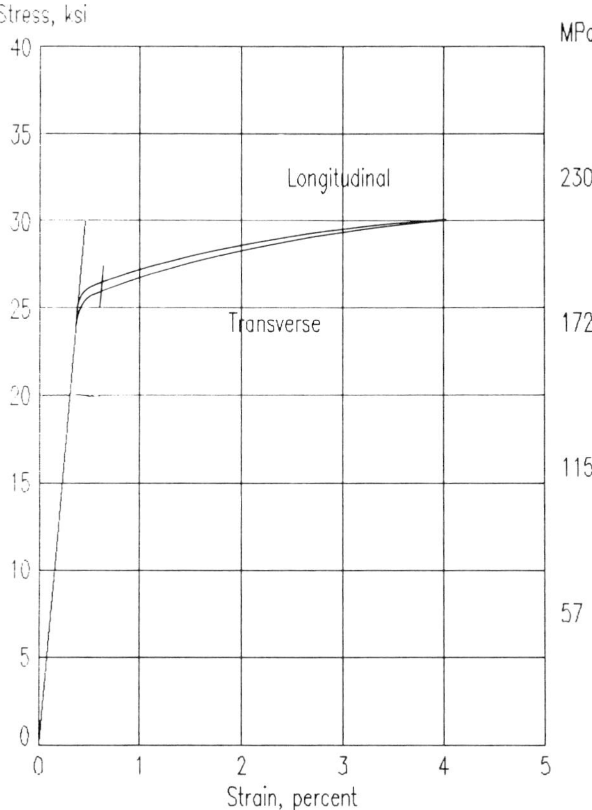

FIGURE 11.4 Compressive stress-strain curves for AZ31B-H24 sheet at 20°C.

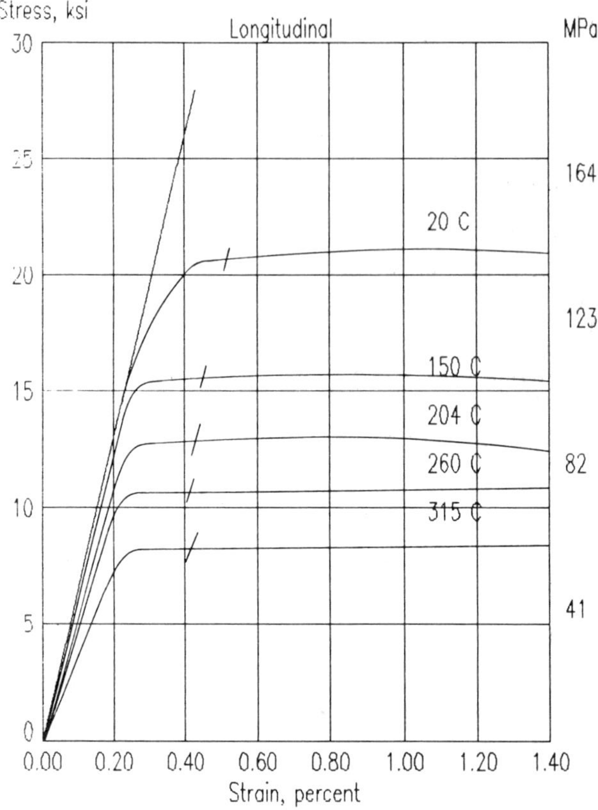

FIGURE 11.5 Tensile stress-strain curves for HK31A-O sheet.

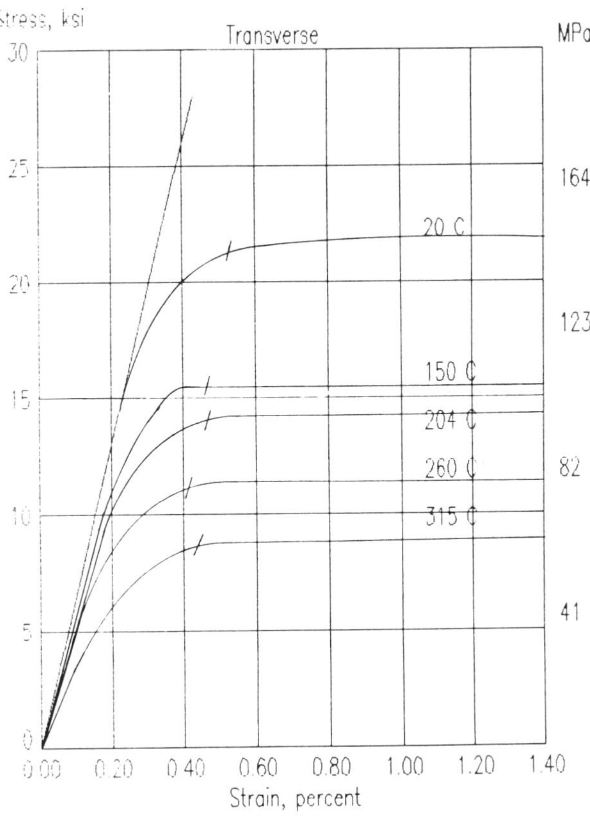

FIGURE 11.6 Tensile stress-strain curves for HK31A-O sheet.

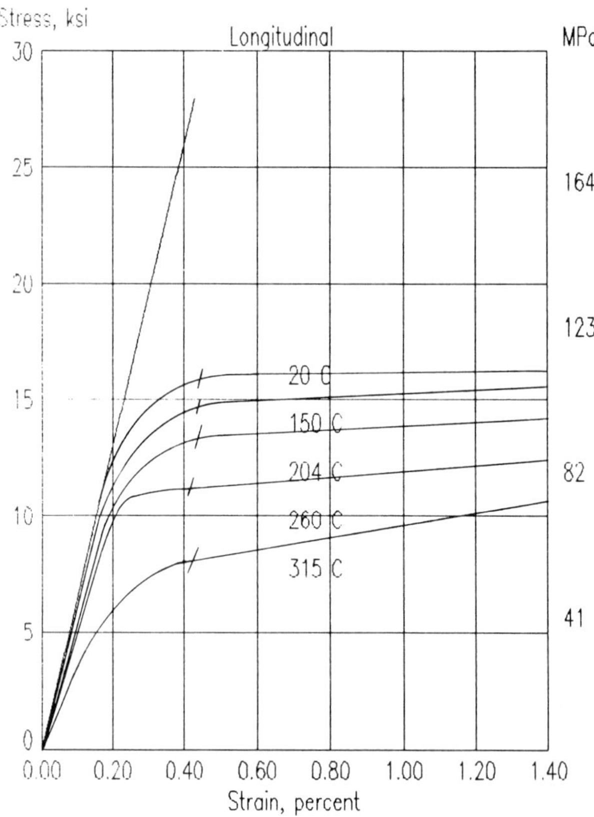

FIGURE 11.7 Compressive stress-strain curves for HK31A-O sheet.

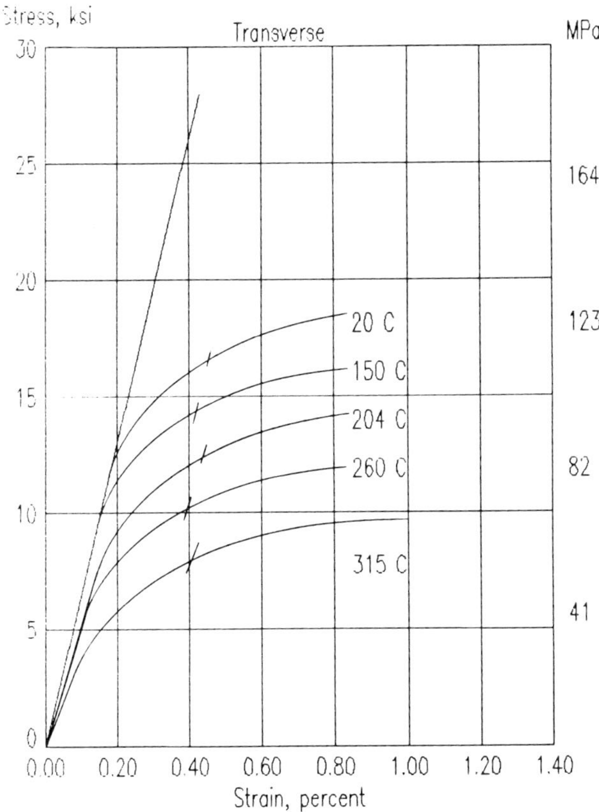

FIGURE 11.8 Compressive stress-strain curves for HK31A-O sheet.

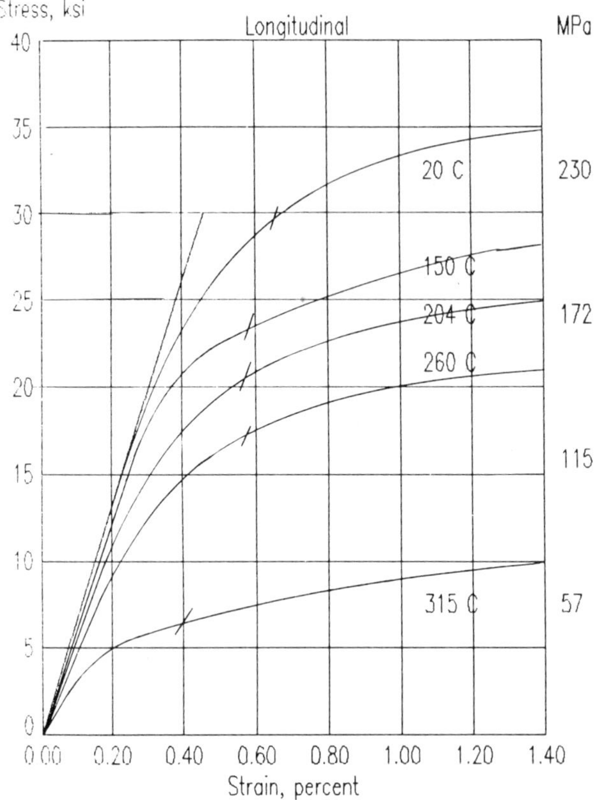

FIGURE 11.9 Tensile stress-strain curves for HK31A-H24 sheet.

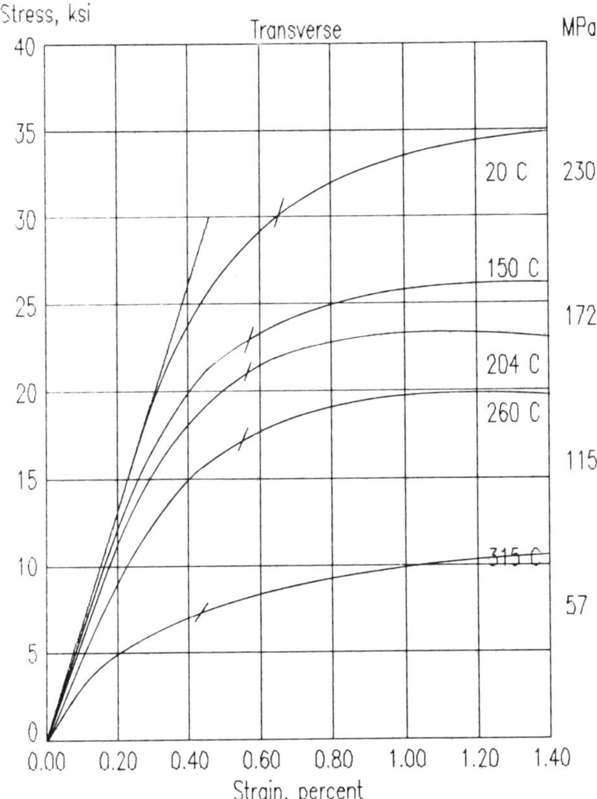

FIGURE 11.10 Tensile stress-strain curves for HK31A-H24 sheet.

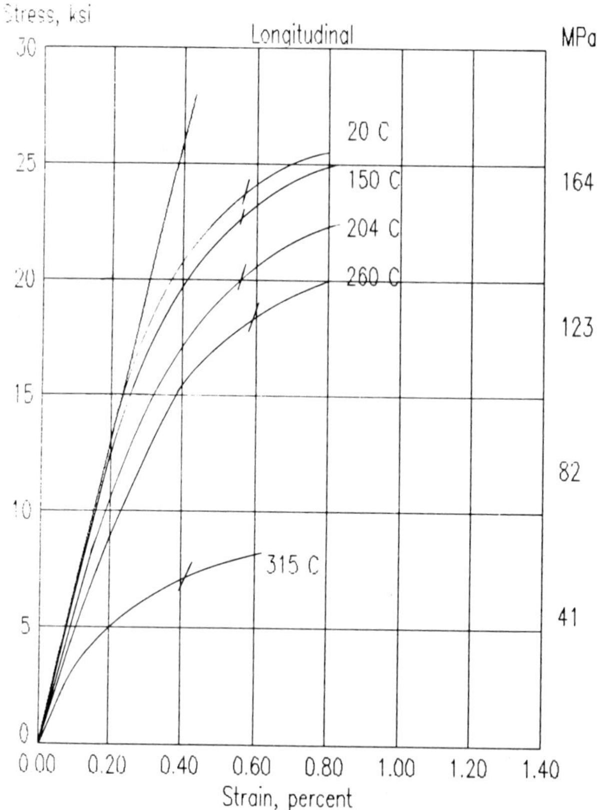

FIGURE 11.11 Compressive stress-strain curves for HK31A-H24 sheet.

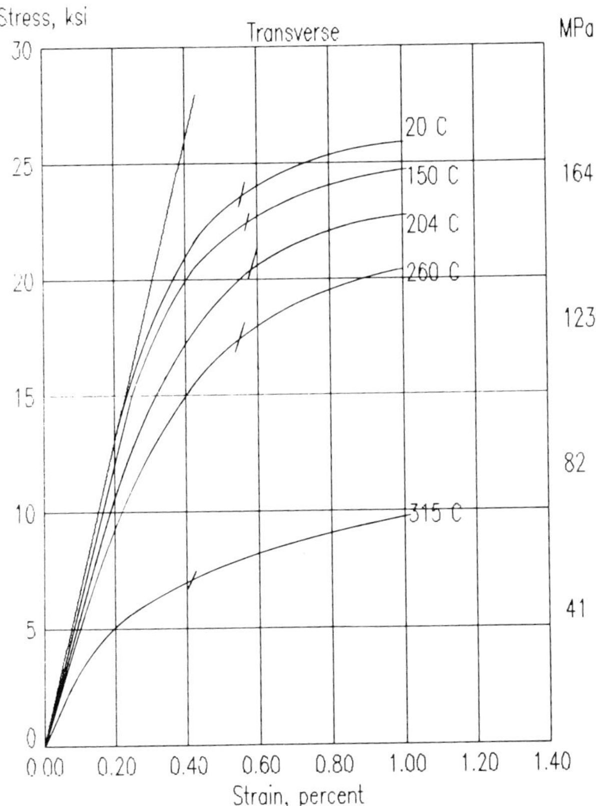

FIGURE 11.12 Compressive stress-strain curves for HK31A-H24 sheet.

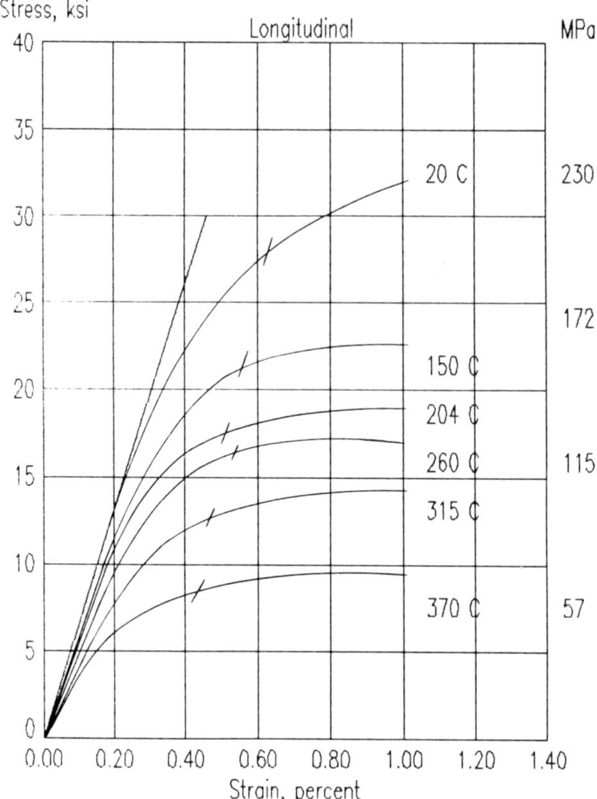

FIGURE 11.13 Tensile stress-strain curves for HM21A-T8 sheet.

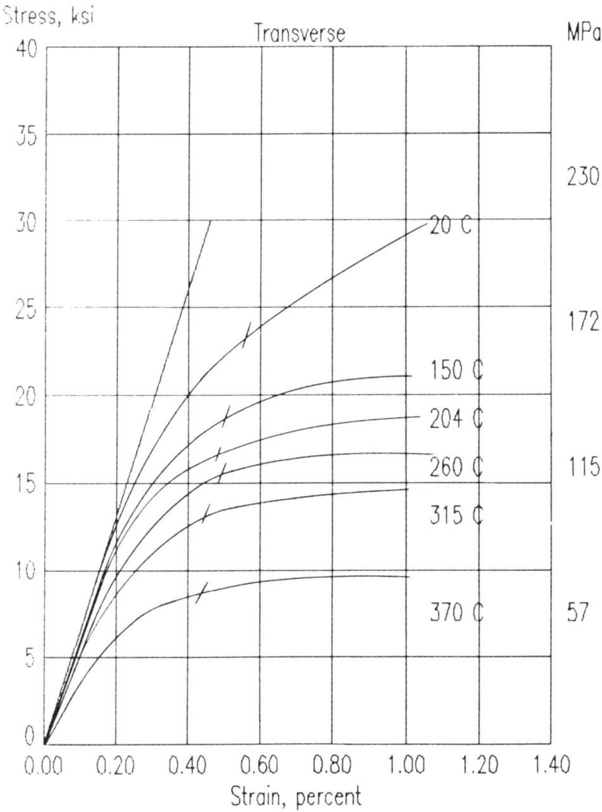

FIGURE 11.14 Tensile stress-strain curves for HM21A-T8 sheet.

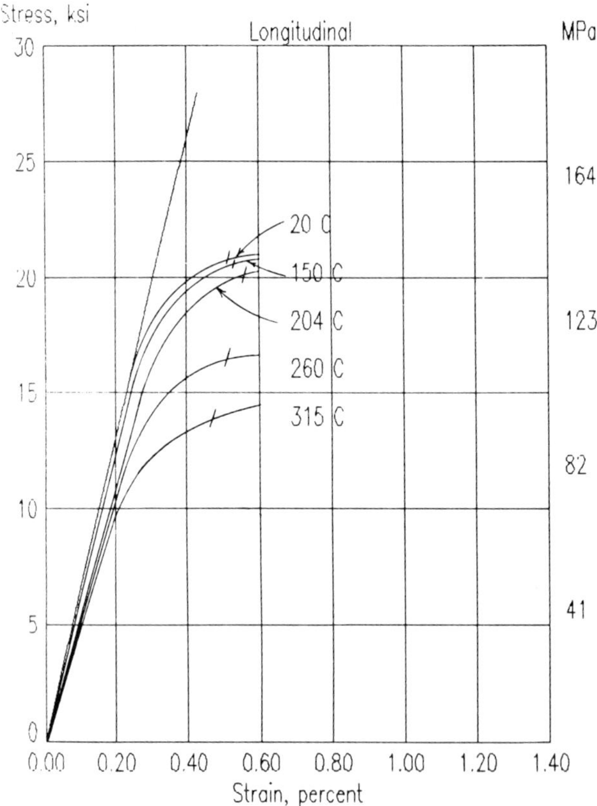

FIGURE 11.15 Compressive stress-strain curves for HM21A-T8 sheet.

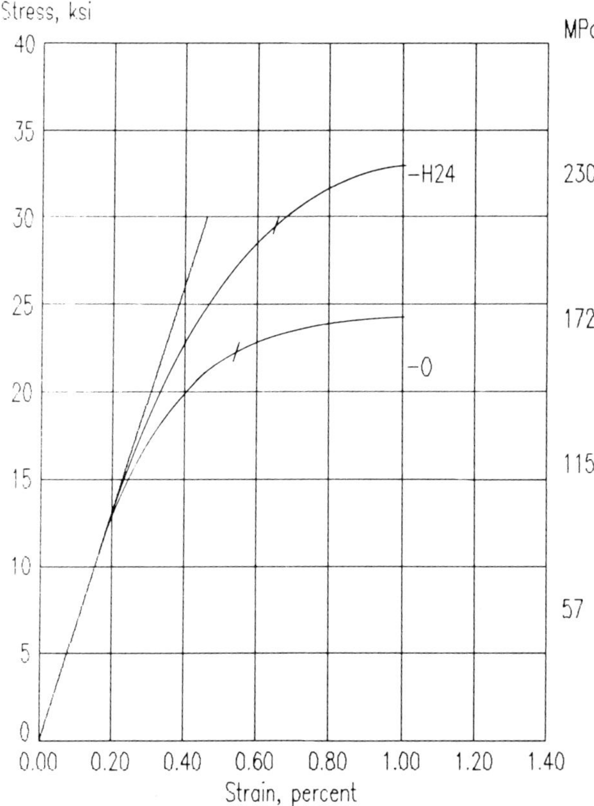

FIGURE 11.16 Longitudinal tensile stress-strain curves for ZE10A sheet.

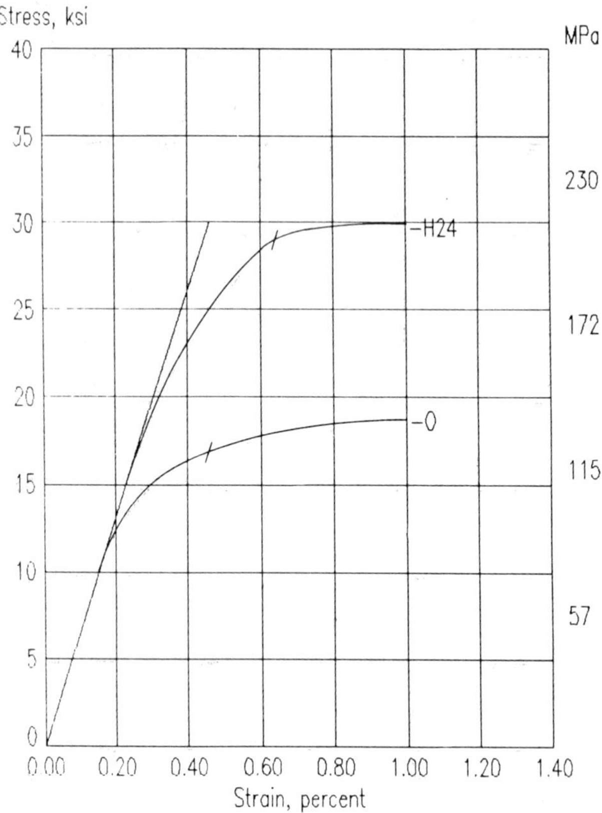

FIGURE 11.17 Longitudinal compressive stress-strain curves for ZE10A sheet.

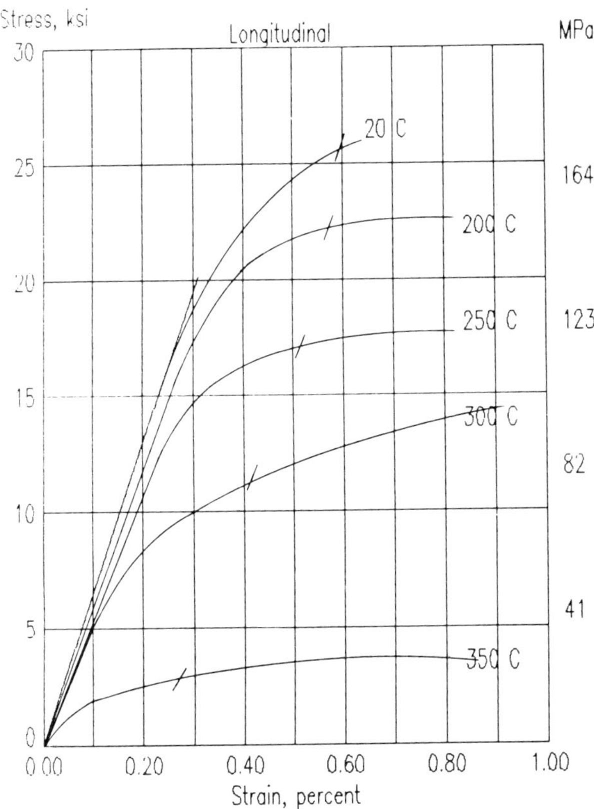

FIGURE 11.18 Tensile stress-strain curves for ZH11A-O sheet.

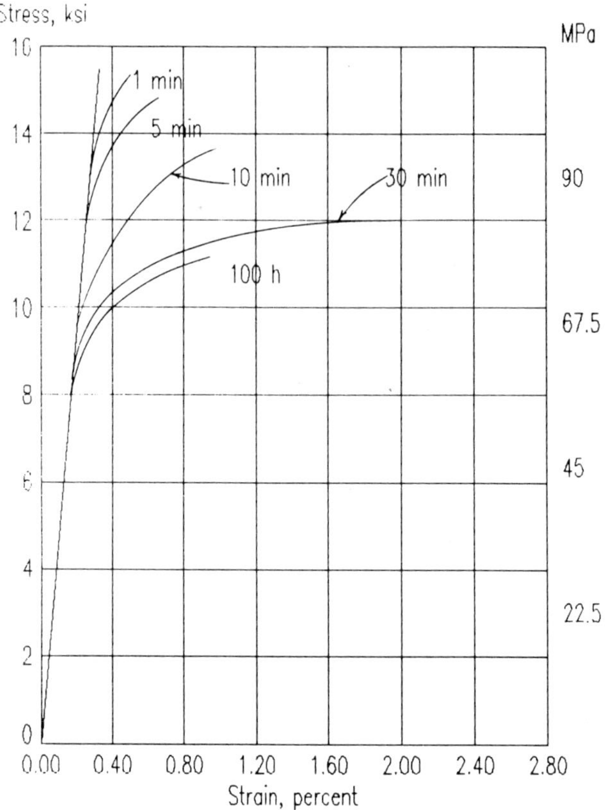

FIGURE 11.19 Isochronous stress-strain curves for HK31A-O sheet at 204°C.

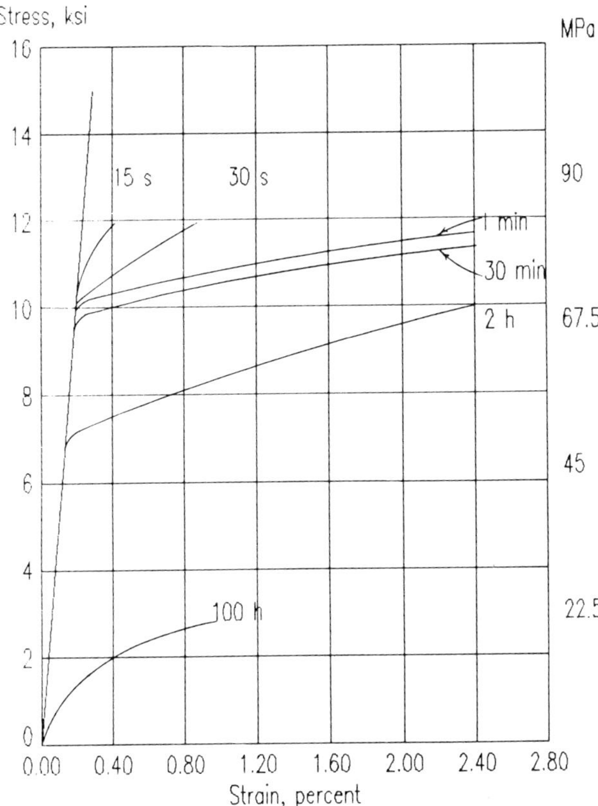

FIGURE 11.20 Isochronous stress-strain curves for HK31A-O sheet at 260°C.

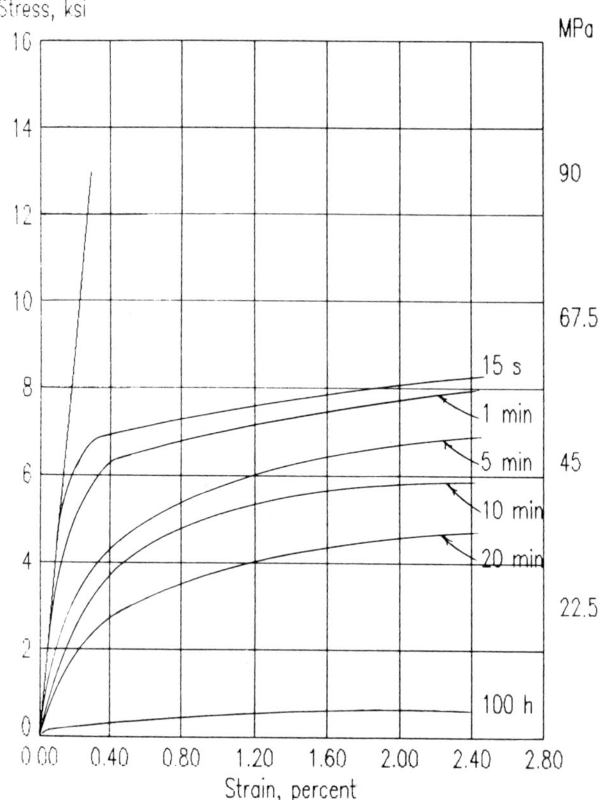

FIGURE 11.21 Isochronous stress-strain curves for HK31A-O sheet at 315°C.

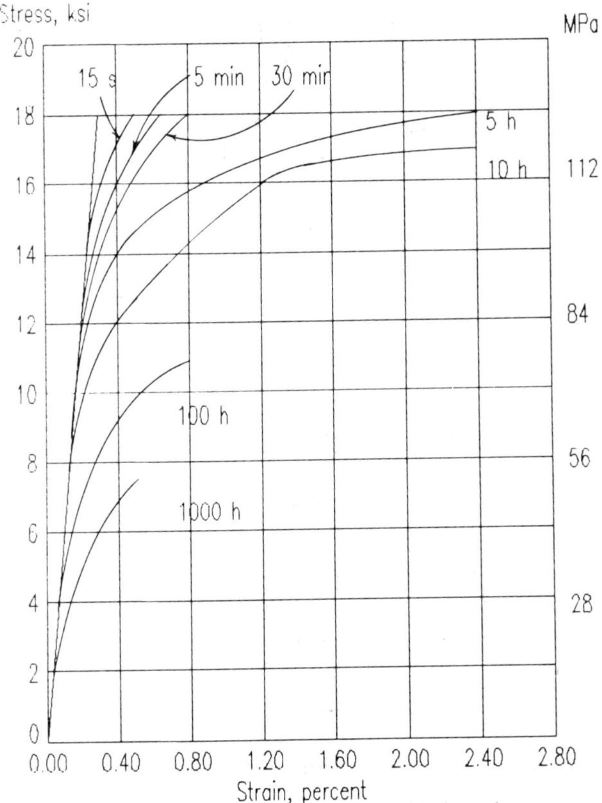

FIGURE 11.22 Isochronous stress-strain curves for HK31A-H24 sheet at 204°C.

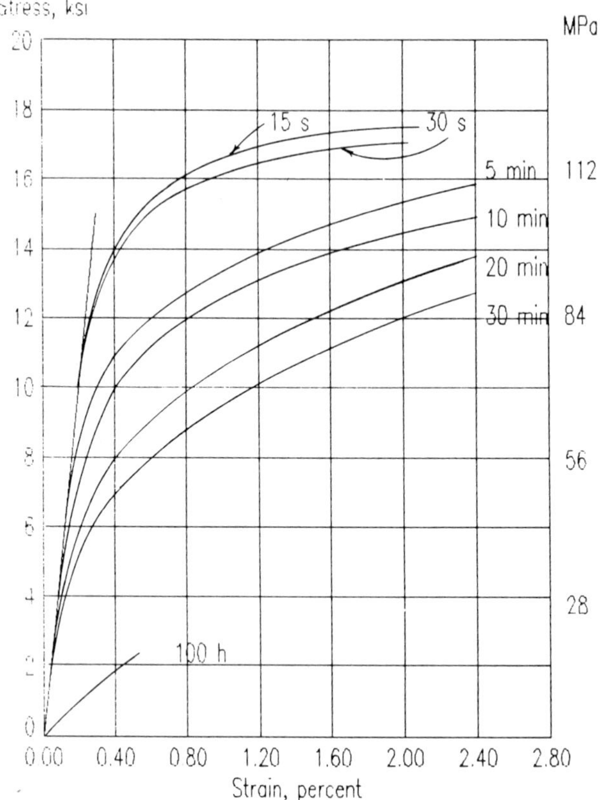

FIGURE 11.23 Isochronous stress-strain curves for HK31A-H24 sheet at 260°C.

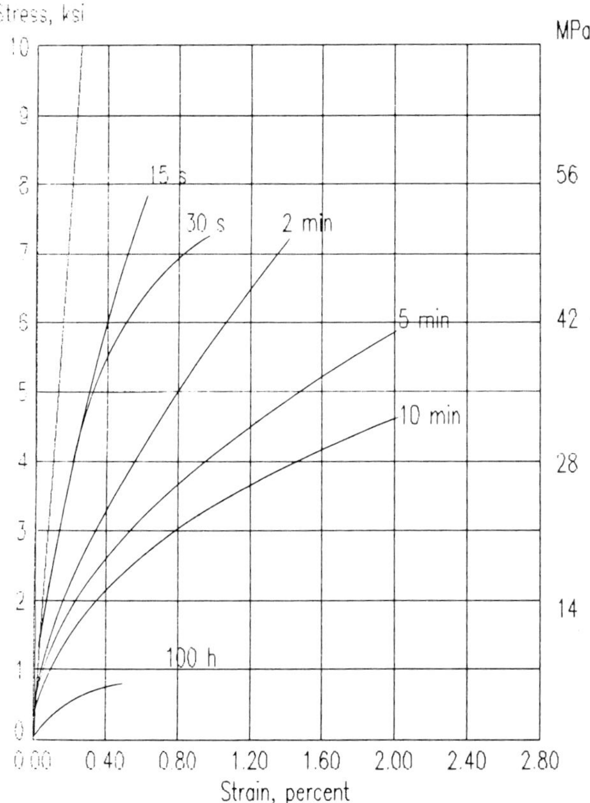

FIGURE 11.24 Isochronous stress-strain curves for HK31A-H24 sheet at 315°C.

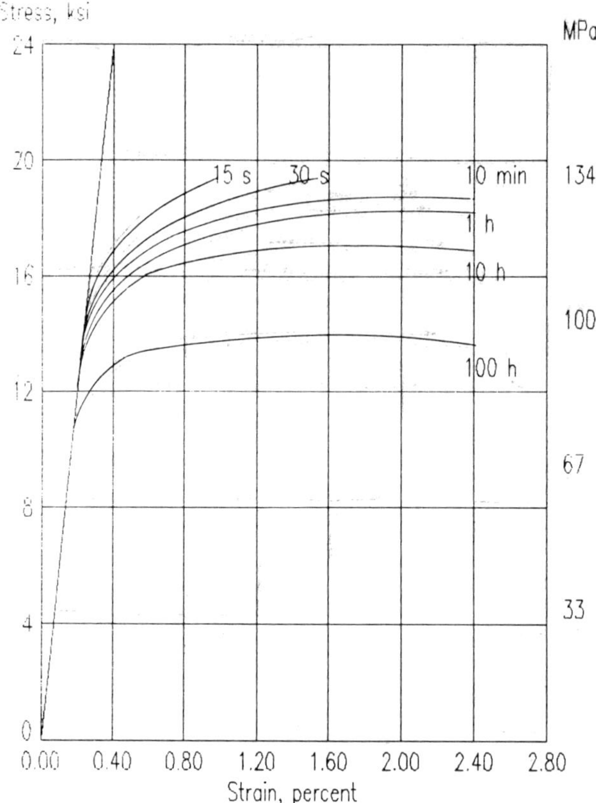

FIGURE 11.25 Isochronous stress-strain curves for HM21A-T8 sheet at 204°C.

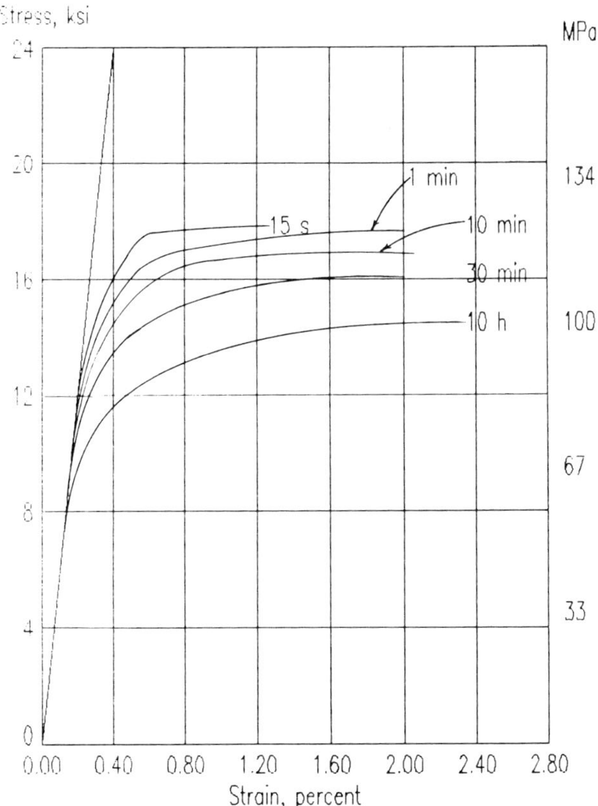

FIGURE 11.26 Isochronous stress-strain curves for HM21A-T8 sheet at 260°C.

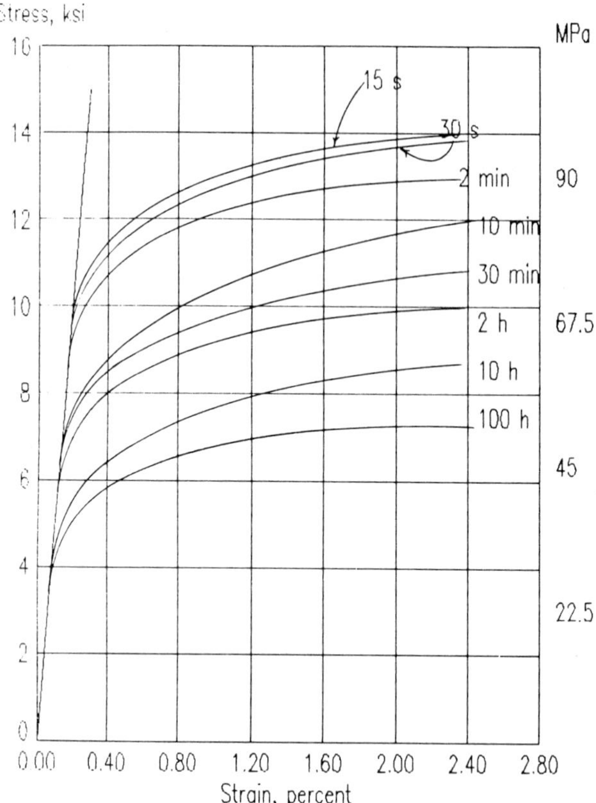

FIGURE 11.27 Isochronous stress-strain curves for HM21A-T8 sheet at 315°C.

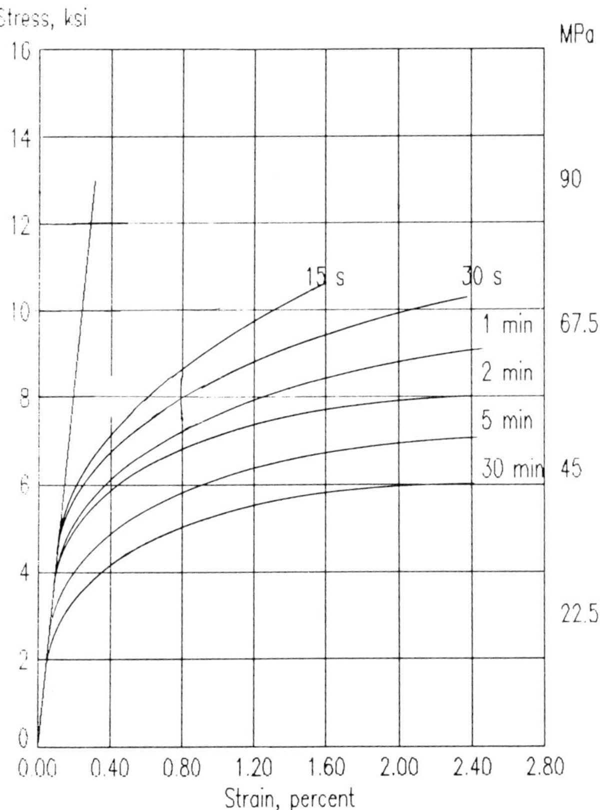

FIGURE 11.28 Isochronous stress-strain curves for HM21A-T8 sheet at 370°C.

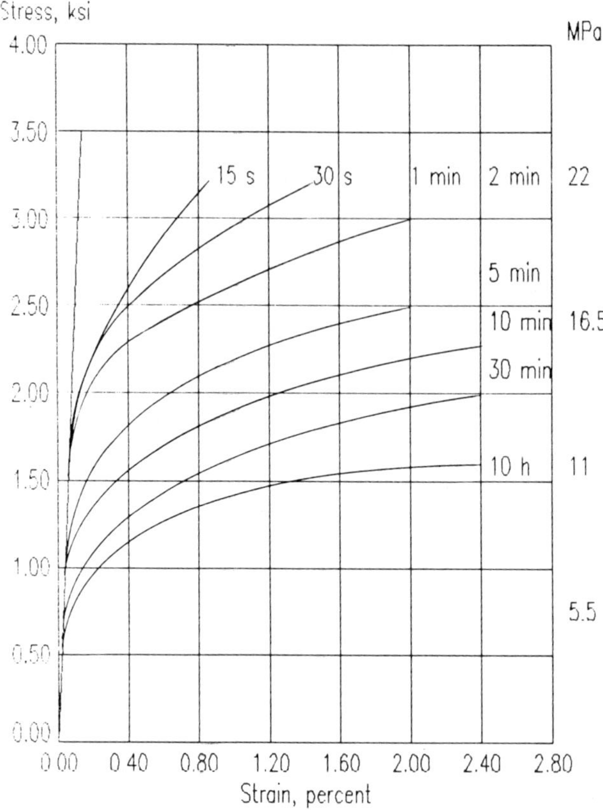

FIGURE 11.29 Isochronous stress-strain curves for HM21A-T8 sheet at 425°C.

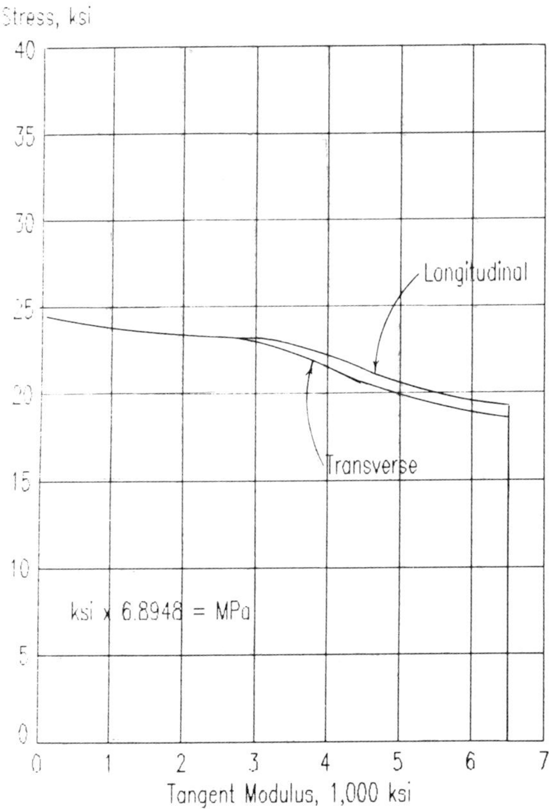

FIGURE 11.30 Compressive tangent modulus curves for AZ31B-H24 at 20°C.

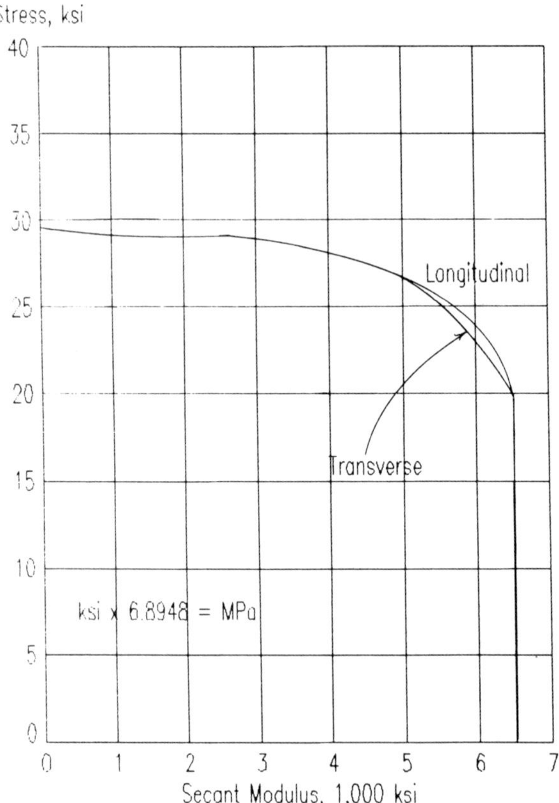

FIGURE 11.31 Compressive secant modulus curves for AZ31B-H24 at 20°C.

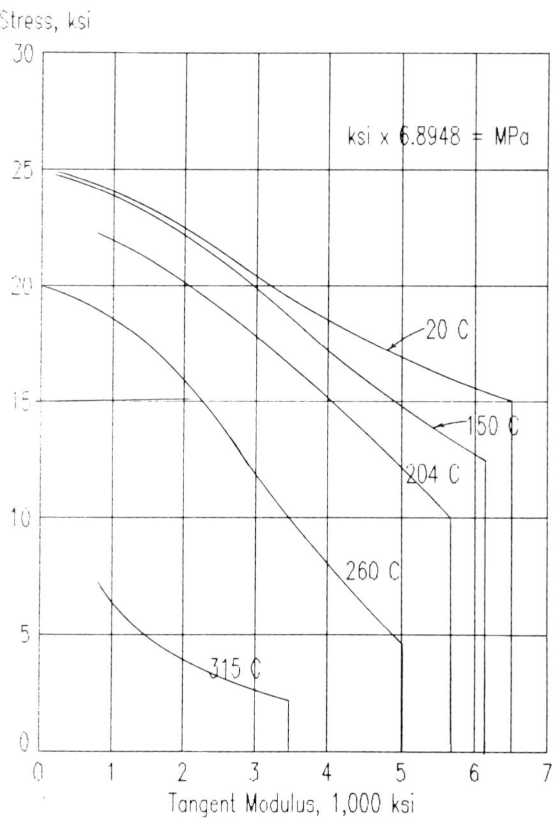

FIGURE 11.32 Compressive tangent modulus curves for HK31A-H24 sheet.

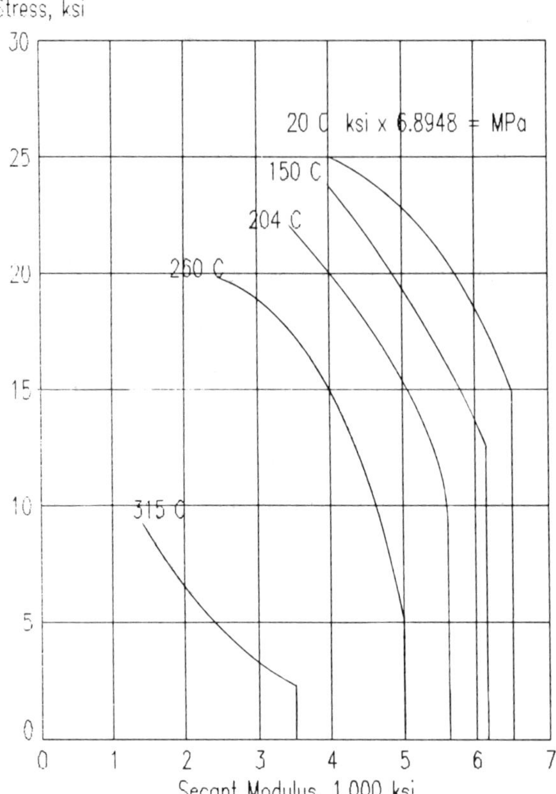

FIGURE 11.33 Compressive secant modulus curves for HK31A-H24 sheet.

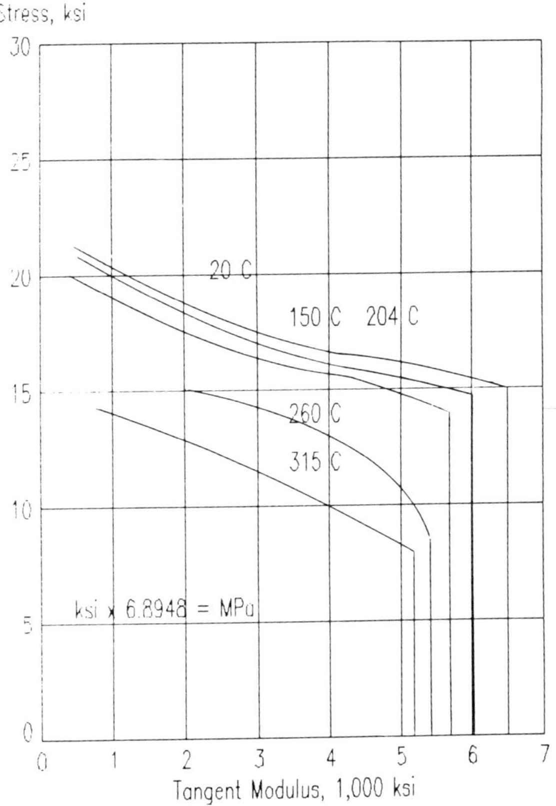

FIGURE 11.34 Compressive tangent modulus curves for HM21A-T8 sheet.

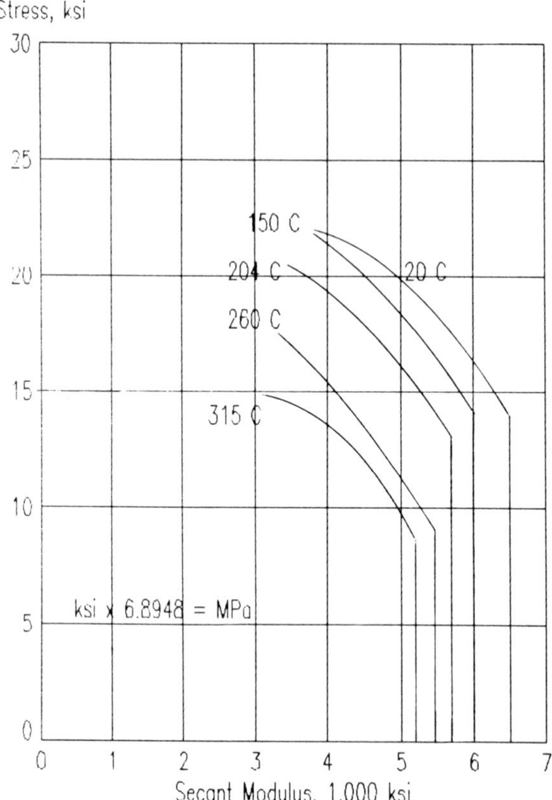

FIGURE 11.35 Compressive secant modulus curves for HM21A-T8 sheet.

11.2 TOUGHNESS

Toughness, as measured by tensile strength in the presence of a sharp notch, by charpy or Izod impact, and by determination of the critical stress intensity factor, is shown in Table 11.6. The ratio of notched to unnotched tensile strength decreases markedly at low temperatures. The values for K_{Ic} are suspiciously high, and probably should be labeled as K_q.

TABLE 11.6 Notch Sensitivity and Toughness of Sheet Alloys (SI Units)

Alloy	Temperature, °C	Tensile strength (MPa)			Notched charpy, joules	K_{Ic} MPa cm$^{1/2}$
		Unnotched	Notched[a]	Ratio		
AZ31B-O	24	260	215	0.83	8	
	-196	400	230	0.53		
AZ31B-H24	24	285	230	0.81		290
	-196	415	165	0.40		
HK31A-O	24	215	185	0.86	5.4	333
	-196	360	205	0.57		
HK31A-H24	24	270	230	0.85	4.1	256
	-196	400	250	0.63		
HM21A-T8	24	245	230	0.94		257
	-78	315	210	0.67		
	-196	375	220	0.59		
ZE10A-O	24	230	200	0.87	9	235
	-78	305	210	0.69		
	-196	365	215	0.59		
ZE10A-H24	24	265	265	1.00		309
	-78	315	260	0.83		
	-196	375	210	0.56		
ZH11A	20				6[b]	

[a]Specimen dimensions: total width: 25.4 mm; notched width: 17.78 mm; thickness: 15.24 mm; 60° V-notch with 0.0076-mm radius.
[b]Izod test.
Source: Refs. 2, 10, 12, 20.

TABLE 11.6a Notch Sensitivity and Toughness of Sheet Alloys (English Units)

Alloy	Temperature, °C	Tensile strength (ksi)			Notched charpy, joules	K_{Ic} ksi $in^{1/2}$
		Unnotched	Notched[a]	Ratio		
AZ31B-O	24	38	31	0.83	5.9	
	-196	58	33	0.53		
AZ31B-H24	24	41	33	0.81		26
	-196	60	24	0.40		
HK31A-O	24	31	27	0.86	4.0	30
	-196	52	30	0.57		
HK31A-H24	24	39	33	0.85	3.0	23
	-196	58	36	0.63		
HM21A-T8	24	36	33	0.94		23
	-78	46	30	0.67		
	-196	54	32	0.59		
ZE10-O	24	33	29	0.87	6.6	21
	-78	44	30	0.69		
	-196	53	31	0.59		
ZE10A-H24	24	38	38	1.00		28
	-78	46	38	0.83		
	-196	54	30	0.56		
ZH11A	20				4.4[b]	

[a]Specimen dimensions: total width: 1 inch; notched width: 0.700 inch; thickness: 0.60 inch; 60° V-notch with 0.0003-inch radius.
[b]Izod test.
Source: Refs. 2, 10, 12, 20.

11.3 BEARING AND SHEAR

Room-temperature bearing and shear strengths are shown in Table 11.7.

TABLE 11.7 Bearing and Shear Strengths of Sheet Alloys (SI Units, MPa)

Alloy	Thickness, mm	Bearing typical		Bearing minimum		Shear typical
		Yield	Ultimate	Yield	Ultimate	
AZ31B-O	0.41-6.35	255	455	200	415	180
	6.36-12.7	235	450	185	415	170
	12.8-50.8	230	450			170
	50.8-76.2	220	450			170
AZ31B-H24	0.41-6.32	325	530	295	470	200
	6.33-9.50	310	495	260	450	195
	9.51-12.7	275	485	235	435	185
	12.8-25.4	270	485			180
	25.5-50.8	240	455			180
	50.9-76.2	230	455			180
AZ31B-H26	6.33-9.50	315	495	275	470	195
	9.51-12.7	305	495	270	450	195
	12.8-19.1	275	495	250	450	195
	19.2-25.4	270	485			195
	25.5-38.0	255	470			185
	38.1-50.8	250	470			180
HK31A-O	0.41-6.35	195	400	165	350	165
	6.36-25.4	185	400	145	350	165
	25.4-76.2	185	395			165
HK31A-H24	0.41-3.18	285	460	235	395	180
	3.19-6.35	285	450	230	395	180
	6.36-25.4	305	470	215	395	185
	25.5-76.2	275	450			180
HM21A-T8	0.41-6.35	255	435	180	365	130
	6.36-12.7	285	460	240	370	
	12.8-25.4	270	435			
	25.6-50.8	260	405			

Source: Ref. 1.

TABLE 11.7a Bearing and Shear Strengths of Sheet Alloys (English Units, ksi)

Alloy	Thickness, in	Bearing typical		Bearing minimum		Shear typical
		Yield	Ultimate	Yield	Ultimate	
AZ31B-O	0.016-0.250	37	66	29	60	26
	0.251-0.500	34	65	27	60	25
	0.501-2.000	33	65			25
	2.001-3.000	32	65			25
AZ31B-H24	0.016-0.249	47	77	43	68	29
	0.250-0.374	45	72	38	65	28
	0.375-0.500	40	70	34	63	27
	0.501-1.000	37	68			26
	1.001-2.000	35	66			26
	2.001-3.000	33	66			26
AZ31B-H26	0.250-0.374	46	72	40	68	28
	0.375-0.500	44	72	39	65	28
	0.501-0.752	40	72	36	65	28
	0.753-1.000	39	70			28
	1.001-1.499	37	68			27
	1.500-2.000	36	68			26
HK31A-O	0.016-0.250	28	58	24	51	24
	0.251-1.000	27	58	21	51	24
	1.001-3.000	27	57			24
HK31A-H24	0.016-0.125	41	67	34	57	26
	0.126-0.250	41	65	33	57	26
	0.251-1.000	44	68	31	57	27
	1.001-3.000	40	65			26
HM21A-T8	0.016-0.250	37	63	26	53	19
	0.251-0.500	41	35	54		
	0.501-1.000	39	63			
	1.001-2.000	38	59			

Source: Ref. 1.

11.4 COLUMNS

See Chapter 10 for a discussion of columns and the use of secant moduli data for arriving at buckling and crippling information. Figures 11.30 through 11.35 show secant modulus and tangent modulus curves; the secant moduli data are used to prepare the curves on buckling and crippling shown in Figures 11.36 through 11.49.

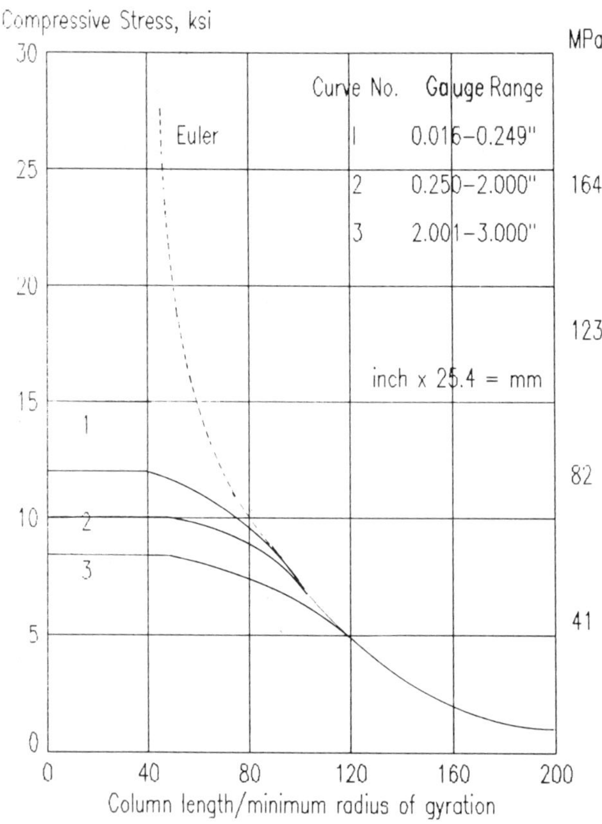

FIGURE 11.36 Column curves at minimum property levels for AZ31B-O sheet and plate.

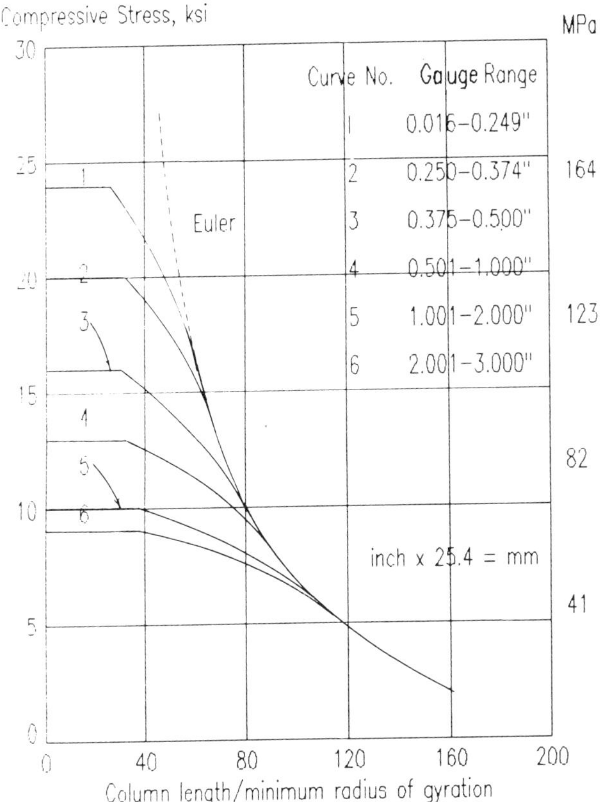

FIGURE 11.37 Column curves at minimum property levels for AZ31B-H24 sheet and plate.

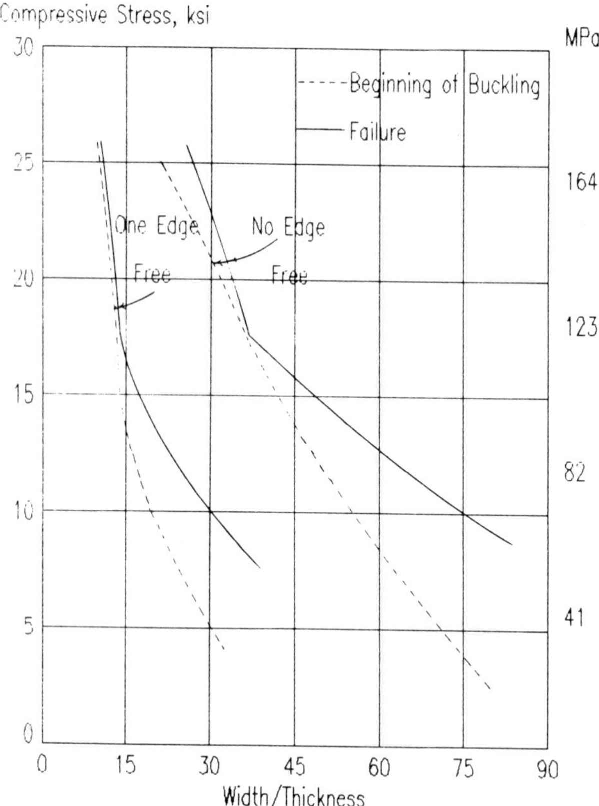

FIGURE 11.38 Crippling curves at minimum property levels for AZ31B-H24 sheet.

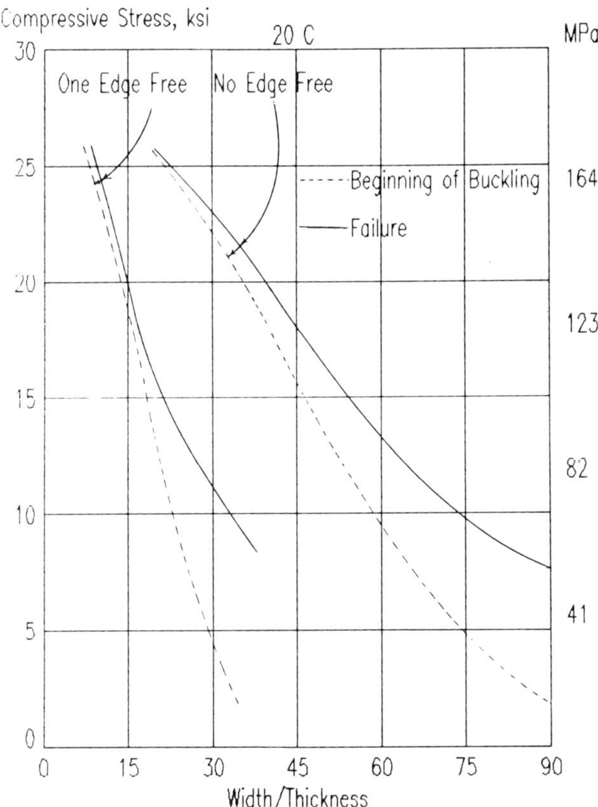

FIGURE 11.39 Buckling and crippling curves for HK31A-H24 sheet.

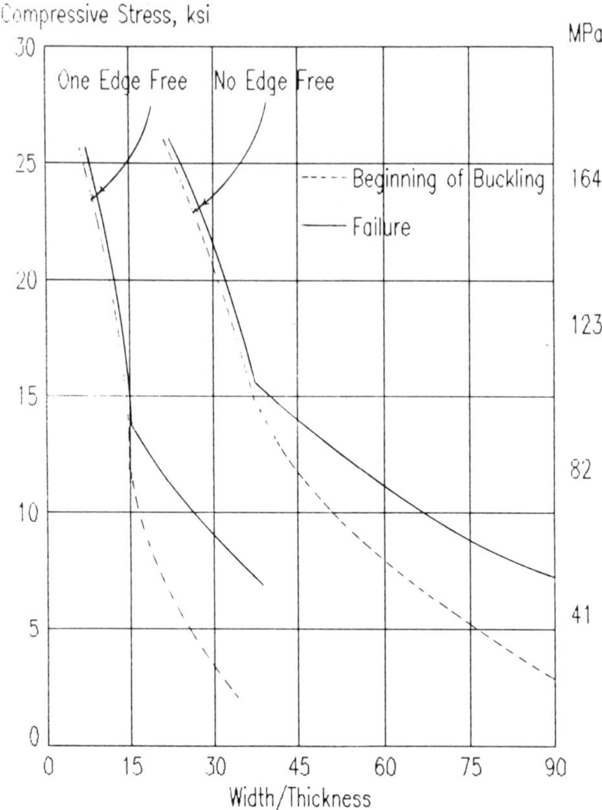

FIGURE 11.40 Buckling and crippling curves for HK31A-H24 sheet at 150°C.

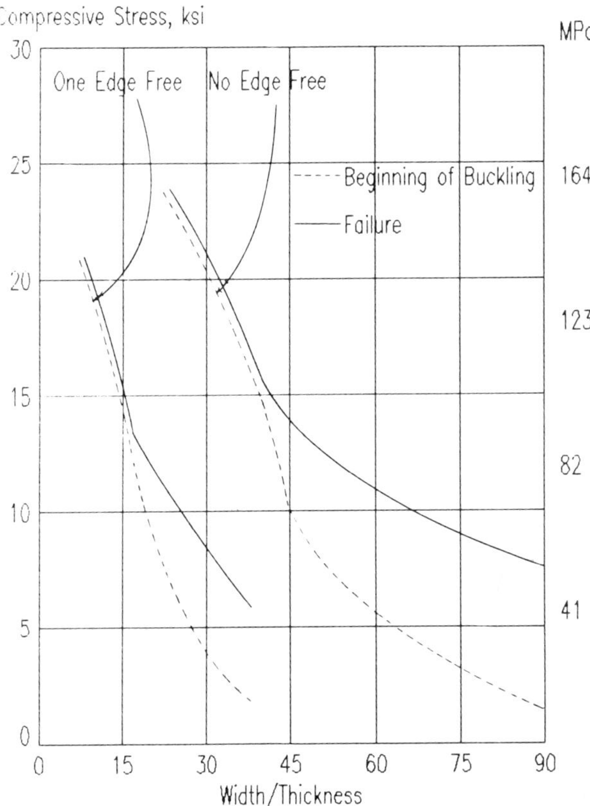

FIGURE 11.41 Buckling and crippling curves for HK31A-H24 sheet at 204°C.

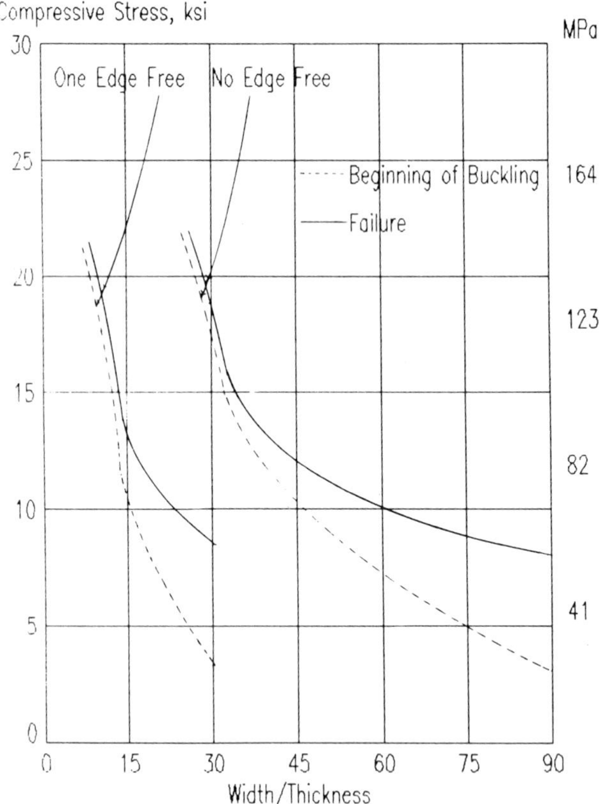

FIGURE 11.42 Buckling and crippling curves for HK31A-H24 sheet at 260°C.

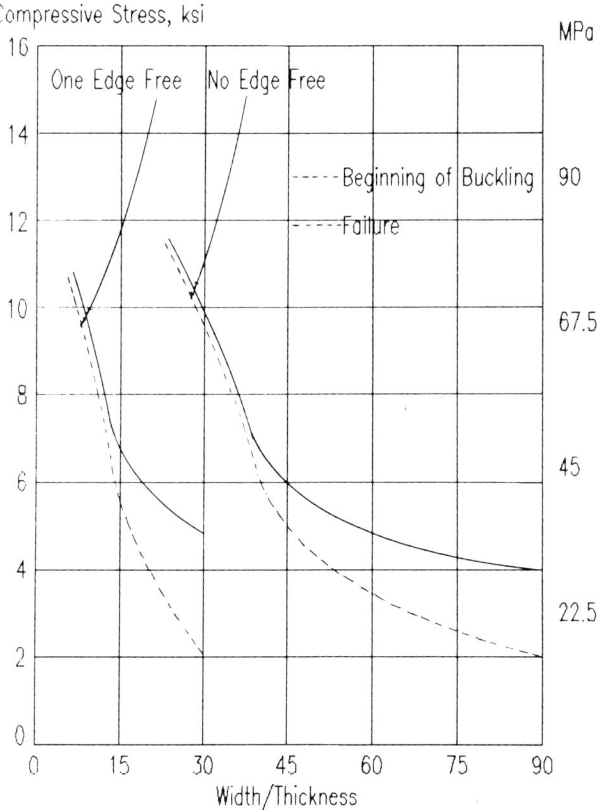

FIGURE 11.43 Buckling and crippling curves for HK31A-H24 sheet at 315°C.

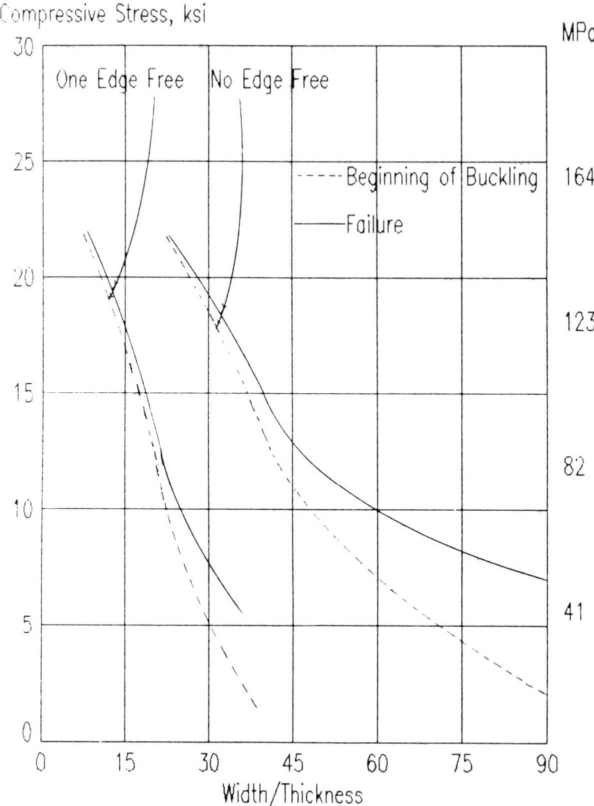

FIGURE 11.44 Buckling and crippling curves for HM21A-T8 sheet at 20°C.

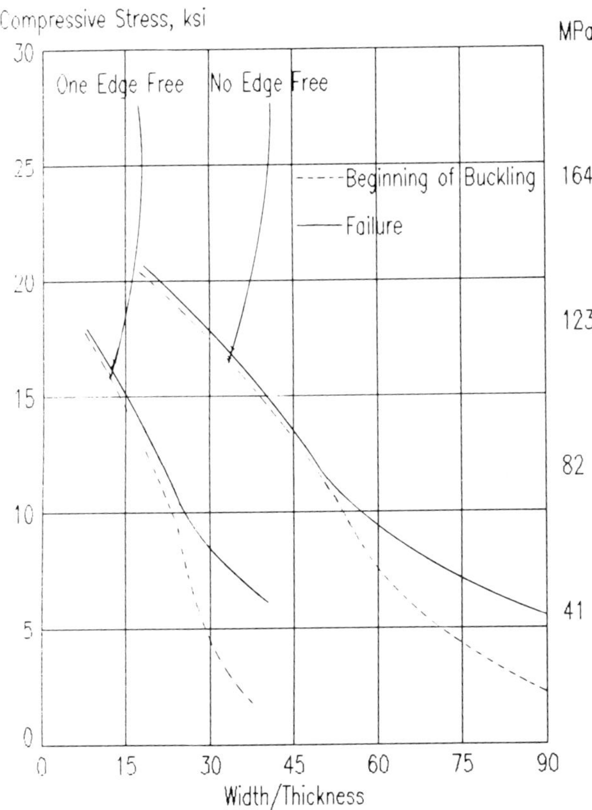

FIGURE 11.45 Buckling and crippling curves for HM21A-T8 sheet at 150°C.

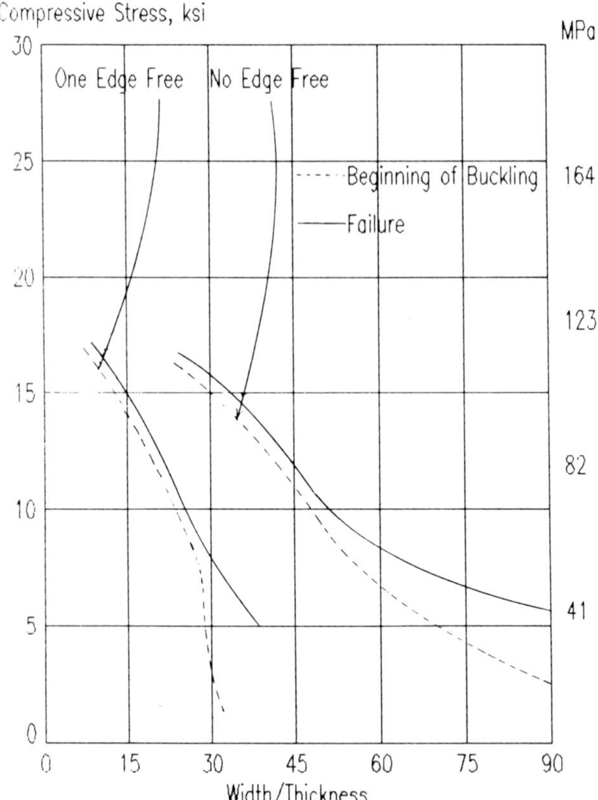

FIGURE 11.46 Buckling and crippling curves for HM21A-T8 sheet at 204°C.

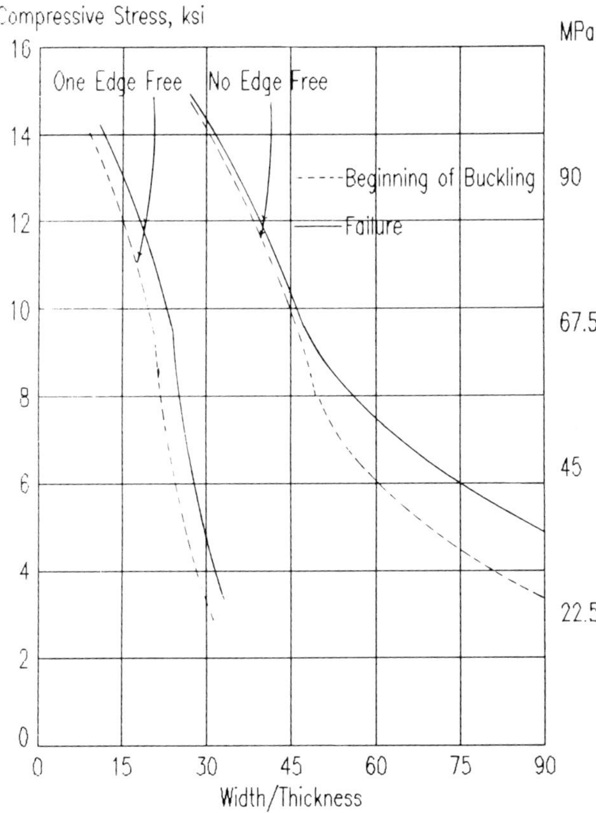

FIGURE 11.47 Buckling and crippling curves for HM21A-T8 sheet at 260°C.

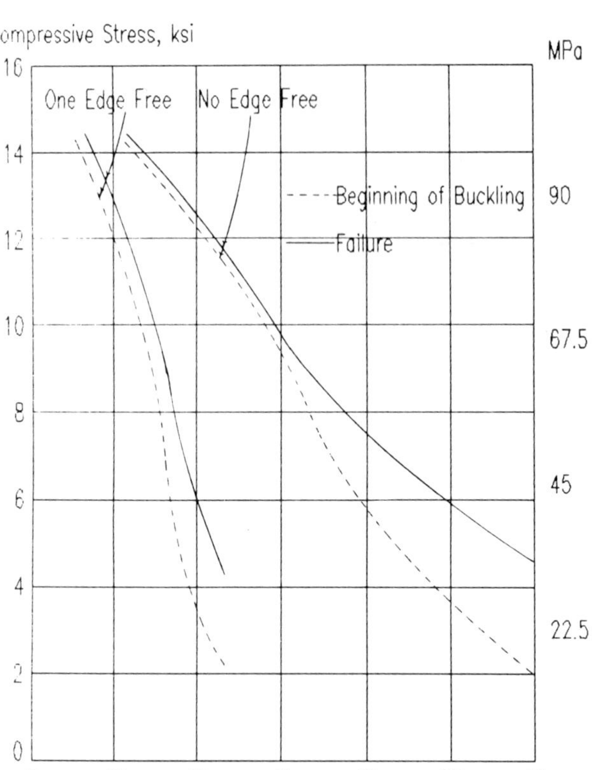

FIGURE 11.48 Buckling and crippling curves for HM21A-T8 sheet at 315°C.

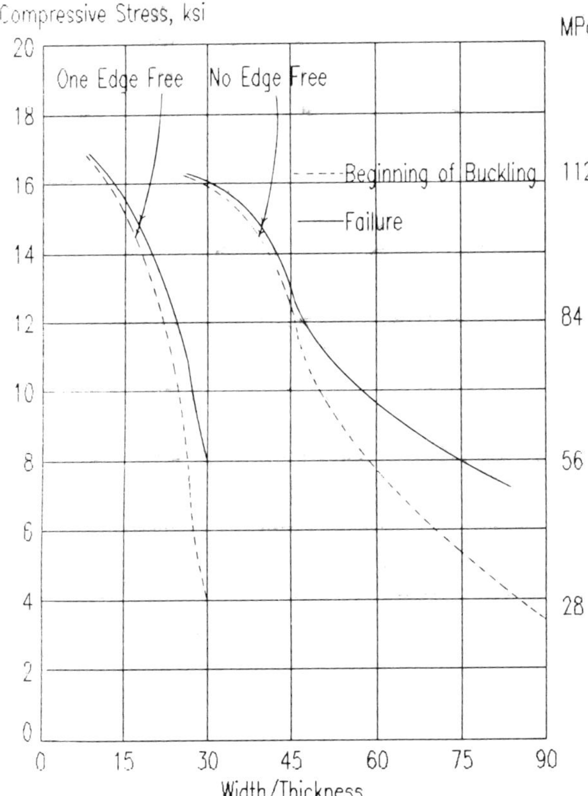

FIGURE 11.49 Buckling and crippling curves for ZE10A-O sheet at 20°C.

11.5 FATIGUE

Graphs showing the scatter bands for the fatigue of sheet and plate are given in Figures 11.50 through 11.58. The effects of R value, of surface conditions, and of temperature are shown. Scatter bands are not available for ZH11A, but values for the alloy are given in Table 11.8 [20].

TABLE 11.8 Fatigue Strength of ZH11A Sheet (Rotating Beam; Unnotched)

Temperature, °C	MPa				ksi			
	5×10^5	10^6	10^7	5×10^7	5×10^5	10^6	10^7	5×10^7
20	117	110	103	100	17	16	15	14.6
150		75				11		
200				59				8.5

Source: Ref. 20.

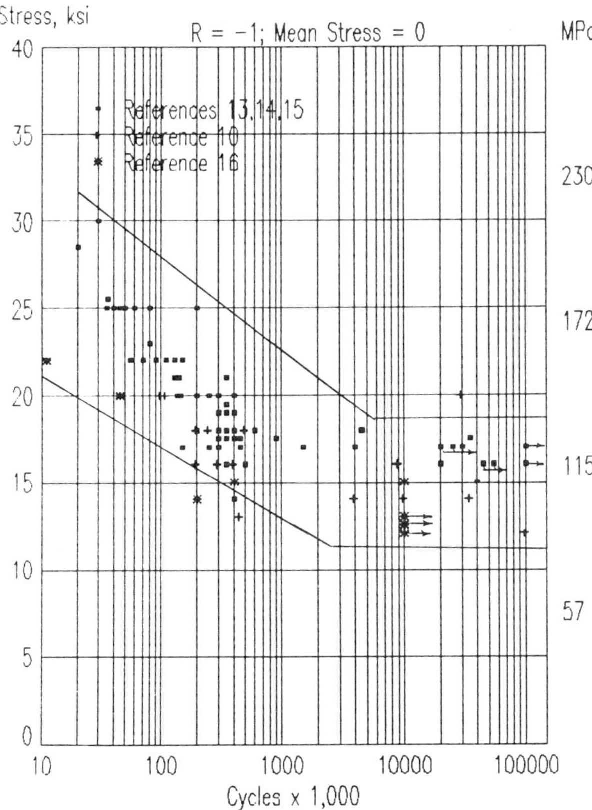

FIGURE 11.50 Cantilever bending fatigue for AZ31B-H24 sheet.

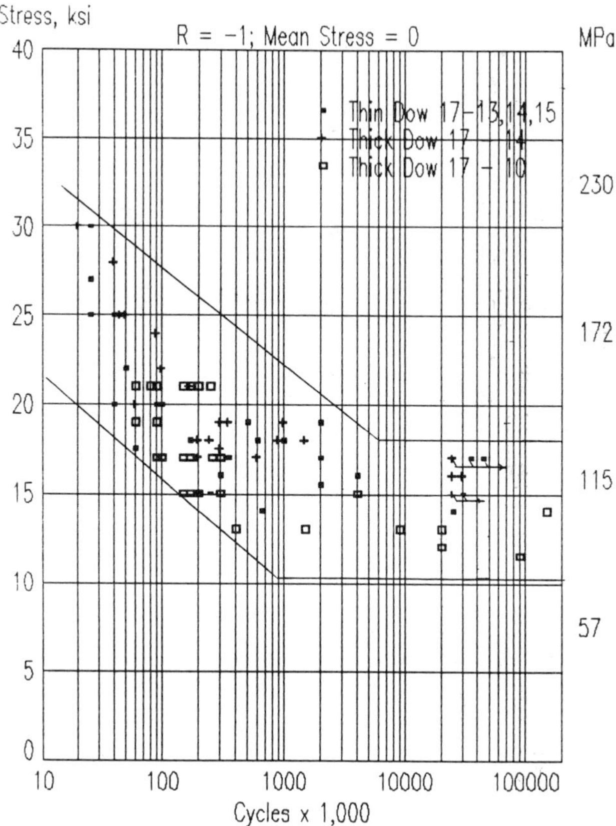

FIGURE 11.51 Cantilever bending fatigue for AZ31B-H24 sheet. (Data from Refs. 10, 13, 14, 15, and 17).

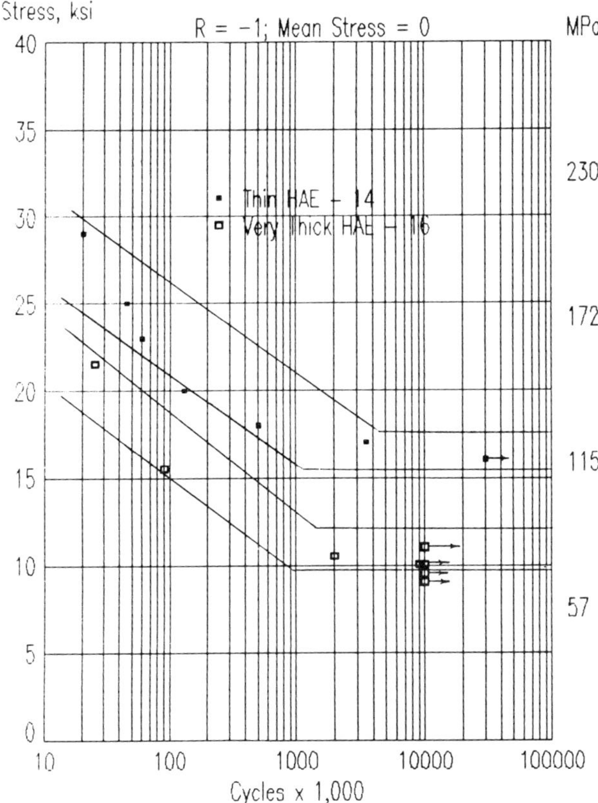

FIGURE 11.52 Cantilever bending fatigue for AZ31B-H24 sheet at 20°C. (Data from Refs. 14 and 16.)

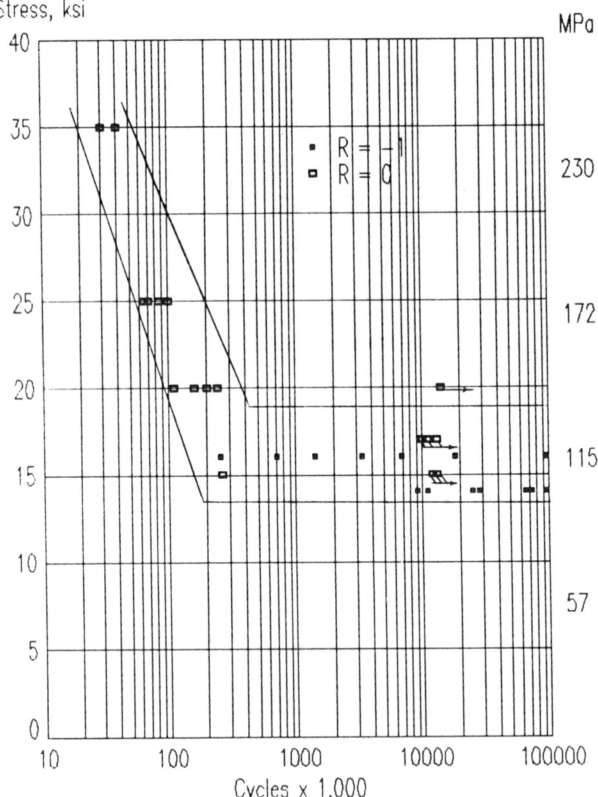

FIGURE 11.53 Cantilever bending fatigue for AZ31B-H24 plate at 20°C.

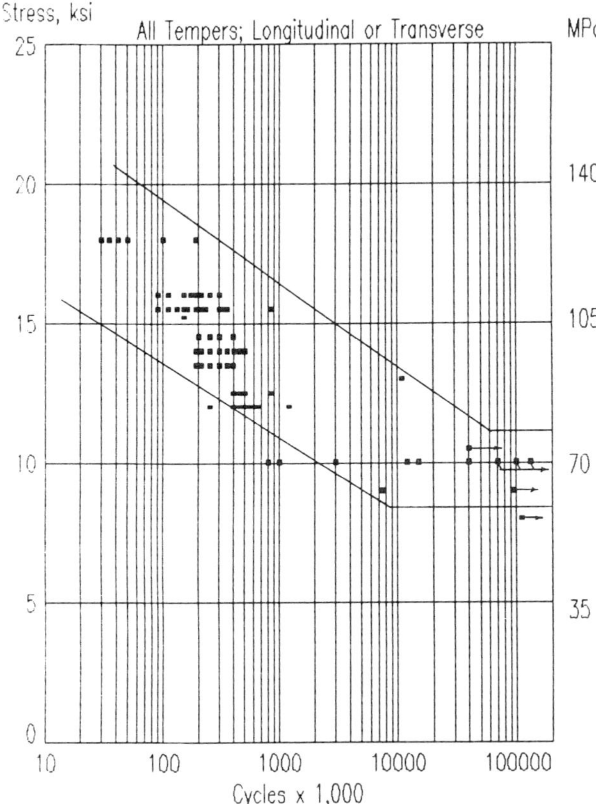

FIGURE 11.54 Cantilever bending fatigue for HK31A sheet at 20°C.

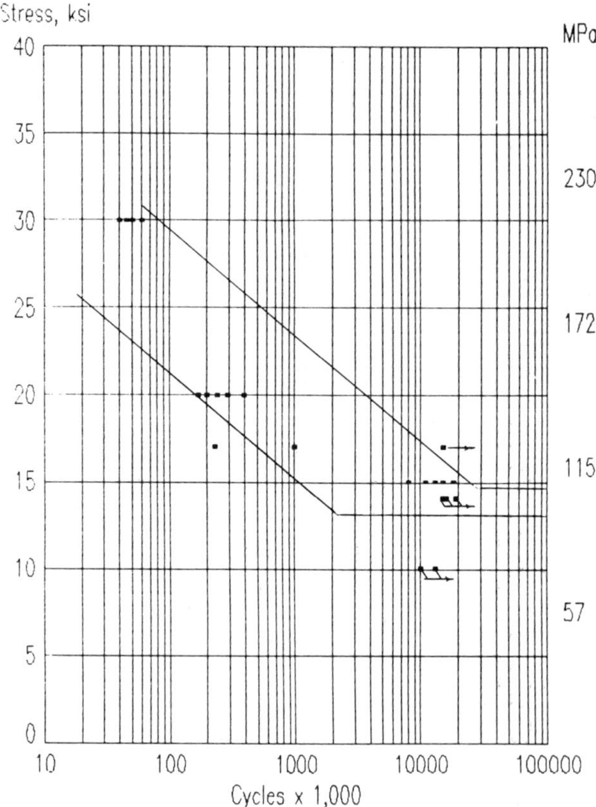

FIGURE 11.55 Cantilever bending fatigue for HK31A-H24 plate at 20°C (R = 0).

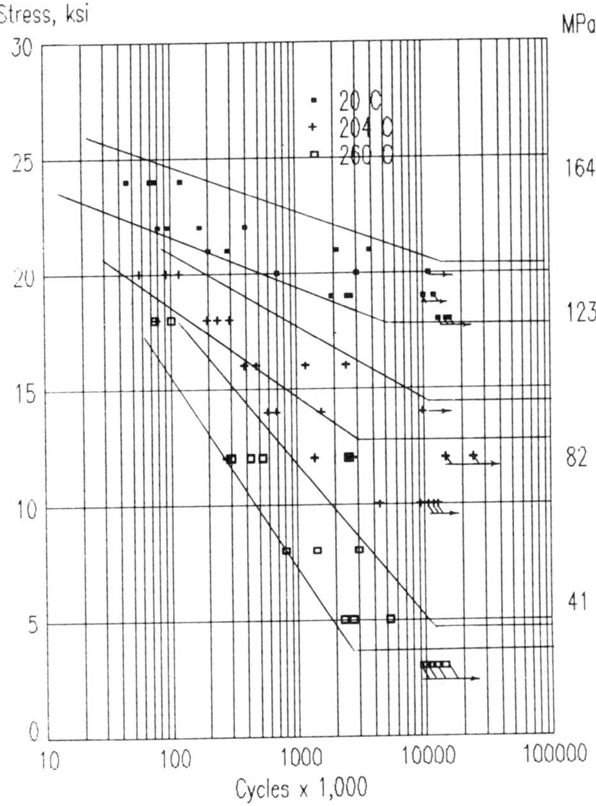

FIGURE 11.56 Axial-load fatigue for HK31A-H24 sheet (R = 0.25).

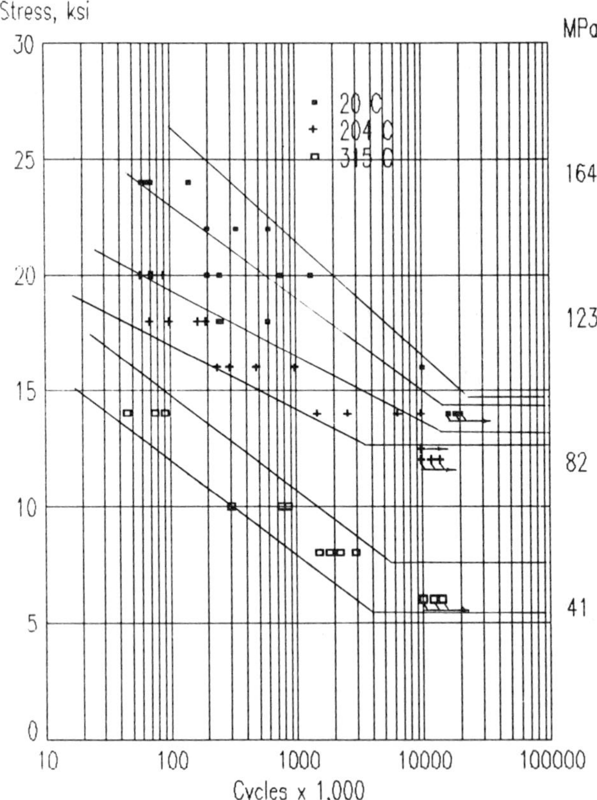

FIGURE 11.57 Axial-load fatigue for HM21A-T8 sheet (R = 0.25).

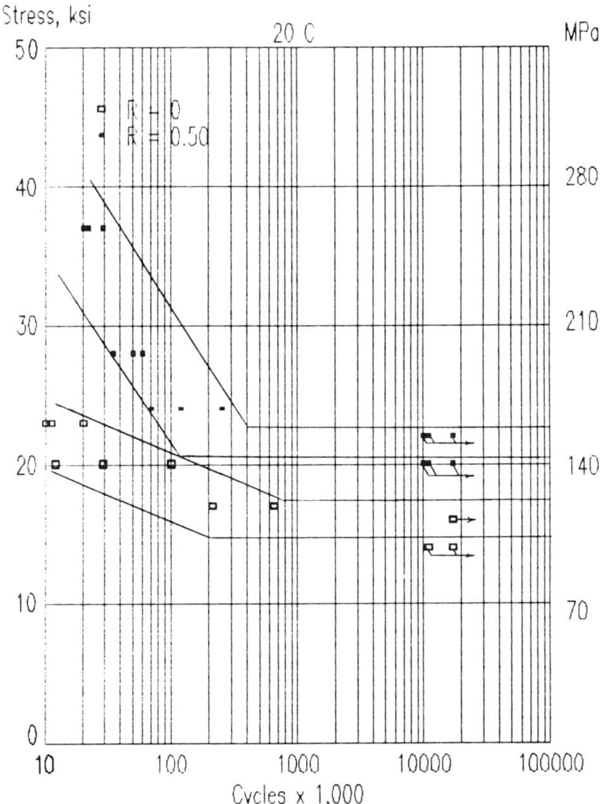

FIGURE 11.58 Axial-load fatigue for ZE10A sheet.

11.6 DENTING

Often a service failure is a dent rather than a break, a typical example being the fender of an automobile. Resistance to denting is influenced by a combination of ability to resist deformation and to restore deformation after removal of the load through elastic and anelastic recovery [17]. The low modulus of elasticity plus, even more important, the high anelastic behavior of magnesium results in a low residual dent after a load is removed. Good resistance to denting and good damping capacity are related through the anelastic behavior of magnesium.

Denting resistance is measured using the equipment shown in Figure 11.61. The tup is dropped from various heights onto the sheet secured in the frame, the resulting depression in the sheet measured, and the results related to various energy levels. The data shown in Figures 11.59 and 11.60, where the higher the ordinate number, the larger the dent, show that magnesium is superior to either aluminum or steel at equal weight. Even at an equal thickness, AZ31B is superior to mild steel and 5052 aluminum alloy and equivalent to 2024 aluminum alloy. Magnesium is very tough in its resistance to denting.

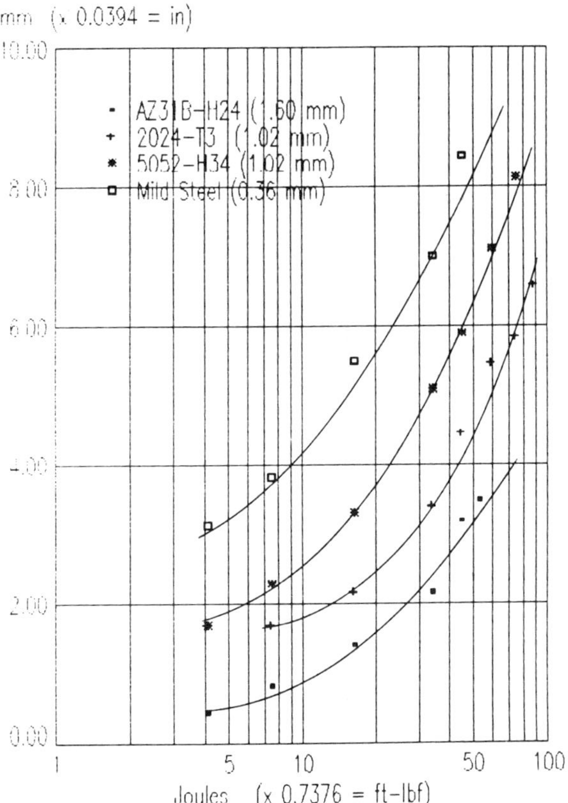

FIGURE 11.59 Equal weight comparison for dent resistance.

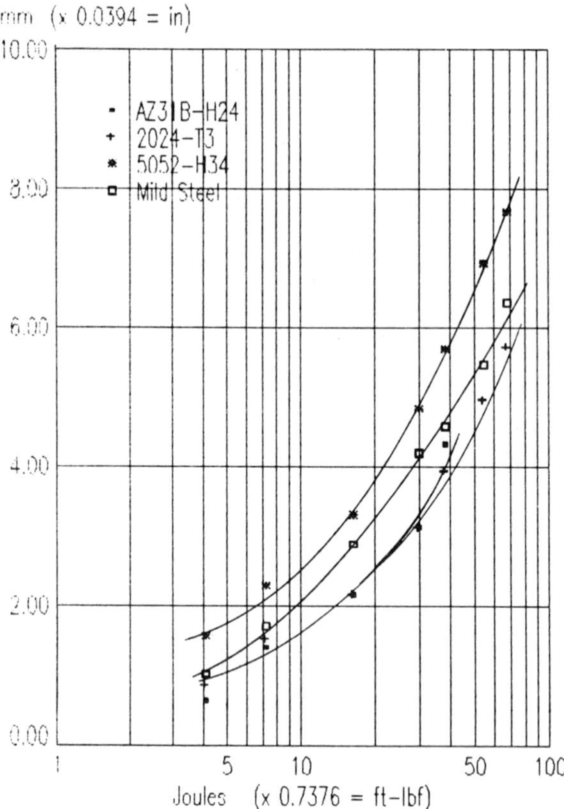

FIGURE 11.60 Equal gauge comparison for dent resistance.

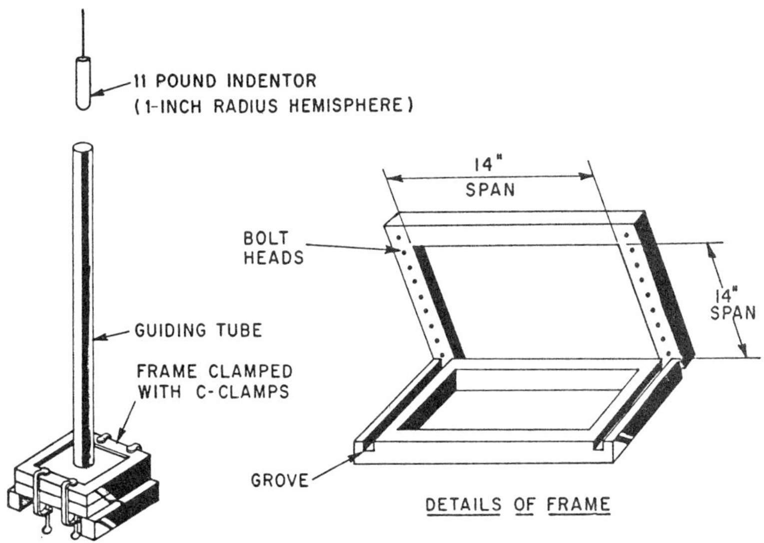

FIGURE 11.61 Dynamic dent test.

11.7 STRESS CORROSION

The stress-corrosion characteristics of magnesium sheet and plate alloys are given in Figures 11.62 through 11.69 [10]. Figure 11.63 is especially interesting in showing that a simple painted surface virtually eliminates stress corrosion. The three bars that did break, did so in the gauge mark, which is a small punch mark put on after painting. It is thus a defect in the painted surface. Even at these points, failure occurred only after three or more years at the highest stresses.

As shown in previous chapters for other forms, alloys containing aluminum can stress corrode under some service conditions. They must be stress relieved to remove residual stresses from operations such as welding. Alloys not containing aluminum do not, in a practical sense, have a stress-corrosion problem.

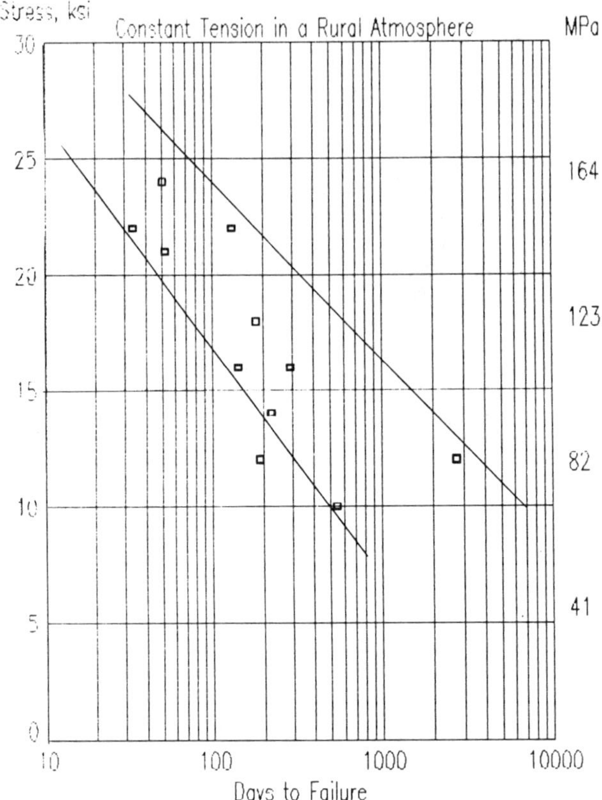

FIGURE 11.62 Stress-corrosion resistance for AZ31B-O sheet.

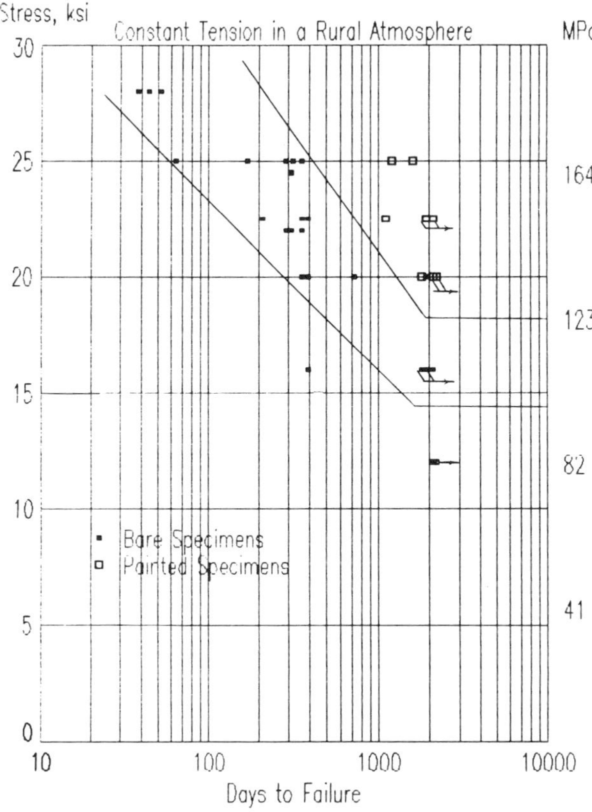

FIGURE 11.63 Stress-corrosion resistance for AZ31B-H14 sheet.

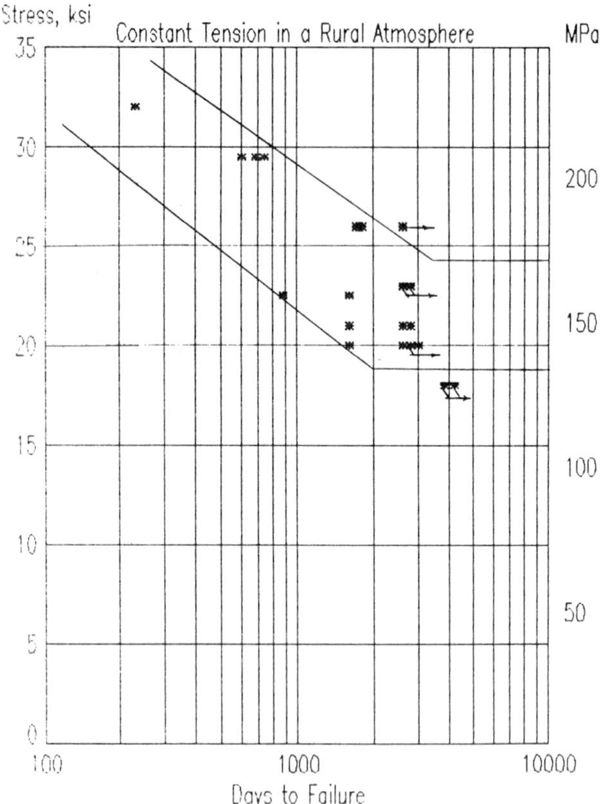

FIGURE 11.64 Stress-corrosion resistance for HK31A-H24 sheet.

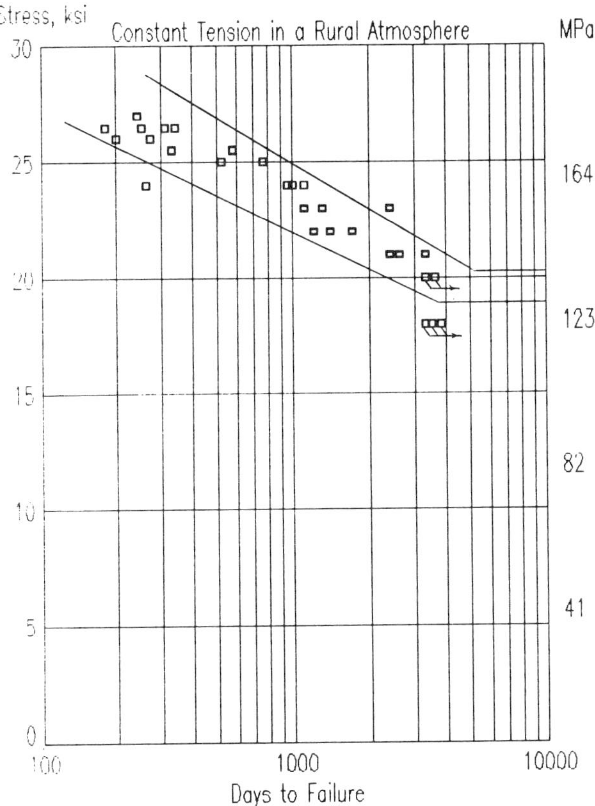

FIGURE 11.65 Stress-corrosion resistance for HM21A-T8 sheet.

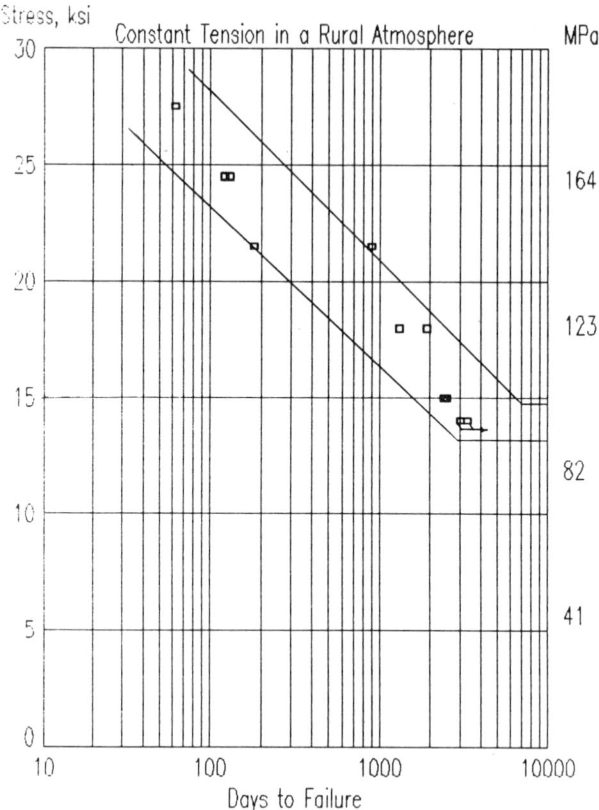

FIGURE 11.66 Stress-corrosion resistance for ZE10A-H24 sheet.

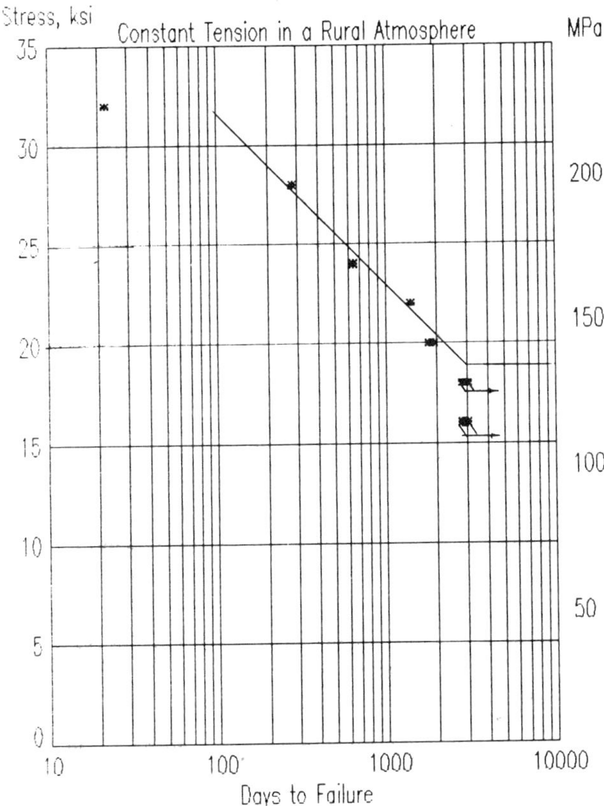

FIGURE 11.67 Stress-corrosion resistance for ZK30A-H24 sheet.

REFERENCES

1. Dow Chemical Company, "Magnesium Mill Products and Alloys," Bulletin Number 141-481, 1983.
2. Dow Chemical Company, "ZE10A Sheet & Plate Magnesium Alloy," Technical Service and Development Letter Enclosure, June 1, 1964.
3. Dow Chemical Company, "Magnesium Tooling Plate," Bulletin Number 141-232, 1983.
4. AMAX Specialty Metals Corporation, *Magnesium Alloys*,
5. Dow Chemical Company, "Magnesium in Design," Bulletin 141-213 , 1967.
6. Dow Chemical Company, "HM21A-T8 Magnesium Alloy Sheet & Plate," TS&D Letter Enclosure, April 24, 1964.
7. Dow Chemical Company, "HK31A-H24 and -O Sheet and Plate," TS&D Letter Enclosure, July 6, 1964.
8. R. W. Fenn, Jr. and L. F. Lockwood, "Low Temperature Properties of Welded Magnesium Alloys," *Welding Journal*, p. 352, American Welding Society, August 1960.
9. R. W. Fenn, Jr., "Low Temperature Properties of Cast and Wrought Magnesium Alloys," *Symposium on Low Temperature Properties of High-Strength Aircraft and Missile Materials*, STP No. 287, p. 61, American Society for Testing Materials, Metals Park, Ohio.
10. Dow Chemical Company, Private communication.
11. Raymond W. Fenn, Jr. and James A. Gusack, "Effect of Strain-Rate and Temperature on the Strength of Magnesium Alloys," *ASTM Proceedings 58*:685 (1958).
12. American Society for Metals, "Properties of Magnesium Alloys," *Metals Handbook*, American Society for Metals, 9th Edition, Vol. 2, 1979.
13. Sara J. Ketcham, "Evaluation of the Dow Chemical Company's New Anodic Treatment for Magnesium," *Report No. AML-NAM-AE-4136*, Aeronautical Materials Laboratory, Naval Air Materials Center, Philadephia, Pennsylvania, July 27, 1954.
14. Sara J. Ketcham, "Anodic Coatings for Magnesium," *Report No. NAMC-AML-AE-1113*, Aeronautical Materials Laboratory, Naval Air Materials Center, Philadelphia, Pennsylvania, February 3, 1960.
15. Sara J. Ketcham, "Investigation of Anodic Coatings for Magnesium Alloys," *Report No. NAMC-AML-1347*, Aeronautical Materials Laboratory, Naval Air Materials Center, Philadelphia, Pennsylvania, January 10, 1962.
16. J. A. Bennett, "The Effect of HAE Coating on the Fatigue Strength of Magnesium Alloy Specimens," *NBS Report 3170*, Metallurgy Division, National Bureau of Standards, March 22, 1954.

17. J. A. Gusack and D. E. Ritzema, "Dent Resistance of Magnesium Alloys Compared to Other Materials," *Proceedings of the International Magnesium Association*, 1961.
18. Colin J. Smithells, *Metal Reference Book*, Butterworths, London and Boston, 1976.
19. Magnesium Elektron Ltd., "ZTY, A Wrought Magnesium Alloy for Use at Elevated Temperature," 1961.
20. "Elektron Magnesium alloy—ZTY," Engineering Alloys Alloy Digest, Digest, Inc, New Jersey, 1968.

12

Mechanical Properties of Forgings

Magnesium forgings are almost always die forgings made on hydraulic presses; hammer forging is seldom used. Alloys AZ31B, AZ61A, AZ80A, ZK30A, and ZK60A are used mainly at room or slightly elevated temperatures. Alloys EK31A, HM21A, and ZH11A are suitable for high-temperature applications.

See Chapter 5 for design considerations.

Mechanical properties are given in Tables 12.1 through 12.7. The properties developed by forgings are dependent upon the alloy composition, the specific forging conditions used, and the orientation of the testing direction to the working direction used during forging. Hence, in the tables, data from specific production forgings are given where possible, including in each case, the orientation of the test specimen with respect to the working direction. Static tensile properties are especially sensitive to the orientation of the test specimen with respect to the forging flow direction. See, for example, Table 12.3 where the tensile yield strength measured in one direction can be as much as double that measured in another direction. Neither creep (Table 12.4) nor fatigue (Table 12.7) are sensitive to direction. The flow patterns developed during forging are dependent not only on the forging practice, but also on the starting material. If an extruded bar is used, for example, as the starting material for forging, the flow patterns already developed in the extrusion affect the flow results in the forging. Selection of the correct starting material can be important.

It is wise to consult with the forger before a design is finalized in order to assure that the optimum properties are obtained in the places and directions that are the most important.

Information on the stress corrosion of forged ZK60A is given in Figure 12.1. The stress-corrosion limit is the same for the -F and the -T6 condition in spite of a 54% higher tensile yield strength in the -T6 condition. However, the limit is high enough that it should seldom if ever be a limiting factor in design.

TABLE 12.1 Longitudinal Tensile Properties of Forgings at Room Temperature (SI Units)

		Typical				Minimum		
			MPa				MPa	
Alloys	Temper	%E	TYS	CYS	TS	%E	TYS	TS
AZ31B	F	9	195	85	260	6	130	235
AZ61A	F	12	180	115	195	6	150	260
AZ80A	F	8	215	170	315	5	180	290
	T5	6	235	195	345	2	195	290
	T6	5	250	185	345			
EK31A	T6	7	185	155	290			
HK31A	T5	21	195	140	260			
HM21A	T5	9	150	110	235	3[a]	170[a]	230[a]
ZH11A	F	13	147		232			
ZK30A	F	8	224	193	309			
ZK60A	T5	16	205	195	305	7	180	290
	T6	11	270	170	325	4	220	295

[a]Applicable only to forgings over 102 mm in thickness.
Source: Refs. 1, 2, 3.

TABLE 12.1a Longitudinal Tensile Properties of Forgings at Room Temperature (English Units)

		Typical				Minimum		
			ksi				ksi	
Alloy	Temper	%E	TYS	CYS	TS	%E	TYS	TS
AZ31B	F	9	29	12	38	6	19	34
AZ61A	F	12	26	17	28	6	22	38
AZ80A	F	8	31	25	46	5	26	42
	T5	6	34	28	50	2	28	42
	T6	5	36	27	50			
EK31A	T6	7	27	22	42			
HK31A	T5	21	28	20	38			
HM21A	T5	9	22	16	34	3[a]	25[a]	33[a]
ZH11A	F	13	21		34			
ZK30A	F	8	32	28	45			
ZK60A	T5	16	30	28	44	7	26	42
	T6	11	39	25	47	4	32	43

[a]Applicable only to forgings over 4 inches in thickness.
Source: Refs. 1, 2, 3, 6.

TABLE 12.2 Effect of Temperature and Direction of Test on Tensile Properties of Forgings (SI Units)

Alloy	Temper	Temperature, °C	Type of forging	Direction	%E	MPa		
						TYS	CYS	TS
AZ80A	F	20	Pancake	Radial	4	195		285
		95			16	160		275
		150			41	140		210
		200			52	90		135
	T5	20			3	220		320
		95			9	180		280
		150			44	135		200
		200			51	95		130
	T6	20			3	250		345
		95			17	210		295
		150			38	150		210
		200			36	100		135
		20		Tangential	2	215		290
		150			24	145		195
		200			32	95		130
EK31A	T6	20	Production	Longitudinal	7	185	155	290
				Transverse	6	220	155	315
				Axial	9	145	130	270
		150		Longitudinal	17	165		215
				Transverse	17	180		230
				Axial	23	140		205

		200		Longitudinal	16	165		200
				Transverse	17	170		210
				Axial	23	135		190
HK31A	T5	20	Pancake	Longitudinal	17	230	185	260
		150			34	195		195
		200			31	165		165
		260			24	145		145
		315			98	55		85
HM21A	T5	20	Production	Longitudinal	14	135	105	230
				Transverse	16	150	115	235
				Axial	14	110	90	210
		200		Longitudinal	30	85		115
				Transverse	37	95		120
				Axial	38	75		90
HM21A	T5	315	Production	Longitudinal	24	70		90
				Transverse	32	80		100
				Axial	27	70		90
		370		Longitudinal	26	65		70
				Transverse	36	65		75
				Axial	33	60		75
ZH11A	F	20		Longitudinal	18	186		290
		100			24	159		220
		150			30	145		186
		200			39	131		152
		250			50	117		131

TABLE 12.2 (*Continued*)

Alloy	Temper	Temperature, °C	Type of forging	Direction	%E	MPa TYS	MPa CYS	MPa TS
ZK 60A	T5	20	Pancake	Radial	13	200		295
		150			49	115		145
		200			51	70		95
	T6	20		Radial	11	240		315
		150			35	150		180
		200			41	105		115
		20		Tangential	10	260		330
		150			33	160		185
		200			37	105		125

Source: Refs. 3, 5, 7.

TABLE 12.2a Effect of Temperature and Direction of Test on Tensile Properties of Forgings (English Units)

Alloy	Temper	Temperature, °C	Type of forging	Direction	%E	ksi		
						TYS	CYS	TS
AZ80A	F	20	Pancake	Radial	4	28		41
		95			16	23		40
		150			41	20		30
		200			52	13		20
	T5	20			3	32		46
		95			9	26		41
		150			44	20		29
		200			51	14		19
	T6	20			3	36		50
		95			17	30		43
		150			38	22		30
		200			36	15		20
		20		Tangential	2	31		42
		150			24	21		28
		200			32	14		19
EK31A	T6	20	Production	Longitudinal	7	27	22	42
				Transverse	6	32	22	46
				Axial	9	21	19	39
		150		Longitudinal	17	24		31
				Transverse	17	26		33
				Axial	23	20		30

TABLE 12.2a (*Continued*)

Alloy	Temper	Temperature °C	Type of forging	Direction	%E	ksi TYS	ksi CYS	ksi TS
		200		Longitudinal	16	24		29
				Transverse	17	25		30
				Axial	23	20		28
HK31A	T5	20	Pancake	Longitudinal	17	33	27	38
		150			34	28		28
		200			31	24		24
		260			24	21		21
		315			98	8		12
HM21A	T5	20	Production	Longitudinal	14	20	15	33
				Transverse	16	22	17	34
				Axial	14	16	13	30
		200		Longitudinal	30	12		17
				Transverse	37	14		17
				Axial	38	11		13
HM21A	T5	315	Production	Longitudinal	24	10		13
				Transverse	32	12		15
				Axial	27	10		13

		370		Longitudinal	26	9	10
				Transverse	36	9	11
				Axial	33	9	11
ZH11A	F	20		Longitudinal	18	27	42
		100			24	23	32
		150			30	21	27
		200			39	19	22
		250			50	17	19
ZK60A	T5	20	Pancake	Radial	13	29	43
		150			49	17	21
		200			51	10	14
	T6	20			11	35	46
		150			35	22	26
		200			41	15	17

Source: Refs. 3, 5, 7.

TABLE 12.3 Effect of Forging and Orientation on Tensile Properties of ZK60A at Room Temperature (SI Units)

			- T5 Temper			- T6 Temper		
				MPa			MPa	
Forging	Location	Orientation	%E	TYS	TS	%E	TYS	TS
Small		Longitudinal	14	270	325	9	365	400
Bracket		Transverse	17	170	285	12	215	305
Impeller		Radial	10	165	275	12	205	305
		Tangential	14	205	295	13	275	330
		Axial	20	145	270	18	170	290
Brake		Tangential	15	205	295	13	260	315
Carrier		Axial	16	165	275	14	185	290
Aircraft	Rim	Tangential	18	215	310	12	315	340
Nose Wheel		Axial	12	180	290	9	220	305
	Hub	Tangential	16	205	310	9	275	315
		Axial	16	125	270	14	205	295
Large	Rim	Tangential	16	195	295	12	270	330
Aircraft		Radial	17	130	260	16	185	290
Wheel		Axial	13	130	260	14	200	295
	Hub	Axial	17	145	285	17	145	290
	Web	Tangential	22	170	285	15	235	295
		Radial	19	185	290	16	240	325
		Axial	15	130	305	14	150	315

Source: Refs. 4, 7.

TABLE 12.3a Effect of Forging and Orientation on Tensile Properties of ZK60A at Room Temperature (English Units)

			-T5 Temper			-T6 Temper		
				ksi			ksi	
Forging	Location	Orientation	%E	TYS	TS	%E	TYS	TS
Small		Longitudinal	14	39	47	9	53	58
Bracket		Transverse	17	25	41	12	31	44
Impeller		Radial	10	24	40	12	30	44
		Tangential	14	30	43	13	40	48
		Axial	20	21	39	18	25	42
Brake		Tangential	15	30	43	13	38	46
Carrier		Axial	16	24	40	13	27	42
Aircraft	Rim	Tangential	18	31	45	12	46	49
Nose Wheel		Axial	12	26	42	9	32	44
	Hub	Tangential	16	30	45	9	40	46
		Axial	16	18	39	14	30	43
Large	Rim	Tangential	16	28	43	12	39	48
Aircraft		Radial	17	19	38	16	27	42
Wheel		Axial	13	19	38	14	29	43
	Hub	Axial	17	21	41	17	21	42
	Web	Tangential	22	25	41	15	34	43
		Radial	19	27	42	16	35	47
		Axial	15	19	44	14	22	46

Source: Refs. 4, 7.

TABLE 12.4 Creep Strengths of Forgings—Stresses for Given Total Extensions after 100 Hours Loading (SI Units)

Alloy	Temper	Forging	Orientation	Temperature, °C	MPa 0.2% T.E.	MPa 0.5% T.E.
AZ80A	F	Pancake	Radial	150	25	45
				200	5	10
	T5			150	25	40
				200	5	15
	T6			150	20	40
				200	5	10
EK31A	T6	Wheel rim	Axial	150	85	140
				200	70	95
			Tangential	150	85	165
				200	65	105
		Impeller	Radial	150	90	140
				200	60	105
HK31A	T5	Pancake	Radial	200	85	130
				260	50	70
				315	10	15
HM21A	T5	Wheel rim	Axial	200	65	85
				315	50	55
				370	20	25
			Tangential	200	65	85
				315	50	55
				370	25	25
		Impeller	Radial	200	75	105
				315	40	45
				370	20	25
ZK60A	T5	Pancake	Radial	150	10	15
				200	2	3
	T6	Pancake	Radial	150	25	45
				200	10	15

Source: Refs. 3, 4, 5.

TABLE 12.4a Creep Strengths of Forgings—Stresses for Given Total Extensions after 100 Hours Loading (English Units)

Alloy	Temper	Forging	Orientation	Temperature, °C	ksi 0.2% T.E.	ksi 0.5% T.E.
AZ80A	F	Pancake	Radial	150	4	7
				200	0.5	1.5
	T5			150	4	6
				200	0.5	2
	T6			150	3	6
				200	0.5	1.5
EK31A	T6	Wheel rim	Axial	150	12	20
				200	10	14
			Tangential	150	12	24
				200	9	15
		Impeller	Radial	150	13	20
				200	9	15
HK31A	T5	Pancake	Radial	200	12	19
				260	7	10
				315	1.5	2
HM21A	T5	Wheel rim	Axial	200	9	12
				315	7	8
				370	3	4
			Tangential	200	9	12
				315	7	8
				370	4	4
		Impeller	Radial	200	11	15
				315	6	7
				370	3	4
ZK60A	T5	Pancake	Radial	150	1.5	2
				200	0.3	0.4
	T6			150	3	7
				200	1.5	2

Source: Refs. 3, 4, 5.

TABLE 12.5 Creep Properties of ZH11A Forgings

Temperature, °C	Stress, MPa	% Creep strain after specified hours				
		50	100	200	500	1000
SI Units						
150	93	0.02	0.02	0.02	0.02	0.025
	108	0.01	0.01	0.02		
200	85	0.02	0.025	0.03	0.03	0.04
	100	0.02	0.03	0.03	0.03	0.04
English Units	Stress, ksi					
150	13.5	0.02	0.02	0.02	0.02	0.025
	15.7	0.01	0.01	0.02		
200	12.4	0.02	0.025	0.03	0.03	0.04
	14.6	0.02	0.03	0.03	0.03	0.04

Source: Ref. 7.

TABLE 12.6 Bearing, Shear, and Toughness of Forgings (SI Units)

Alloy	Temper	Forging	Orientation	Temperature, °C	MPa Bearing Yield	MPa Bearing Ultimate	MPa Shear	Notch charpy, joules
AZ31B	F	Pancake	Longitudinal	20	250	485	130	
AZ61A	F	Pancake	Longitudinal	20	285	495	140	
AZ80A	F	Pancake	Longitudinal	20			50	
	T5					550	160	
EK31A	T6	Production	Tangential	20	315	430	160	2
				150	275	370	130	
				200	270	350	125	
			Axial	20	285	345	145	4
				150	270	360	125	
				200	275	340	115	
HM21A	T5	Production	Tangential	20	150	235	115	7
				200	110	195	75	
				315	110	165	60	
			Axial	370	160	250	125	7
				20	160	250	125	
				200	125	200	75	
				315	115	185	60	
				370	105	130	55	

TABLE 12.6 (*Continued*)

					MPa			
					Bearing			
Alloy	Temper	Forging	Orientation	Temperature, °C	Yield	Ultimate	Shear	Notch charpy, joules
ZK60A	T5	Wheel rim	Tangential	20	285	420	165	6
				200	145	185	45	
			Axial	20	290	380	165	6
				200	150	185	50	
			Radial	20				6
	T6	Wheel rim	Tangential	20	340	475	180	3
				200	180	240	50	
			Axial	20	305	420	170	3
				200	195	260	170	
			Radial	20				5

Source: Refs. 2, 3, 4.

TABLE 12.6a Bearing, Shear, and Toughness of Forgings (English Units)

Alloy	Temper	Forging	Orientation	Temperature, °C	ksi: Bearing Yield	ksi: Bearing Ultimate	ksi: Shear	Notch charpy, ft-lbf
AZ31B	F	Pancake	Longitudinal	20	36	70	10	
AZ61A	F	Pancake	Longitudinal	20	41	72	20	
AZ80A	F	Pancake	Longitudinal	20			22	
	T5					80	23	
EK31A	T6	Production	Tangential	20	46	62	23	0.7
				150	40	54	19	
				200	39	51	18	
			Axial	20	41	50	21	3.2
				150	39	52	18	
				200	40	49	17	
HM21A	T5	Production	Tangential	20	22	34	17	5.2
				200	16	28	11	
				315	16	24	9	
				370	14	19	8	
			Axial	20	23	36	18	5.2
				200	18	29	11	
				315	17	27	9	
				370	15	19	8	

TABLE 12.6a (*Continued*)

Alloy	Temper	Forging	Orientation	Temperature, °C	ksi: Bearing Yield	ksi: Bearing Ultimate	ksi: Shear	Notch charpy, ft-lbf
ZK60A	T5	Wheel rim	Tangential	20	41	61	24	4.4
				200	21	27	7	
			Axial	20	42	55	24	4.4
				200	22	27	7	
			Radial	20				4.4
	T6		Tangential	20	49	69	26	2.2
				200	26	35	7	
			Axial	20	44	61	25	2.2
				200	28	38	25	
			Radial	20				3.7

Source: Refs. 2, 3, 4.

TABLE 12.7 Fatigue Properties of Forgings (SI Units)

Alloy	Temper	Forging	Orientation	Fatigue Test	Surface	Stress (MPa) at cycles 10^5	10^6	10^7	10^8
AZ61A	F	Pancake	Longitudinal	Rotating Beam	Polished	180	150	145	140
				Flexure[a]	As forged	110	85	75	
AZ81A	F	Pancake	Longitudinal	Rotating	Polished	200	170	145	130
	T5			Beam		195	160	140	125
						170	140	125	105
	T6			Flexure[a]	As forged	125	95	95	
EK31A	T6	Wheel rim	Tangential	Rotating	Polished	130	95	90	90
				Beam	Notched[b]	95	70	60	60
				Axial[c]	Polished	200	165	150	
		(tested at 200°C)				180	145	120	
HM21A	T5	Wheel rim	Tangential	Rotating	Polished	110	85	60	60
				Beam	Notched[b]	70	55	40	35
				Axial[c]	Polished	145	115	95	
		(tested at 200°C)				115	110	105	
		(tested at 315°C)				95	85	70	

TABLE 12.7 (*Continued*)

Alloy	Temper	Forging	Orientation	Fatigue Test	Surface	Stress (MPa) at cycles 10^5	10^6	10^7	10^8
ZK 60A	T5	Wheel rim	Tangential	Rotating	Polished	160	140	125	125
			Axial	Beam	Polished	160	140	125	125
					Notched[b]	95	85	70	60
			Axial	Flexure[a]	Polished	130	95	90	85
	T6	Wheel rim	Tangential	Rotating	Polished	185	150	130	115
			Axial	Beam	Polished	185	150	130	115
					Notched[b]	115	90	70	60
			Axial	Flexure[a]	Polished	140	125	110	105

[a] $R = -1$
[b] Stress concentration = 2.0
[c] $R = 0.25$
Source: Refs. 2, 3, 4.

TABLE 12.7a Fatigue Properties of Forgings (English Units)

Alloy	Temper	Forgings	Orientation	Fatigue Test	Surface	Stress (ksi) at cycles			
						10^5	10^6	10^7	10^8
AZ61A	F	Pancake	Longitudinal	Rotating Beam	Polished	26	22	21	20
				Flexure[a]	As forged	16	12	11	
AZ81A	F	Pancake	Longitudinal	Rotating	Polished	29	25	21	19
	T5			Beam		28	23	20	18
	T6					25	20	18	15
				Flexure[a]	As forged	18	14	14	
EK31A	T6	Wheel rim	Tangential	Rotating	Polished	19	14	13	13
				Beam	Notched[b]	14	10	9	9
				Axial[c]	Polished	29	24	22	
		(tested at 200°C)				26	21	17	
HM21A	T5	Wheel rim	Tangential	Rotating	Polished	16	12	9	9
				Beam	Notched[b]	10	8	6	5
				Axial[b]	Polished	21	17	14	
		(tested at 200°C)				17	16	15	
		(tested at 315°C)				14	12	10	

TABLE 12.7a (*Continued*)

Alloy	Temper	Forgings	Orientation	Fatigue Test	Surface	Stress (ksi) at cycles			
						10^5	10^6	10^7	10^8
ZK 60A	T 5	Wheel rim	Tangential	Rotating	Polished	23	20	18	18
			Axial	Beam	Polished	23	20	18	18
					Notched[b]	14	12	10	9
			Axial	Flexure[a]	Polished	19	14	13	12
	T 6	Wheel rim	Tangential	Rotating	Polished	27	22	19	17
			Axial	Beam	Polished	27	22	19	17
					Notched[b]	17	13	10	9
			Axial	Flexure[a]	Polished	20	18	16	15

[a] R = -1
[b] Stress concentration = 2.0
[c] R = 0.25
Source: Refs. 2, 3, 4.

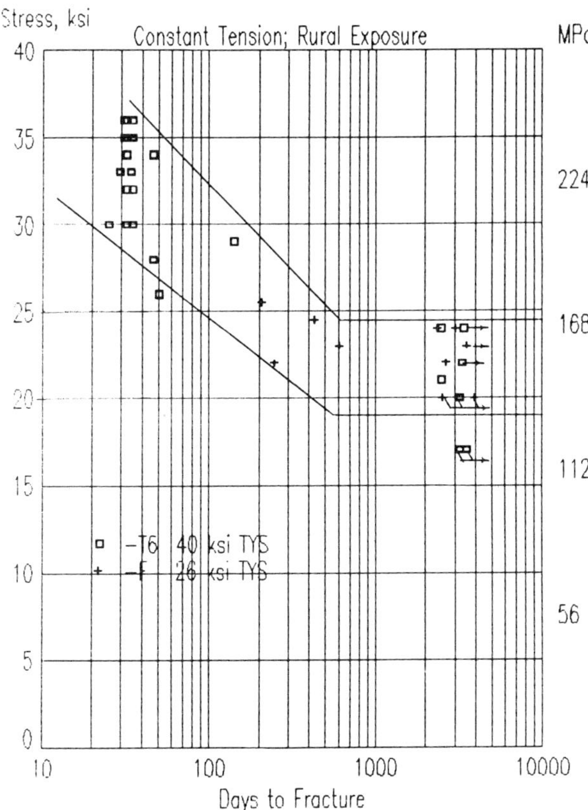

FIGURE 12.1 Stress corrosion of forged ZK60A.

REFERENCES

1. ASTM Standard B91, "Standard Specification for Magnesium Alloy Forging," *ASTM Annual Book of Standards*, Vol. 02.02, Sec. 2, 1984.
2. Dow Chemical Company, "Magnesium in Design," Bulletin 141-213, 1967.
3. Dow Chemical Company, "Magnesium Forging Alloys for Elevated Temperature Service," Technical Service and Development Letter Enclosure, April 24, 1962.
4. Dow Chemical Company, "Magnesium ZK60A-T6 Forging Alloy," TS&D Letter Enclosure, September 14, 1959.
5. Dow Chemical Company, Private communication.
6. Colin J. Smithells, *Metals Reference Book*, Butterworths, London and Boston, 1976.
7. Alloy Digest, "Elektron Magnesium Alloy–ZTY," Engineering Alloys Digest, Inc., New Jersey, 1968.

IV

Corrosion and Finishing

Corrosion behavior depends upon the metal in question and upon the environment to which it is exposed. In general, magnesium alloys have about the same corrosion resistance to normal exposures as mild steel, but are less resistant than aluminum alloys.

For many applications, the corrosion resistance is adequate for service without any protective treatments, although finishing may be desired for appearance. If the service conditions are expected to be at all severe, then finishing systems can be applied that will provide protection, the required degree of protection dictating the selection made.

The two chapters in this section give details on the corrosion behavior of magnesium and magnesium alloys in various environments and of finishing systems that can be applied for various objectives.

13 Corrosion

Corrosion can take many forms. It can be simply a slight tarnishing, an overall loss of material from the surface, or pitting of the surface Intergranular corrosion, a corrosive penetration of the grain boundaries, common in some metals, is not seen in magnesium. The corroding medium is anything that reacts with the metal, and can be a solid, a liquid, or a gas.

13.1 THE FILMS THAT FORM ON MAGNESIUM

Thermodynamically, magnesium, which is about as active as sodium, should react completely with the oxygen in the air and vigorously with water. The fact that it does neither is a tribute to the very protective films that form on the surface. The story of the corrosion behavior of magnesium is the story of the behavior of the films and their protective qualities in various environments.

Protection varies from perfection to allowing slight tarnishing to significant loss of weight and dimensions, depending upon the corrosion conditions and the composition and metallurgical state of the alloy under consideration.

In the absence of condensed water, no corrosion occurs in the atmosphere. An inadvertent experiment in the Dow Chemical Company laboratories provided a very striking demonstration of this. Dr. J. D. Hanawalt, then director of the laboratory, had had a tree of pure magnesium crystals condensed from the vapor on his desk for many years under the cover of an unsealed bell jar.

The crystals were quite beautiful, being about a centimeter on a side, with very flat and specular faces. Speculating one day on what would happen, we removed the bell jar cover. One day later, the crystals were tarnished to a light grey, and after several days were quite dark. Since the air in the office was free to enter the bell jar even with the cover on, the only difference was the settling of dust on the metal. Dust is deliquescent, so that for the first time in many years the crystals were in contact with liquid moisture. Without the water on the surface, they remained bright; with water, they tarnished rapidly. Since it is rare that moisture will not form on the surface, the usual reaction of magnesium to an indoor or mild outdoor exposure is a tarnishing to a light grey.

In a more controlled experiment, the corrosion rate during exposure to humid air was found to be less than 10^{-5} mm/year (4×10^{-7} in/year), provided that no condensation occurred. If the humidity was raised to allow condensation, the rate became greater than 4×10^{-5} mm/day (16×10^{-7} in/day) [1]. In the first case, the film was amorphous, but probably related to magnesium oxide; in the latter, the film was magnesium hydroxide, the reaction product between magnesium oxide and water.

When a freshly-prepared surface of magnesium is exposed to the atmosphere at room temperature, the first reaction is formation of magnesium oxide. If no liquid water is present, the reaction stops with the formation of only a very thin film of oxide and protection is essentially perfect. If moisture is present, and can form as liquid water on the surface of the magnesium, the magnesium oxide is converted to magnesium hydroxide. This compound has an equilibrium pH of 10.4, and is therefore stable in the presence of bases. However, it is not stable in the presence of acids, and will rapidly break down when these are present. Strong acids, such as hydrochloric, destroy the film and rapidly dissolve magnesium. The carbon dioxide in the atmosphere dissolves in any condensed water to form carbonic acid which reacts with magnesium hydroxide to form carbonates. The same is true of other acid-producing compounds, such as sulphur dioxide in industrial atmospheres, the latter forming sulphates. Some films that have been found on magnesium exposed to the atmosphere are [1,2,3,4]:

$Mg(OH)_2$ (magnesium hydroxide)
$3MgCO_3 \cdot Mg(OH)_2 \cdot 3H_2O$ (hydromagnesite)
$MgCO_3 \cdot 3H_2O$ (nesquehonite)
$MgCO_3 \cdot 5H_2O$ (lansfordite)
$MgCO_3 \cdot 5Mg(OH)_2 \cdot 2Al(OH)_3 \cdot 4H_2O$ (hydrotalcite), if aluminum is present
$MgSO_4$ (magnesium sulphate)
$MgSO_4 \cdot 7H_2O$ (epsomite)

While these films and others like them are remarkably protective, they are slightly soluble in water and will therefore not provide complete protection over long periods. They are also subject to breakdown in the presence of certain activating ions such as chloride, bromide, sulphate, and chlorate. It has been suggested that these types of ion are carried by electrochemical transport to anodic sites on the metal surface, where they form the magnesium salt which is acidic to the hydroxide film and therefore damaging [5]. Salt water, which contains chlorides, is therefore a harsh environment for magnesium. The corrosion information in later sections of this chapter provide quantitative data on the corrosion behavior of magnesium alloys in various environments.

Two other anions, namely chromate and fluoride, form very protective films on magnesium. Thus, a small quantity of sodium fluoride in water produces a film of magnesium fluoride, which is far more protective than the normally formed magnesium hydroxide. Magnesium has even been used for constructing handling equipment for concentrated HF. Chromic acid is the basis for many protective treatments applied to magnesium as is detailed in Chapter 14.

13.2 EFFECT OF COMPOSITION AND STRUCTURE

Magnesium is anodic to almost all other metals, and the potential difference between magnesium and other metals in saline solutions is large enough that the films present on the magnesium will not prevent the flow of current with accompanying corrosion. Galvanic corrosion is therefore a potential problem that must be dealt with, and this is discussed fully in Chapter 14. Hanawalt, Nelson, and Peloubet [6] were the first to point out the quantitative relation between corrosion of alloys in salt water and the presence of phases in the alloy cathodic to magnesium. Figure 13.1, taken from their work, shows that there is a definite "tolerance limit" for iron in magnesium, above which the corrosion rate in salt water increases dramatically. When the iron is above the tolerance limit, the corrosion seen is due to galvanic corrosion between the magnesium and the iron-containing phase. Similar limits were found for all elements investigated, and Figure 13.2 gives information on these. Some interactions among elements present in magnesium were also noted by the same authors. In particular, the presence of aluminum lowers the tolerance limit for iron by an order of magnitude. If zinc is also present, the tolerance limit again rises. If manganese is present, the tolerance limit for iron increases.

Loose [7] noted that, in addition to the chemical effects pointed out by Hanawalt et al., the size and distribution of the cathodic phases as influenced by cooling rate or heat treatment, is important. Aune

[8] has recently expanded this point, showing that processing parameters have a large influence on the corrosion rate with constant impurity levels.

Recently, the subject of the effect of impurity levels and distribution on the corrosion of AZ91 has been under intensive study [8,9,10]. Some very detailed and careful work by Hillis and co-workers [10] has shown that the tolerance limits in cast AZ91 for the most important elements (Fe, Cu, and Ni) are influenced by cooling rate and by the amount of manganese present. Their results can be summaried by stating that the tolerance limits for corrosion in salt water are as follows:

For high-pressure die castings (rapid freezing rate):

Fe%, maximum = 0.032 × Mn%
Ni%, maximum = 0.005%
Cu%, maximum = 0.070%

For gravity castings (slow freezing rate):

Fe%, maximum = 0.032 × Mn%
Ni%, maximum = 0.001%
Cu%, maximum = 0.040%

AZ91 alloy conforming to these limits is labeled AZ91D for the die-cast application or AZ91E for the gravity-cast application.

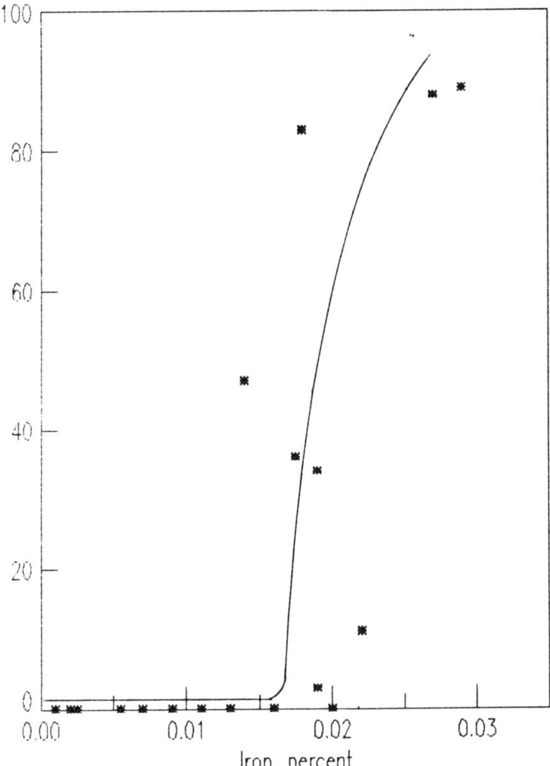

FIGURE 13.1 Effect of iron on corrosion of pure magnesium (alternate immersion in 3% NaCl). From Ref. 6.

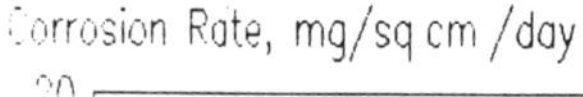

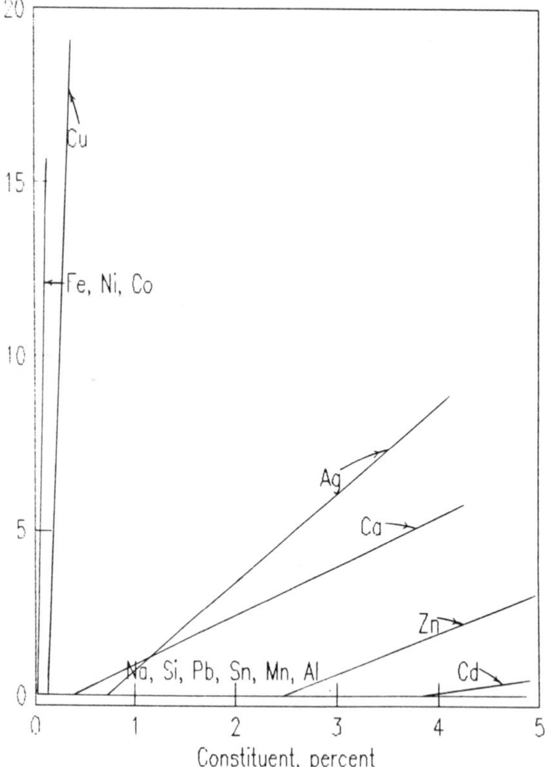

FIGURE 13.2 Corrosion of magnesium binary alloys (alternate immersion in 3% NaCl). From Ref. 6.

13.3 ATMOSPHERIC CORROSION

Information on the behavior of metals during exposure to the atmosphere is obtained by exposing panels of the materials to various rural, industrial, or marine environments. Data are gathered on loss of metal from the surface, expressed as microns/year (mils/year), and on change of tensile strength with time. The latter is calculated by taking the load sustained at fracture of the corroded sample divided by the original, uncorroded area. The tensile strength so calculated therefore includes loss of cross-sectional area and any notch effects introduced by, for example, pitting. It is a measure of the continuing load-carrying ability of a structure.

Because of the presence of hydrotalcite in the film, magnesium alloys that contain aluminum as an alloying element are more tarnish resistant in the atmosphere than are those free of aluminum.

Table 13.1 gives some information on the effect of exposure of magnesium alloys, one aluminum, and one steel alloy to marine, industrial, and rural atmospheres [11]. The rural exposure site is near Midland, Michigan where the average sulphur dioxide concentration is 7 ppb. The industrial site is within the Dow Chemical Company property in Midland, Michigan, where the average sulphur dioxide content is 20 to 25 ppb. In general, the marine site is harsher than the rural, but the industrial is clearly the harshest. While aluminum is the least affected by exposure, magnesium is considerably superior to steel in a marine environment, and about equivalent to steel in either industrial or rural exposure. The good performance of AZ91D die casting is especially noteworthy.

Tables 13.2 and 13.3 [12] give information on the corrosion of magnesium alloys exposed 80 feet (24 m) and 800 feet (240 m), respectively, from the ocean at Kure Beach, North Carolina. At 800 feet, the panels are free of ocean spray and have only a very little salt-laden air to contend with. At 80 feet, salt spray is frequent.

While there are some trends that can be seen in the data, the overall conclusion is that the corrosion rate is independent of time of exposure. There are some noteworthy differences among alloy types, which can best be seen by simply averaging all the data for each type. When this is done, the following results:

	microns/year		mils/year	
	24 m	240 m	80 ft	800 ft
All wrought AZ alloys	18.6	10.8	0.73	0.43
ZK60 Extrusions	27.7	14.9	1.09	0.59
AZ91D High-pressure die casting	8.4	4.6	0.33	0.18

It is clear that those alloys that contain aluminum are superior to those that do not, as represented by ZK60A; that the exposure at 80 feet is more severe than the one at 800 feet; and that the high-purity version of the high-pressure die-cast alloy is in a class by itself. In fact, the latter is not too different from the aluminum alloy cited in Table 13.1 for similar conditions of exposure. To put the numbers in perspective, it should be realized that the total loss of section thickness, in millimeters, after ten years of exposure, is obtained by dividing the numbers above by 100. Thus, for the worst case of ZK60A exposed 80 feet from the ocean, the section thickness decreases after ten years by 0.3 mm (12 mils). In the best case, that of die-cast AZ91D at 800 feet from the ocean, the decrease after ten years is only 0.05 mm (2 mils).

As pointed out above, the tensile strength of the exposed panels is obtained by dividing the load needed to fracture the specimen by the thickness of the unexposed panel. The original panels were all roughly 1600 microns (0.064 in) in thickness. The tensile strength thus should be less than the original strength, due simply to loss of section, by the factor obtained by calculating the ratio of the exposed thickness to 1600. This does not account for the numbers given in the tables. Thus, for AZ31B-O exposed for 8.5 years, the factor is 0.92. When the original strength, 275 MPa, is multiplied by this, the result is 253 MPa, which is slightly higher than the observed 225 MPa. The difference is small since magnesium does not corrode intergranularly, but what there is, is due to pitting that has occurred during exposure which reduces the cross section more than the average number and also results in some stress concentration. The difference arrived at by this calculation is a rough measure of the pitting tendency and of the sensitivity to stress concentration.

TABLE 13.1 Corrosion Rates During Atmospheric Exposure for 2.5 to 3.0 Years (Loss/year)

		Exposure site					
		Rural		Industrial		Marine[a]	
Alloy	Form	Microns	Mils	Microns	Mils	Microns	Mils
AZ31B-H24	Sheet	13.2	0.52	25.4	1.0	17.5	0.69
HK31A-H24		185	0.73	30.5	1.2	16.3	0.64
HM21A-T8		20.3	0.80	33.0	1.3	22.4	0.88
ZE10A-O		22.4	0.88	30.5	1.2	27.9	1.1
2024-T3 (Aluminum		0.1	0.004	2	0.08	2	0.08
Steel (0.27% Cu)		15	0.59	25.4	1.0	150	5.9
AZ31B-F	Extrusion	13.5	0.53	25.4	1.0	19.6	0.77
HM31A-F		17.8	0.70	35.6	1.4	20.3	0.80
ZK60A-T5		16.8	0.66	33.0	1.3	25.4	1.0
AZ63A	Casting	8.6	0.34	22.4	0.88	19.3	0.76
AZ91A-T6		4.3	0.17	15.8	0.62	22.4	0.88
AZ91D-T6		2.8	0.11	14.5	0.57	6.4	0.25
AZ92A-T6		9.4	0.37	20.3	0.80	25.4	1.0
EZ33A-T6		20.1	0.79	40.6	1.6	27.9	0.11
HK31A-T6		17.0	0.67	35.6	1.4	27.9	0.11
HZ32A-T5		15.5	0.61	38.1	1.5	27.9	0.11
ZH62A-T5		14.7	0.58	40.6	1.6	40.6	1.6
ZK51A-T5		14.5	0.57	35.6	1.4	25.4	1.0

[a] 80 feet from the ocean at Kure Beach, N. C.

Source: Refs. 1, 11. 12

TABLE 13.2 Atmospheric Exposure at 80 Feet from the Ocean (Bare Magnesium)

Alloy	Temper	Time, years	Corrosion rate Microns/yr	Corrosion rate Mils/yr	Tensile strength MPa	Tensile strength ksi
AZ31B Sheet	O	0			275	40
		1.0	21.3	0.84	275	40
		2.0	19.7	0.78	275	40
		4.5	20.5	0.81		
		8.5	14.4	0.57	225	33
		9.0	14.6	0.57		
AZ61A Sheet	O	0			285	41
		0.5	11.8	0.46	270	39
		1.0	14.8	0.58	260	38
		2.2	9.5	0.37	265	38
		4.1	11.5	0.45	250	36
		8.1	14.9	0.59	225	33
		12.0	12.0	0.47	215	31
	H24	0			320	46
		0.5	11.8	0.46	295	43
		1.0	13.6	0.54	285	41
		2.0	12.1	0.48	280	41
		4.9	17.8	0.70	240	35
		5.5	13.4	0.53	235	34
		8.0	16.0	0.63	235	34
AZ61A Extruded	F	0			310	45
		0.5	23.4	0.92	295	43
		1.0	23.9	0.94	290	42
		2.0	28.8	1.13	290	42
		7.0	25.1	0.99	275	40
		8.0	20.0	0.79	275	40
		11.0	25.1	0.99	270	39
		12.0	21.3	0.84	260	38
AZ80A Extruded	F	0			315	46
		0.5	24.1	0.95	310	45
		1.0	21.6	0.85	285	41
		2.2	27.0	1.11	295	43
		5.5	22.5	0.89	280	41
		7.0	23.3	0.92	265	38
		8.0	17.5	0.69	280	41
		11.0	21.7	0.85	270	39
		12.0	19.3	0.76	250	36

TABLE 13.2 (*Continued*)

Alloy	Temper	Time, years	Corrosion rate Microns/Yr	Corrosion rate Mils/yr	Tensile strength MPa	Tensile strength ksi
AZ91D Die cast	F	0.5	9.8	0.39		
		1.0	6.9	0.27		
		2.0	9.4	0.37		
		5.0	10.2	0.40		
		8.0	7.2	0.28		
		12.0	7.1	0.28		
M1A Sheet	O	0			245	36
		0.5	34.4	1.35	245	36
		1.0	29.2	1.15	230	33
		2.2	24.6	0.97	230	33
		5.5	15.8	0.62	235	34
		7.0	15.8	0.62	225	33
		8.0	12.5	0.49	225	32
		12.0	11.6	0.46	215	31
ZK60A Extruded	F	0			330	48
		1.4	34.5	1.36	300	44
		2.0	35.5	1.40	305	44
		4.0	24.3	0.96	275	40
		7.0	20.4	0.80	290	42
		8.4	20.1	0.79	275	40
	T5	0			365	53
		1.4	35.5	1.40	320	46
		2.0	37.5	1.48	320	46
		4.0	26.4	1.04	295	43
		7.0	21.3	0.84	290	42
		8.4	21.1	0.83	260	38

Source: Ref. 12.

TABLE 13.3 Atmospheric Exposure at 800 Feet from the Ocean (Bare Magnesium)

Alloy	Temper	Time, years	Corrosion rate		Tensile strength	
			Microns/yr	Mils/yr	MPa	ksi
AZ31A Sheet	O	0			255	37
		0.5	15.7	0.62	250	36
		1.0	19.3	0.76	250	36
		2.0	16.4	0.65	245	36
		5.0	13.8	0.54	225	33
		8.0	11.1	0.44	215	31
		10.0	7.8	0.31		
	H24	0			285	41
		0.5	17.1	0.67	285	41
		1.0	19.6	0.77	280	41
		2.0	16.3	0.64	275	40
		5.0	15.3	0.60	255	37
		8.0	13.6	0.54	245	36
AZ61A Sheet	O	0			290	42
		0.5	8.6	0.34	280	41
		1.0	7.4	0.29	280	41
		2.0	7.4	0.29	275	40
		4.0	6.2	0.24	260	38
		5.0	11.9	0.47	240	35
		8.0	14.2	0.56	225	33
		12.0	11.5	0.45	210	30
	H24	0			320	46
		0.5	5.5	0.22	300	44
		1.0	5.5	0.22	300	44
		2.0	6.3	0.25	290	42
		4.0	8.2	0.32	270	39
		5.0	9.4	0.37	260	38
		8.0	11.0	0.43	225	33
		10.0	12.8	0.50	205	30
		12.0	11.0	0.43	210	30
AZ61A Extruded	F	0			310	45
		0.5	7.4	0.29	300	44
		1.0	6.2	0.24	300	44
		2.0	9.2	0.36	295	43
		4.0	8.3	0.33	285	41
		5.0	14.3	0.56	290	42
		8.0	14.8	0.58	290	42
		12.0	11.8	0.46	270	39

TABLE 13.3 (*Continued*)

Alloy	Temper	years	Corrosion rate		Tensile strength	
			Microns/yr	Mils/yr	MPa	ksi
AZ80A	F	0			315	46
Extruded		0.5	7.8	0.31	305	44
		1.0	5.4	0.21	305	44
		2.0	6.9	0.27	310	45
		4.0	6.7	0.26	295	43
		5.0	8.7	0.34	300	44
		8.0	10.0	0.39	290	42
		12.0			280	41
AZ91D	F	0.5	3.0	0.12		
Die cast		1.0	4.6	0.18		
		2.0	4.9	0.19		
		4.0	5.0	0.20		
		5.0	4.1	0.16		
		8.0	4.6	0.18		
		12.0	4.6	0.18		
ZK60A	F	0			325	47
Extruded		1.4	19.4	0.76	320	46
		4.0	11.2	0.44	315	46
		7.0	13.1	0.52	310	45
		9.8	12.7	0.50	310	45
	T5	0			365	53
		1.4	20.5	0.81	330	48
		4.0	15.8	0.62	310	45
		7.0	13.4	0.53	305	44
		9.8	12.9	0.51	295	43

Source: Ref. 12.

13.4 CORROSION IN FRESH WATER

Magnesium immersed in stagnant distilled water protected from absorption of carbon dioxide will corrode at the start and then the rate will decrease to a very low value. The first reaction is production of magnesium hydroxide on the surface. As this forms, the pH of the system rises due to the dissolution of hydroxide until the point is reached where further dissolution of the hydroxide is inhibited and corrosion essentially stops (see Figure 13.3 [1]). If the water is replenished in any way so that the pH is prevented from rising, the dissolution of magnesium hydroxide is not inhibited, and corrosion will continue at the initial higher rate.

If the water can absorb carbon dioxide from the atmosphere, carbonic acid is produced, the film of magnesium hydroxide that has formed is attacked, and the corrosion rate rises. See Figure 13.3 for this effect also. As shown in Figure 13.3, if aluminum is present in the alloy (AZ92A), the effect of carbon dioxide is much decreased. This is due to the presence of hydrocalcite on the surface if aluminum is present and is the same effect that was noted for the atmospheric exposure data in the previous section.

As the temperature of the water is raised, the corrosion rate increases as is shown in Figure 13.4 [1]. Table 13.4 shows corrosion rates of several alloys in boiling de-ionized water [1]. There is a clear effect of alloy content on the rates. If fluoride is added to the water, the film on the metal changes from magnesium hydroxide to magnesium fluoride, and protection is greatly increased, as is shown in Table 13.5 [12]. In the presence of fluoride, even pure magnesium has a relatively low rate of corrosion in boiling water.

TABLE 13.4 Corrosion in Boiling De-ionized Water

	Corrosion rate		
Alloy	Microns/year	Mils/year	Inches/year
Pure Mg	16,200	638	0.638
AZ31B	800	31.5	0.032
AZ92A	400	15.8	0.016
HK31A	300	11.8	0.012
M1A	2,000	78.7	0.079
ZK60A	25,400	1000	1.000

Source: Ref. 1.

TABLE 13.5 Corrosion in Boiling Water Inhibited with 0.025N NaF

Alloy	Corrosion rate Microns/year	Mils/year	Inches/year
Pure Mg	55.9	2.20	0.0022
AZ31B	25.4	1.00	0.0010
AZ92A[a]	81.3	3.20	0.0032

[a]Die casting with the as-cast surface.
Source: Ref. 1.

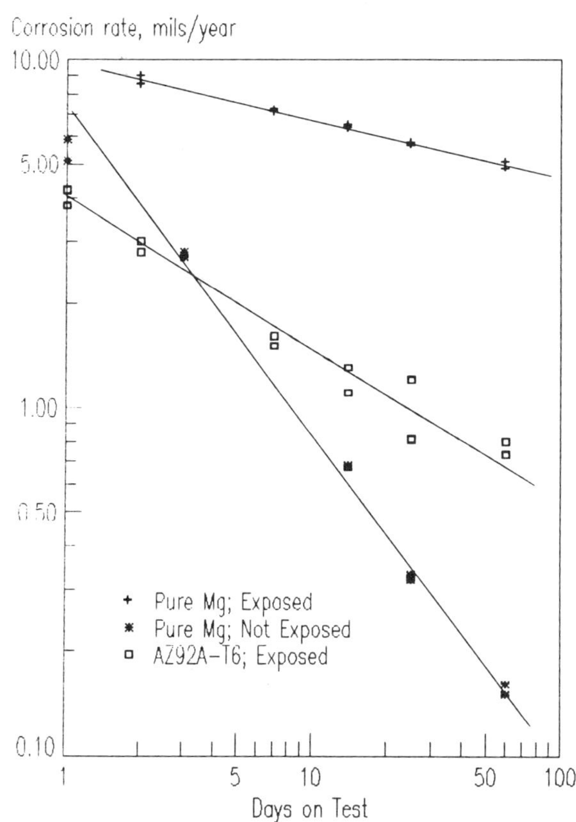

FIGURE 13.3 Corrosion rate in distilled water. The effect of CO_2 in water on the corrosion rate of pure magnesium and AZ92A-T6. Those samples "exposed" were free to absorb CO_2 from the air. Those "un-exposed" could not.

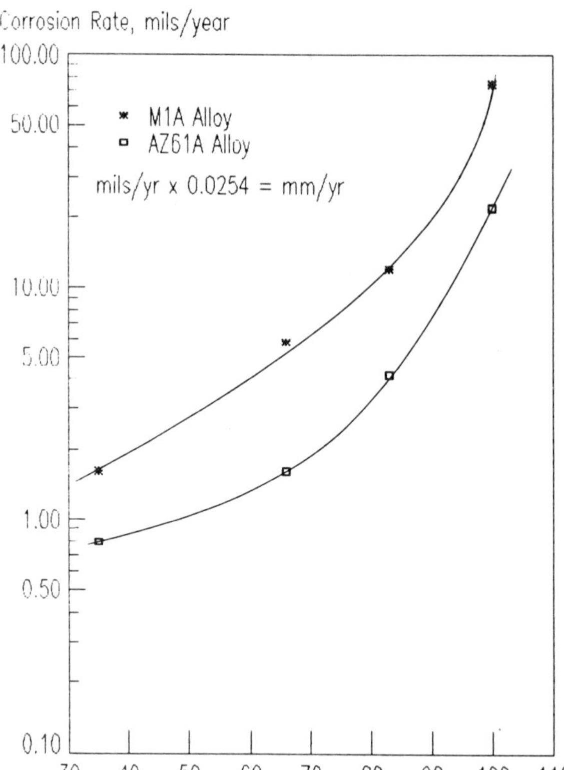

FIGURE 13.4 Corrosion of magnesium alloys in distilled water.

13.5 CORROSION IN SALT WATER

Because the chloride ion in salt water attacks the stability of the film that forms naturally on the metal, the rates are higher than in the absence of chloride. The rates are governed almost entirely by the presence of cathodic phases, as discussed in Section 13.2. If these phases are below the tolerance limits, the next most important compositional consideration is the presence of aluminum, which is beneficial. Typical salt-water corrosion rates of metal controlled for cathodic impurities are:

ZK60A 1.5 mm/year; 59 mils/year
AZ31B 0.6 mm/year; 24 mils/year
AZ91D 0.4 mm/year; 16 mils/year

All three alloys have about the same level of cathodic impurities. AZ31B contains both aluminum and manganese and hence the rate is lower than for ZK60A; AZ91D contains aluminum and manganese, and the freezing rate is rapid enough to affect the distribution of the cathodic phases that are present.

Without control of the cathodic impurities, rates can be as high as 200 mm/year (7.9 in/year).

13.6 CORROSION IN SELECTED MEDIA

Table 13.6 [13] lists a large number of substances with general comments on the suitability of magnesium in contact with them. In the table, concentration refers to concentrations in water, with 100 being anhydrous. A "yes" in the table means that laboratory and even commercial experience indicates stability in the material. Testing for the specific application is warranted. The recommendations in Table 13.6 are for room temperature. If the application involves elevated temperatures, testing should be undertaken before commitment to use.

A good rule of thumb is that materials that are basic, neutral, or contain fluorine cause little or no corrosion, while materials that are acidic do cause corrosion. Thus, while magnesium is stable in the presence of most organic compounds, methyl alcohol, which is an acid, attacks it rapidly. Ethyl alcohol, which is only slightly ionized, attacks the metal only slightly. Dry alkyl halides cause little corrosion even up to their boiling points. Magnesium can be safely and effectively cleaned by use of chlorinated solvents.

Magnesium is inert in the presence of halogenated refrigerants at room temperature. However, because of its high affinity for fluorine, use of magnesium in contact with fluoride-containing compounds at high temperatures must be approached with caution. An extremely rapid reaction, even explosive in some cases, can take place.

TABLE 13.6 Suitability of Testing Magnesium in Various Substances

CHEMICAL	CONCENTRATION (%)	SERVICE TEST WARRANTED
Acetaldehyde	Any	No
Acetic Acid	Any	No
Acetone	Any	Yes
Acetylene	100	Yes
Alcohol, Butyl	100	Yes
Alcohol, Ethyl	100	Yes
Alcohol, Isopropyl	100	Yes
Alcohol, Methyl	100	No
Alcohol, Propyl	100	Yes
Ammonia (Gas or Liquid)	100	Yes
Ammonium Salts (Most)	Any	No
Ammonium Hydroxide	Any	Yes
Aniline	100	Yes
Anthracene	100	Yes
Arsenates (Most)	Any	Yes
Benzaldehyde	Any	No
Benzene	100	Yes
Bichromates	Any	Yes
Boric Acid	1-5	No
Brake Fluids (Most)	100	Yes
Bromides (Most)	Any	No
Bromobenzene	100	Yes
Butter	100	No
Butylphenols	100	Yes
Calcium Arsenate	Any	Yes
Calcium Carbonate	100	Yes
Calcium Chloride	Any	No
Calcium Hydroxide	100	Yes
Camphor	100	Yes
Carbon Bisulphide	100	Yes
Carbon Dioxide (Dry)	100	Yes
Carbon Monoxide	100	Yes
Carbon Tetrachloride	100	Yes
Carbonated Water	Any	No
Castor Oil	100	Yes
Cellulose	100	Yes
Cement	100	Yes
Chlorides (Most)	Any	No
Chlorine	100	No
Chlorobenzenes	100	Yes
Chloroform	100	Yes
Chlorophenols	Any	No
Chlorophenylphenol	100	Yes
Chromates (Most)	Any	Yes
Chromic Acid	Any	Yes
Citronella Oil	100	Yes
Cod Liver Oil (Crude)	—	Yes
Copals	100	Yes
Coumarin	100	Yes
Cresol	100	Yes
Cyanides (Most)	Any	Yes
Dichlorohydrin	100	Yes
Dichlorophenol	100	Yes
Dichromates (See Bichromates)		
Diethanolamine	100	Yes
Diethyl Aniline	100	Yes
Diethyl Benzene	100	Yes
Diethylene Glycol Solutions	Any	Yes, may need inhibitors
Diphenyl	100	Yes
Diphenylamine	100	Yes
Diphenylmethane	100	Yes
Diphenyl Oxide	100	Yes
Dipropylene Glycol	100	Yes
Divinylbenzene	100	Yes
Dry Cleaning Fluids	100	Yes
Ethers	100	Yes
Ethanolamine (Mono)	100	Yes
Ethyl Acetate	100	Yes
Ethyl Benzene	100	Yes
Ethyl Bromide	100	No
Ethylcellulose	100	Yes
Ethyl Chloride	100	Yes
Ethyl Salicylate	100	Yes
Ethylene (Gas)	100	Yes
Ethylene Dibromide	100	Yes
Ethylene Glycol Solutions	Any	Yes, may need inhibitors
Fats, Cooking (Acid-free)	100	Yes
Fatty Acids	Any	No
Ferric Chloride	Any	No
Fluorides (Most)	Any	Yes
Fluosilicic Acid	Any	No
Formaldehyde	Any	Yes
Fruit Juices and Acids	Any	No
Fuel Oil	100	Yes
Gasohol (10% Ethanol)	100	Yes, if inhibited
Gasohol (10% Methanol)	100	Yes, if inhibited
Gasoline (Lead-free)	100	Yes, if inhibited
Gasoline (Leaded)	100	Yes, if inhibited
Gelatine	Any	Yes
Glycerine C.P.	100	Yes
Grease (Acid-free)	100	Yes
Heavy Metal Salts (Most)	Any	No
Hexamine	3	Yes
Hydrochloric Acid	Any	No
Hydrofluoric Acid	5-60	Yes
Hydrogen Peroxide	Any	No
Hydrogen Sulfide	100	Yes
Iodides	Any	No
Iodine Crystals (Dry)	100	Yes
Isopropyl Acetate	100	Yes
Isopropyl Benzene	100	Yes
Isopropyl Bromide	Any	No
Kerosene	100	Yes
Lanolin	100	Yes
Lard	100	Yes
Lead Arsenate	Any	Yes
Lead Oxide	Any	No
Linseed Oil	100	Yes
Magnesium Arsenate	Any	Yes
Magnesium Carbonate	100	Yes
Magnesium Chloride	Any	No
Mercury Salts	Any	No
Methane (Gas)	100	Yes
Methyl Bromide	Any	No
Methyl Cellulose	100	Yes
Methyl Chloride	100	Yes
Methylene Chloride	100	Yes
Methyl Salicylate	100	Yes
Milk (Fresh and Sour)	100	No
Mineral Acids	Any	No
Monobromobenzene	100	Yes
Monochlorobenzene	100	Yes
Naphtha	100	Yes
Naphthalene	100	Yes
Nicotine Sulphate	40	Yes
Nitrates (All)	Any	No
Nitrous Gases	100	No
Nitric Acid	Any	No
Nitroglycerin	Any	No
Oil, Animal — (Acid- and Chloride-free)	Any	Yes
Oil, Mineral (Chloride-free)	100	Yes
Oil, Vegetable (Chloride-free)	100	Yes
Oleic Acid	100	Yes
Olive Oil	100	Yes
Organic Acids (Most)	Any	No
Orthochlorophenol	100	No
Orthodichlorobenzene	100	Yes
Orthophenylphenol	100	Yes
Oxygen	100	Yes
Paraphenylphenol	100	Yes
Paradichlorobenzene	100	Yes
Pentachlorophenol	100	Yes
Perchloroethylene	100	Yes
Permanganates (Most)	Any	Yes
Phenol	100	Yes
Phenyl Ethyl Acetate	100	Yes
Phenylphenols	100	Yes
Phosphates (Most)	Any	Yes
Phosphoric Acid	Any	No
Polypropylene Glycols	100	Yes
Potassium Fluoride	Any	Yes
Potassium Hydroxide	Any	Yes
Potassium Nitrite	Any	No
Potassium Permanganate	Any	Yes
Propylene Glycol U.S.P.	100	Yes
Propylene Oxide	100	Yes, may need inhibitors
Pyridine (Acid Free)	100	Yes
Pyrogallol	Any	No
Rubber & Rubber Cements	100	Yes
Sea Water	100	No
Sodium Bromate	Any	No
Sodium Bromide	Any	No
Sodium Carbonate	Any	Yes
Sodium Chloride	Any	No
Sodium Cyanide	Any	Yes
Sodium Dichromate	Any	Yes
Sodium Fluoride	Any	Yes
Sodium Hydroxide	Any	Yes
Sodium Phosphate (Tribasic)	Any	Yes
Sodium Silicate	Any	Yes
Sodium Sulfide	3	Yes
Sodium Tetraborate	3	Yes
Steam	100	No
Stearic Acid (Dry)	100	Yes
Styrene Polymer	100	Yes
Sugar Solutions (Acid-Free)	Any	Yes
Sulphates (Most)	Any	No
Sulphur	100	Yes
Sulphur Dioxide (Dry)	100	Yes
Sulphur Chloride	Any	No
Sulphuric Acid	Any	No
Sulphurous Acid	Any	No
Tannic Acid	3	No
Tanning Solutions	Any	No
Tar, Crude & Its Fractions	100	Yes
Tartaric Acid	Any	No
Tetrahydro-naphthalene	100	Yes
Titanium Tetrachloride	100	Yes
Toluene (Toluol)	100	Yes
Trichlorbenzene	100	Yes
Trichloroethylene	100	Yes
Trichlorophenol	100	Yes
Tung Oil	100	Yes
Turpentine	100	Yes
Urea	100	Yes
Urea in Aqueous Solution (Cold)	Any	Yes
Urea in Aqueous Solution (Warm)	Any	No
Vinegar	Any	No
Vinylidine Chloride	100	Yes
Vinyl Toluene	100	Yes
Water, Boiling	100	No
Water, Distilled	100	Yes
Water, Rain	100	Yes
Waxes (Acid-free)	100	Yes
Xylol	100	Yes

Source: Ref. 13.

REFERENCES

1. Hugh P. Godard, W. B. Jepson, M. R. Bothwell, and Robert L. Kane, *The Corrosion of Light Metals*, John Wiley & Sons, Inc., New York, 1967.
2. E. I. Levitina, *Zhur. Obschei Khim* (24), 216-218 (1954).
3. G. D. Bengough and L. Whitby, *Transactions of the Institute of Chemical Engineering* (11), 176-189 (1933).
4. J. J. Casey and A. R. Maupin, "Outdoor Storage of Magnesium," a talk presented to the American Institute of Mining, Metallurgical, and Petroleum Engineers, February 1957.
5. J. L. Robinson and P. F. King, *Journal of the Electrochemical Society* (108), 36-41 (1961).
6. J. D. Hanawalt, C. E. Nelson, and J. A. Peloubet, "Corrosion Studies of Magnesium and Its Alloys," *Transactions of the American Institute of Mining and Metallurgical Engineers* (147), 273-299 (1942).
7. W. S. Loose, "Magnesium Alloys," *Corrosion Handbook*, John Wiley and Sons, Inc., New York, 1948.
8. T. Kr. Aune, "Minimizing Base Metal Corrosion on Magnesium Products. The Effect of Element Distribution (Structure) on Corrosion Behavior," *Proceedings of the International Magnesium Association* (1983).
9. James E. Hillis, "The Effects of Heavy Metal Contamination on Magnesium Corrosion Performance," *SAE Technical Paper Number 830523* (1983).
10. K. N. Reichek, K. J. Clark, and J. E. Hillis, "Controlling the Salt Water Corrosion Performance of Magnesium AZ91 Alloy," *SAE Technical Paper Number 850417*, 1985.
11. American Society for Metals, "Properties and Selection: Nonferrous Alloys and Pure Metals," *Metals Handbook*, (2) American Society for Metals, Metals Park, Ohio, 1979.
12. Dow Chemical Company, Private communication.
13. Dow Chemical Company, "Magnesium: Designing Around Corrosion," *Bulletin 141-396*, 1982.

BIBLIOGRAPHY

1. J. C. Fuggle, L. M. Watson, and D. J. Fabian, "X-Ray Photoelectron Studies of the Reaction of Clean Metals (Mg, Al, Cr, Mn) with Oxygen and Water Vapour," *Surface Science 49*, North-Holland Publishing Company, 1975.

14

Corrosion Protection and Finishing

As pointed out in Chapter 13, the corrosion behavior of magnesium alloys depends upon the innate film-forming characteristics of the alloy being used, the environment in which it must serve, and the presence or absence of galvanic couples. The principle deleterious environmental factors are acidic water and water containing chlorides. This chapter deals with methods of combatting corrosive conditions that may arise from a variety of causes. The methods can vary from nothing at all for service in mild conditions to elaborate systems that will protect the metal from the severest conditions.

Designing for corrosion protection means considering all factors that are of importance, including proper physical design of the part, proper cleaning of the surface of the part, proper treatment of the surface for the expected service conditions, and proper accounting for any dimensional changes that occur because of the treatments used. These factors are all covered in the following sections of this chapter.

14.1 GALVANIC CORROSION

When two dissimilar metals are connected electrically both by a metallic connection and by a liquid electrolyte, there is an electrical potential difference between the two pieces such that a current can flow from one to the other through both the electrolyte and the metallic connection. The current through the electrolyte is accomplished by a flow of ions, which means that one of the metals must dissolve in the electrolyte to provide the ions. The dissolving metal is the one with the greater negative voltage and is known as the anode. The other metal, with a less negative voltage, known as the cathode, does not dissolve. The specific voltage that is characteristic of any two connected metals is dependent upon the electrolyte; even the order of voltages can change with changing electrolyte. However, since salt water is the electrolyte most generally encountered, the most useful electromotive series is based on salt water. Table 14.1 shows this series for a 3-6% NaCl solution. In this table, a metal with a greater negative voltage is anodic in salt water to a metal with a smaller negative voltage. It is apparent that magnesium is anodic to all other metals in the table, and will be the corroding partner in a couple while protecting the other from any corrosion.

While the potential difference shown in Table 14.1 is a rough guide to the amount of galvanic corrosion to be expected from any given couple, it is not the only factor. In general, the greater the potential difference, the greater will be the corrosion. Thus a magnesium-steel couple is more serious than a magnesium-aluminum couple. However, the voltages shown are for the bare metals. In practice, films build both on the magnesium and on the other metal, which changes the effective potentials significantly. Thus, tin, which should be considerably more damaging than steel, is less harmful than would be expected when connected to magnesium because the surface of the tin coats with a film, or "cathodically polarizes." Tin coating is, in fact, a way of alleviating galvanic attack on magnesium by steel.

Since the electromotive series depends upon the electrolyte, two pieces of the same metal connected metallically will exhibit a potential difference if in contact with electrolytes of different composition. The more anodic of the two pieces will dissolve while the more cathodic will be protected. This situation typically arises if electrolyte penetrates between two faying surfaces. As the solution deep in the gap reacts with the metals of the surface its composition changes. If the gap is narrow, replenishment of the solution from the outside does not take place, so that a compositional difference between the electrolyte in the gap and the electrolyte external to the gap persists. The potentials that arise result in corrosion that can be severe. This particular type of galvanic corrosion, known as crevice corrosion, must be guarded against by prevention of seepage of the electrolyte between gaps such as those created by two faying surfaces. The sketch in the lower left corner of Figure 14.1 depicts exactly this situation. Note that the presence of two different metals is not needed for this type of galvanic corrosion to occur.

Galvanic corrosion can be eliminated by eliminating at least one of the factors that produce it: the foreign metal in the couple can be removed, the metallic connection can be broken, or the electrolyte connection can be broken. All three methods are used in practice.

The most common source of galvanic corrosion is the coupling of magnesium to another metal as by use of fasteners, or by joining a magnesium part to a part made of another metal as in an automobile engine. A second source that is sometimes overlooked is surface contamination with small particles of a dissimilar metal. Protective measures must be taken in both cases to prevent the galvanic corrosion that may result.

TABLE 14.1 Electromotive Series in 3-6% NaCl Solution

Metal	Volts vs. 0.1N calomel electrode
Mg	-1.73
Mg alloys	-1.67
Galvanized steel	-1.14
Zn	-1.05
Cadmium plated steel	-0.86
99,99 % Al	-0.85
Al + 12% Si	-0.83
Steel	-0.78
Grey cast iron	-0.78
Pb	-0.55
Sn	-0.50
Chrome steel, active	-0.43
60/40 Brass	-0.22
Ni	-0.14
Chrome steel, passive	-0.13
Ag	-0.05
Au	+0.18

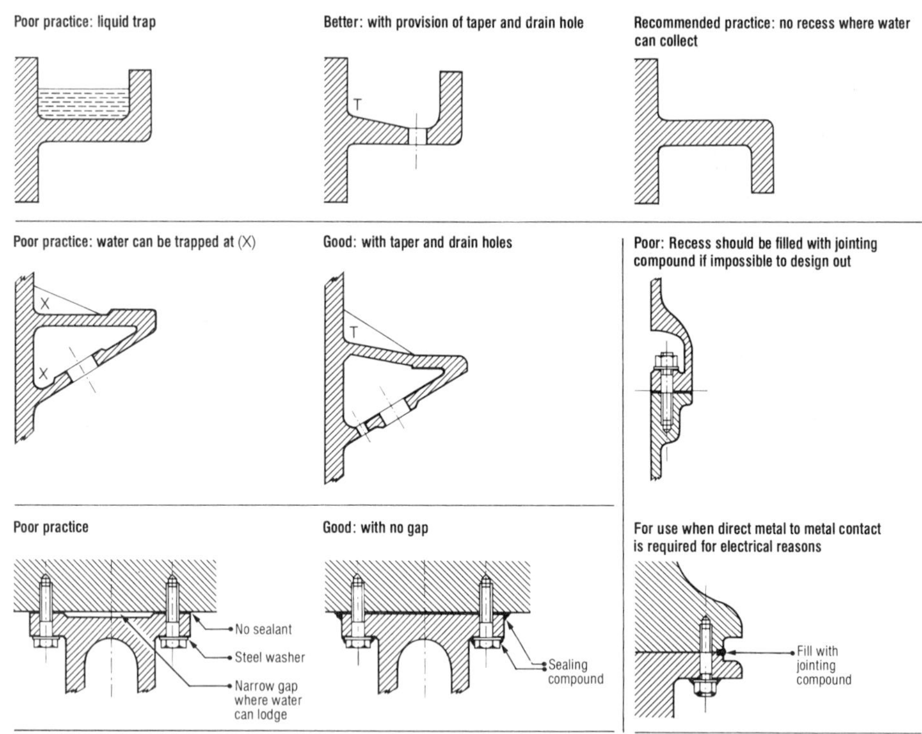

FIGURE 14.1 Design principles for corrosion prevention.

14.2 SURFACE CONTAMINATION

The surface of magnesium is of course important to the corrosion behavior. If particles of another metal become embedded in the surface, galvanic corrosion will result if an electrolyte becomes available. If salts, especially sodium chloride, are left on the surface, the naturally protective films on the magnesium are damaged and corrosion will be more severe than in their absence. The solutions for both types of contamination are to prevent them happening or to clean the surface if contamination occurs. A few sources of surface contamination should be noted.

If the surface is cleaned by blasting with iron shot, particles of iron will be left on the surface and must be removed to prevent a serious corrosion problem. Even blasting with sand, glass, or aluminum oxide can be deleterious since the material used is usually not pure enough to prevent any contamination of the surface, and the cold-worked surface can be anodic to unworked areas leading to galvanic corrosion even in the absence of other cathodic particles. The solution is acid etching to remove at least 0.05 mm (0.002 in) from the surface or fluoride anodizing.

If a lubricant containing oil, graphite, or molybdenum disulfide is used during forming operations, the graphite, the molybdenum disulfide, or any reduced carbon from the oil, will be left on the surface. Although not included in Table 14.1, both molybdenum disulfide and carbon are cathodic to magnesium, and galvanic corrosion can occur. They must be cleaned off the surface. If the magnesium is pickled in an acid that contains heavy-metal salts, the magnesium reduces the salts to the elemental metals which then plate on the surface of the magnesium. The particles must be cleaned off if they occur. In this case, the easiest solution is to assure that no pickling solutions contain enough heavy-metal contamination to be a problem.

If the surface of the magnesium contains a chromate coating from a previous treatment, and the surface is heated above about 300°C, some of the chromate will be reduced to metallic chromium and a galvanic couple results. Refraining from heating such a surface to that high a temperature is the solution.

Sodium chloride is ubiquitous in this world, so that its presence on the surface can almost be taken for granted. Even handling with the fingers deposits salt. The problem is not serious even with bare magnesium if the exposure is mild. If, however, a clear coating is applied to a surface not first cleaned of any salt contamination, filiform corrosion will occur even upon very mild exposure. This is not deleterious to either dimensions or mechanical properties but is quite unsightly. The solution is proper cleaning before application of the clear coating. The ferric nitrate bright pickle is one solution designed to clean the surface of chlorides while leaving the original metallic appearance. This should always be used if a clear coating is desired over a metallic-appearing surface. The system of ferric nitrate bright pickle plus a clear coating is suitable only for mild exposures.

In general, contamination is so hard to avoid completely, that cleaning the surface prior to further protective treatment is a universal requirement. For those applications where magnesium is used without any surface treatment, severe surface contamination must be removed.

14.3 PART DESIGN

Proper protection from corrosion starts with good part design, which must take into account drainage of water from the part during service, selection of the appropriate finishing systems to be used, allowance for dimensional changes that will occur during finishing, and proper joint design to prevent galvanic corrosion during service.

It was pointed out in Chapter 13 that stagnant water free to absorb carbon dioxide from the atmosphere can result in serious corrosion. It is consequently important to design parts so that no water collects in pockets. Figure 14.1 [1] illustrates the principles involved. These should be carefully followed.

The finishing system to be used is a compromise between the service conditions expected for the part and the cost of the finishing. A summary of choices that are available is given in the tables of this chapter.

When paint is planned as a coating, the tendency for paint to be too thin at sharp edges should be countered by rounding all edges. A radius of 0.75 mm (0.03 in) should be the minimum accepted.

Many of the treatments described later in this chapter change the dimensions of the part. The amount of this change must be known and accounted for in the original design of the part. The changes to be expected are given with each treatment that is described.

Measures for prevention of galvanic corrosion must be a part of the original design. From the standpoint of galvanic corrosion alone, fasteners should be of aluminum, preferably of those alloys that contain substantial quantities of magnesium and little or no copper, such as the 5000 series. These alloys cathodically polarize sufficiently when connected with magnesium that galvanic corrosion is minimized. If steel fasteners must be used for strength purposes, they should be coated with aluminum, zinc, cadmium, or tin, since these metals also cathodically polarize. In all cases of joining, the two metals should be separated by insulators, and access of any electrolyte should be denied. Figure 14.2 shows good joint designs [2]. Note again that denial of electrolyte access is as important when joining two pieces of magnesium as when joining magnesium to a dissimilar metal.

When painting joints of dissimilar metals, both the anode and the cathode should be covered, but it is most important to cover the cathode. The anode must supply the ions for the electrolytic current. If these ions must come from a small area to service a large cathodic area, it is clear that severe pitting will occur. If, on the other hand, the ions can come from a large anode area to service only a small cathode area, the corrosion may be only superficial. When magnesium corrodes in a galvanic situation, the solution becomes basic. Any aluminum present is attacked by the basic solution, sometimes severely. Therefore, poor protection of an assembly containing both magnesium and aluminum can lead not only to corrosion of the magnesium member but also corrosion, and even destruction, of the aluminum member. This precaution also applies to paint coatings, providing an additional reason for complete and high-quality covering of both the anode and the cathode.

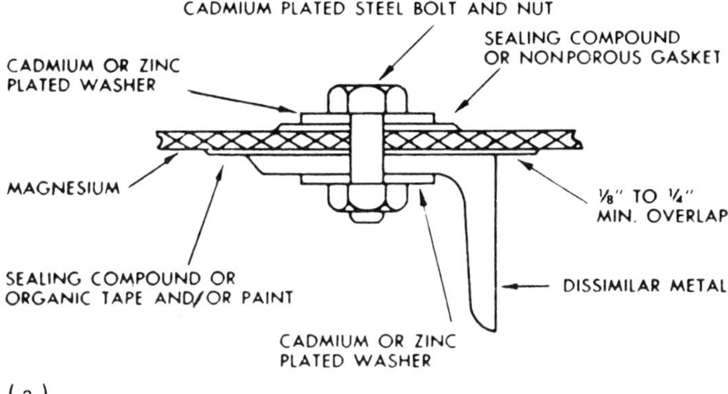

(a)

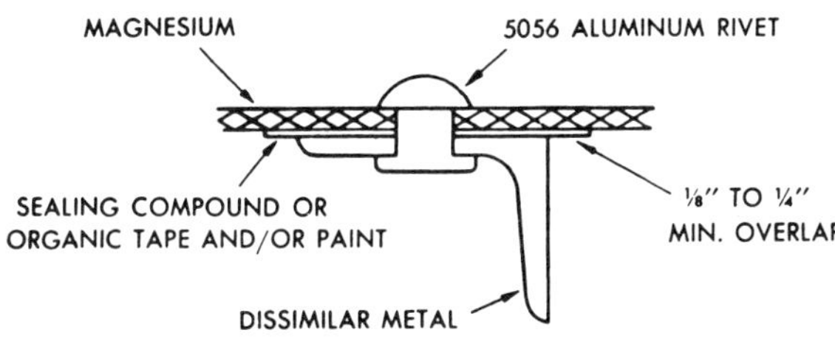

(b)

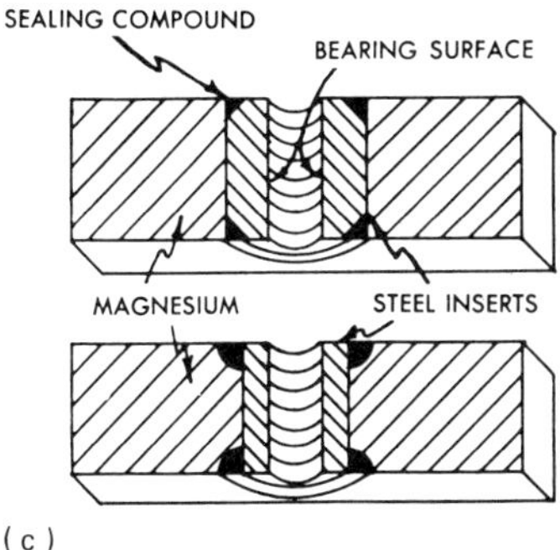

(c)

FIGURE 14.2 (a) Alternate methods of protecting magnesium bolted to dissimilar metal or wood. (b) Magnesium riveted to dissimilar metal. (c) Methods of avoiding contact corrosion with inserts.

14.4 CLEANING

Cleaning of magnesium surfaces can be done mechanically, by acid etching, by anodically cleaning, or by solvent or alkaline cleaning for degreasing. If mechanical cleaning, such as by shot or sand blasting, is used, the surface must be etched subsequently to a depth of at least 0.05 mm (0.002 in) to remove contamination and worked metal. Most acid etches remove metal and the dimensional changes must be known if tolerances are important for the part. Chromic acid, which removes no magnesium, is an exception.

Table 14.2 lists a number of cleaning etching baths together with their characteristics. Table 14.3 gives details on bath compositions and procedures for use.

TABLE 14.2 Cleaning Solution Characteristics

Name	Metal removal	Uses
Acetic-nitrate	0.01-0.02 mm	Removes mill scale from wrought products.
Acid fluoride	Nil	Activation bath for the dichromate coating.
Cathodic degreasing	Nil	Decreased time for degreasing.
Caustic soak	Nil	Removes grease, oil, and chromate coatings.
Chromic acid	Nil	Removes flux, oxides, corrosion products, chemical treatments.
Chromic nitrate	0.01 mm	Removes mill scale, oxide, and burned-on graphite from wrought products.
Chromic-nitrate-hydrofluoric	0.01-0.02 mm	Removes surface segregation from castings.
Chromic-sulfuric	0.002 mm	Preparation for spot welding on wrought products.
Fluoride anodize[a]	Nil	Cleans all alloys and forms. Especially useful for sand castings with contaminated surfaces.
Heavy duty alkaline	Nil	Removes grease, oil, and chromate coatings.
Hydrofluoric acid	Nil	Activating prior to chemical treatment. Removal of powdery coatings that can occur after the chrome pickle.
Hydrofluoric-sulfuric	0.002 mm	Brightening of castings.
Mild etch	0.002-0.005 mm/5 min	Mild cleaning.
Solvent	Nil	Removes grease and oil.
Sulfuric acid	0.05 mm	Removes the effects of sand blasting on castings.

[a]Patented by MEL.

Conversion factor: mm × 0.0394 = in.

Note: See Table 14.3 for compositions and procedures.

TABLE 14.3 Compositions and Procedures for Cleaning Solutions

Name	Composition (aqueous)	Procedures
Acetic-nitrate	200 g/l CH_3COOH 50 g/l $NaNO_3$	0.5-1 min at 20-30°C. Use a tank lined with 3003 aluminum, ceramic, lead or rubber.
Acid fluoride	47 g/l $NaHF_2$, KHF_2, NH_4HF_2	Immerse 5 minutes at 20°C.
Cathodic degreasing	30 g/l Na_3PO_4 30 g/l $Na_2CO_3 \cdot 10H_2O$ 1 g/l wetting agent	Cathodic degreasing for 0.5-3 min. 1-4 A/dm^2 current density; 4-6 volts.
Caustic soak	100 g/l NaOH	Immerse for 10-20 min at 90-100°C. Rinse thoroughly.
Chromic acid	180 g/l CrO_3	Immerse for 1-15 min at 20-100°C. If etching of the magnesium occurs, check bath for chloride contamination. Use tanks of lead, stainless steel, or 1100 aluminum.
Chromic-nitrate	180 g/l CrO_3 30 g/l $NaNO_3$	Thoroughly rinse in cold water. Immerse for 3 min at 20-30°C. Agitate during immersion. Use stainless steel tanks or tanks lined with ceramic, lead, rubber or vinyl.
Chromic-nitrate-hydrofluoric	280 g/l CrO_3 25 ml/l of 70% HNO_3 8 ml/l of 60% HF	Immerse for 0.5-2 min at 20-30°C. Use tanks lined with rubber or vinyl.
Chromic-sulfuric	180 g/l CrO_3 0.5 ml/l H_2SO_4	Use only for cleaning parts to be spot welded. Immerse 3 min at 20-30°C. Use stainless steel, ceramic, rubber or 1100 aluminum.
Fluoride anodize	15-25% NH_4HF_2	Use alternating current, starting at a low voltage and raising gradually to 120 V. Continue for 10-15 min or until the current falls below 50 A/m^2. Use tanks, connec-

TABLE 14.3 (*Continued*)

Name	Composition (aqueous)	Procedures
		tors, and clamps of magnesium. The fluoride coating produced must be removed by the chromic acid cleaner before further treatments.
Heavy duty alkaline	15-60 g/l NaOH 10 g/l $Na_3PO_4 \cdot 12H_2O$ 1 g/l wetting agent	Immerse for 3-10 min at 90-100°C. Rinse thoroughly.
Hydrofloric acid	11 volume % HF in water	Immerse for 0.5-5 min at a bath temperature of 20-30°C.
Hydrofluoric-sulfuric	15-20 Volume % HF 5 Volume % H_2SO_4	Immerse 2-5 min at 20°C.
Mild etch	30 g/l $Na_4P_2O_7$ 65 g/l $Na_2B_4O_7 \cdot 10H_2O$ 7 g/l NaF	Immerse for 2-5 min at 75-80°C.
Solvent	Standard materials	Immersion or vapor degrease.
Sulfuric acid	30 ml/l H_2SO_4	10-15 sec immersion at 20-30°C, or until 0.05 mm of surface is removed.

Conversion factor: 1 g/l = 0.134 oz/gal.

14.5 CHEMICAL CONVERSION COATINGS

Coatings are applied to supplement or replace the natural protective films, or to cover them up with something else. The purpose of any coating is protection or appearance, or both. They can vary from a simple wax or oil through chemical conversion coatings to elaborate organic or inorganic coatings.

As pointed out in Chapter 13, the natural films that form on magnesium, while quite protective, do not serve perfectly in all environments. In addition they are basic, with a typical pH or 10.5, making it difficult to apply a paint directly. Chemical conversion coatings are therefore applied to provide somewhat more protection than that provided by the natural films as well as a better foundation for subsequently applied paint. Table 14.4 lists a number of conversion baths that are available together with their characteristics. Table 14.5 gives details on bath compositions and procedures.

14.5.2 Anodic Coatings

Anodic coatings are a special form of chemical conversion coating. In general, they are tougher, harder, and more abrasion resistant than chemical conversion coatings, and are excellent bases for painting. The most corrosion resistant finishes available generally start with an anodic coating. For maximum corrosion resistance, anodic coatings should be sealed with unpigmented organic resin. Table 14.6 lists anodic coatings that are used, together with their compositions and procedures.

14.5.3 Metallic Coatings

All metals that are commonly electroplated can be put on magnesium and provide a combination of good appearance and resistance to corrosion and abrasion. All plating schemes for magnesium start by putting on a zinc coating by chemical conversion. Once this is on, and a copper strike applied, the surface to be plated is copper and normal plating techniques can be used. Rather surprisingly, a high-quality copper-nickel-chromium plating is protective even if the plate is damaged. When a scribe mark is put on such a plated magnesium panel and then the panel subjected to salt spray, there is very little or no undercutting of the plate. Thus, even though one would expect a very bad galvanic corrosion problem if the metallic plate and the magnesium become connected by an electrolyte, such does not seem to be the case. Electroplating can provide excellent protection in areas where the plating is not broken. Details of plating solutions and procedures are not included since this is beyond the scope of this book.

14.5.4 Organic Coatings

Organic coatings can vary from simple oils or waxes to multi-paint coatings. The primer must be resistant to alkalis if service is expected to be at all severe. Vinyl, epoxy, polyvinylbuteral and epichlorhydrin-bisphenol resins are satisfactory; alkyds and nitrocellulose should be avoided. Baking the primer usually increases alkali resistance and thus adhesion of the primer to the magnesium. If the service is expected to be severe, the primer should also contain inhibitive chromates.

The finish coat can be any desired that is compatible with the primer used and meets the service requirements of appearance and protection desired.

14.5.5 Special Coatings

A wide variety of special coatings are available and most of these can be applied to magnesium. These include rubbers and ceramics.

TABLE 14.4 Chemical Conversion Coating Characteristics

Name	Metal removal	Uses
Chrome pickle (Dow #1)	0.015 mm	Some protection during shipment and storage. A good paint base. Used on all forms.
Chrome-alum (Dow #4)	Nil	A black coating used primarily for die castings, but suitable for all alloys and forms except those alloys containing thorium.
Dichromate (Dow #7)	Nil	Better protection than provided by the chrome pickle as well as a better paint base. Because there is no dimensional change, can be applied after machining. Not used on alloys containing thorium, or M2A.
Galvanic anodize (Dow #9)	Nil	A dark brown to black coating often used on optical equipment such as cameras. Can be used on all alloys and forms.
Ammonium phosphate (Dow #18)	Nil	A paint base, not as protective as those based on chromic acid, but less toxic.
Dilute chromic acid (Dow #19)	Nil	A less expensive paint base treatment than the chrome pickle. Used on all alloys and forms. Best used for mild and indoor applications.

TABLE 14.4 (*Continued*)

Name	Metal removal	Uses
Modified chrome pickle (Dow #20)	0.013 mm/min	Develops a more uniform coating than the chrome pickle. A good paint base, used on all alloys and forms.
Ferric nitrate bright pickle (Dow #21)	Nil	A bright metallic finish that will protect during storage or mild use for up to six months. Especially useful as the preparation for clear paints.
Chromate manganate (Dow #22)	Nil	Some storage protection and a paint base. Not as protective nor as good a base as the chrome pickle, but much less expensive for the materials.
Stannate immersion[a] (Dow #23)	Nil	Deposits a coating of tin on any steel parts, eliminating galvanic corrosion. Not suitable if aluminum is present. Because of good electrical conductivity, a good RF grounding system. As good a paint base as the dichromate or the #17 anodize.

[a]Patented by the Dow Chemical Company.
Conversion factor: mm × 0.0394 = in.
Note: See Table 14.4 for compositions and procedures.

TABLE 14.5 Compositions and Procedures for Conversion Coatings

Name	Composition (aqueous)	Procedures
Chrome pickle	180 g/l $Na_2Cr_2O_7 \cdot H_2O$ 190 ml/l 65% HNO_3	Immerse 0.5-2 min at 20-40°C using mild agitation. Hold parts 5 sec above tank before cold-water rinsing.
Chrom-alum	30 g/l $K_2Cr_2(SO_4)_4 \cdot 24H_2O$ 100 g/l $Na_2Cr_2O_7 \cdot H_2O$	Immerse 2-5 min at 100°C. Use aluminum, ceramic or stainless steel tanks.
Dichromate	120-180 g/l $Na_2Cr_2O_7 \cdot 2H_2O$ 2.5 g/l CaF_2, or MgF_2	Acid fluoride pickle, rinse thoroughly, then boil for 30 min. Minimum temperature is 95°C.
Galvanic anodize	30 g/l $(NH_4)_2SO_4$ 30 g/l $Na_2Cr_2O_7 \cdot 2H_2O$ 2.5 ml/l NH_4OH	Immerse 10-30 min at 50-60°C. Connect work to a steel cathode.
Ammonium phosphate	120 g/l $NH_4H_2PO_4$ 30 g/l $(NH_4)_2SO_3 \cdot H_2O$ 15 ml/l 30% NH_3	Immerse for 1.5-2 min at 20°C with mild agitation. Neutralize by immersing in 120 g/l NaOH at 70-90°C for 1-2 min with mild agitation.
Dilute chromic acid	10 g/l CrO_3 7.5 g/l $CaSO_42H_2O$	In making the bath, add the chemicals in the order shown and stir vigorously for 15 min. Immerse parts 0.5-2 min at 20°C. Can be brushed or sprayed on until a brassy finish is obtained.
Modified chrome pickle	15 g/l $NaHF_2$ 180 g/l $Na_2Cr_2O_7 \cdot 2H_2O$ 10 g/l $(Al_2(SO_4)_3 \cdot 14H_2O$ 125 ml/l 70% HNO_3	5 sec to 2 min immersion at 20°C followed by a 5 sec air drain.

TABLE 14.5 (*Continued*)

Name	Composition (aqueous)	Procedures
Ferric nitrate bright pickle	180 g/l CrO_3 40 g/l $Fe(NO_3)_3 \cdot 9H_2O$ 3.5 g/l KF	Immerse for 0.25-3 min at 15-40°C with agitation. May also be sprayed on parts.
Chromate manganate	180 g/l $Na_2Cr_2O_7 \cdot 2H_2O$ 5 g/l $KMnO_4$ 2.6 ml/l H_2SO_4	15-30 sec at 20-40°C.
Stannate immersion	10 g/l NaOH 50 g/l $K_2SnO_3 \cdot 3H_2O$ 10 g/l $NaC_2H_3O_2 \cdot 3H_2O$ 50 g/l $Na_4P_2O_7$	Immerse for 20 min at 80°C. Insulate workpiece from the tank and heating coil. Prepare the bath by adding the chemicals steadily and in the order given, agitating to assure each chemical is dissolved before adding the next. Add the $Na_4P_2O_7$ at 60-65°C.

Conversion factor: 1 g/l = 0.134 oz/gal.

TABLE 14.6 Baths and Procedures for Anodic Coatings

Name	Composition (aqueous)	Procedures
HAE	35 g/l KF 35 g/l Na_3PO_4 165 g/l KOH 35 g/l $Al(OH)_3$ 20 g/l K_2MnO_4 or $KMnO_4$	Use alternating current with a current density of 1.5-2.5 A/dm^2. Maintain bath temperature below 20°C, with a maximum of 30°C. For the thin coating, run for 7-10 min at 65-70 volts; for the thick coating, run for 60-90 min at 80-90 volts. The thin coating will increase the dimensions by 5 microns/surface. The thick coating will increase dimensions by 20 microns/surface. Metal is removed, so the film produced is twice the build up.
Dow 17	225-450 g/l NH_4HF_2 50-120 g/l $Na_2Cr_2O_7 \cdot 2H_2O$ 50-110 ml/l 85% H_3PO_4	Use alternating current with a current density of 0.5-5 A/dm^2. Maintain bath temperature at 70-80°C. For the thin coating, run for 4-5 min at 60-75 volts; for the thick coating, run for 25 min at 90-100 volts. The thin coating will increase the dimensions by 5 microns/surface. The thick coating will increase the dimensions by 25 microns/surface. Metal is also removed, so the film produced is twice the build up.

Conversion factor: 1 g/l = 0.134 oz/gal.

14.6 GENERAL FINISHING SYSTEMS

Because the wide choice of surface preparation schemes can lead to confusion if not frustration, Figure 14.3 is included to indicate the general order of preparation for the major forms of cast and wrought products [3]. How far one proceeds along any chain depends upon the service that will be demanded of the part.

General recommendations on the degree of protection appropriate to various conditions, and good paths for achieving each degree are given in Table 14.7. This is to be considered as a guide, not as a firm recommendation. In each case, it is wise to perform actual service tests before selecting a final system for production.

TABLE 14.7 A Guide for Finishing

Type of service	Chemical treatment	Primer	Finish paint
Mild indoor	None, ferric nitrate	None	None.
Indoor	Chrome pickle, modified chrome pickle, ferric nitrate, or chromate manganate	None, or One coat	One coat, or one or more coats compatible with the primer.
Normal inland	Chrome pickle, dichromate, anodize, modified chrome pickle, or chromate manganate	One coat	Two or more coats compatible with the primer.
High humidity	Chrome pickle, Dichromate, anodize, or modified chrome pickle	One coat	Two or more coats of baked alkyd enamel, air-drying vinyl lacquer, or baked epoxy.
Marine or other severe exposures	Dichromate or sealed anodize	One or two coats	Two or more coats depending on number of primer coats used. Primer and finish preferably baked.

Source: Ref. 3.

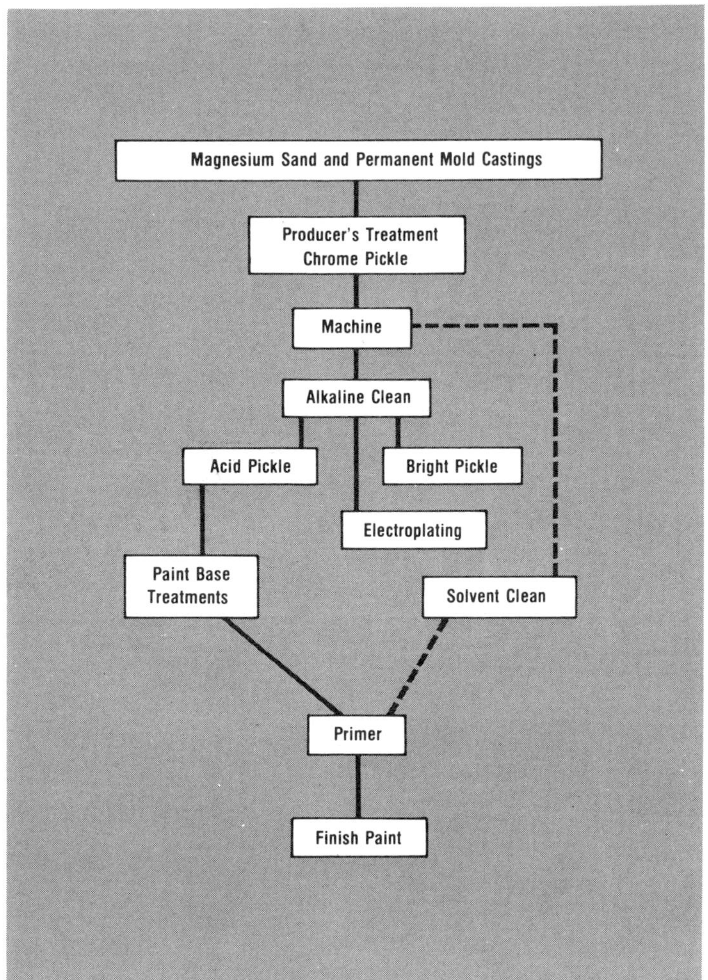

FIGURE 14.3 Alternate finishing schemes. From Ref. 3.

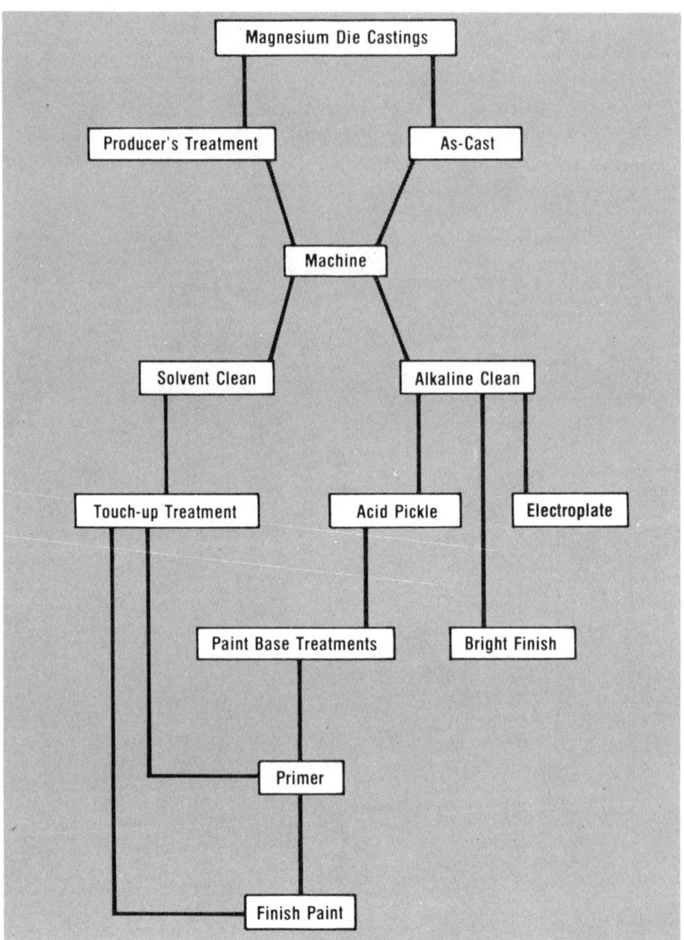

FIGURE 14.3 (*Continued*)

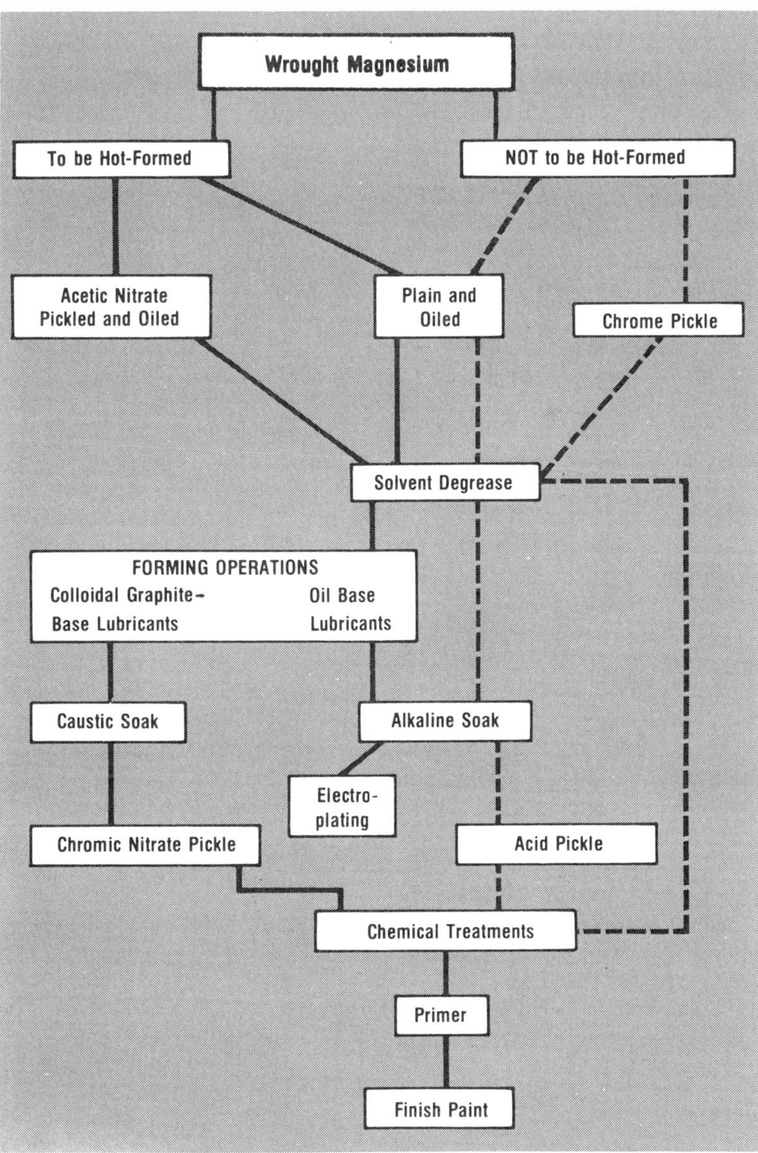

FIGURE 14.3 *(Continued)*

REFERENCES

1. Magnesium Elektron Ltd., "Surface Treatment for magnesium Alloys in Aerospace and Defense," Bulletin.
2. Dow Chemical Company, "Magnesium Finishing," Bulletin 141-229 (1963).
3. Dow Chemical Company, "Operations in Matnesium Finishing," 1982.

BIBLIOGRAPHY

1. Norsk Hydro a.s., "Normag Magnesium Pure and Alloys,"
2. AMAX Specialty Metals Corporation, Magnesium Division, "Magnesium, a Look at Magnesium Die Casting,"
3. E. L. White and F. N. Fink, "Corrosion Protection of Magnesium and Magnesium Alloys," Memorandum 205, Defense Metals Information Center, Battelle Memorial Institute, Columbus, Ohio (1965).
4. American Society for Metals, *Metals Handbook*, "Properties and Selections: Nonferrous Alloys and Pure Metals," 9th edition, Vol. 2, American Society for Metals, Metals Park, Ohio, 1979.
5. W. Unsworth and J. F. King, "The Realities of Magnesium Corrosion and Protection," *Proceedings of the International Magnesium Association*, 1984.
6. Magnesium Elektron Ltd., "Fluoride Anodizing of Magnesium Alloy Components," Bulletin 202, February 1977.
7. Magnesium Elektron Ltd., "Chromate Treatments for Magnesium Alloys," Bulletin 203, June 1980.
8. Magnesium Elektron Ltd., "Protection and Surface Treatment of Magnesium Alloys," Bulletin 252, August 1984.
9. Magnesium Elektron Ltd., "Surface Sealing of Magnesium Alloy Components," Bulletin 200, February 1977.
10. Magnesium Elektron Ltd., "H. A. E. Hard Anodizing of Magnesium Alloy Components," Bulletin 201, February 1977.

Index